Quantum Topology

SERIES ON KNOTS AND EVERYTHING

Editor-in-charge: Louis H. Kauffman

Published:

Vol. 1: Knots and Physics
L. H. Kauffman

Vol. 2: How Surfaces Intersect in Space
J. S. Carter

Vol. 3: Quantum Topology
edited by L. H. Kauffman & R. A. Baadhio

Vol. 4: Gauge Fields, Knots and Gravity
J. Baez & J. P. Muniain

Vol. 5: Gems, Computers and Attractors for 3-Manifolds
S. Lins

Vol. 6: Knots and Applications
edited by L. H. Kauffman

Vol. 7: Random Knotting and Linking
edited by K. C. Millett & D. W. Sumners

Vol. 8: Symmetric Bends: How to Join Two Lengths of Cord
R. E. Miles

Vol. 9: Combinatorial Physics
T. Bastin & C. W. Kilmister

Vol. 10: Nonstandard Logics and Nonstandard Metrics in Physics
W. M. Honig

Forthcoming:

Vol. 11: History and Science of Knots
edited by J. C. Turner & P. van de Griend

K&E Series on Knots and Everything — Vol. 3

Quantum Topology

Louis H. Kauffman
*Department of Mathematics,
Statistics and Computer Science
University of Illinois at Chicago*

Randy A. Baadhio
*Theory Group, Physics Division,
Lawrence Berkeley Laboratory
and
Department of Physics
University of California at Berkeley*

World Scientific
Singapore • New Jersey • London • Hong Kong

Published by

World Scientific Publishing Co. Pte. Ltd.
5 Toh Tuck Link, Singapore 596224
USA office: 27 Warren Street, Suite 401-402, Hackensack, NJ 07601
UK office: 57 Shelton Street, Covent Garden, London WC2H 9HE

British Library Cataloguing-in-Publication Data
A catalogue record for this book is available from the British Library.

About the cover:
West African Artistic Representation of Cosmogony

Courtesy of S. R. N'Djimbi-Tchikaya

Series on Knots and Everything — Vol. 3
QUANTUM TOPOLOGY

ISBN-13 978-981-02-1544-6
ISBN-10 981-02-1544-4
ISBN-13 978-981-02-2575-9 (pbk)
ISBN-10 981-02-2575-X (pbk)

To the Memory of
Our Fathers,

Abraham Joseph Kauffman
and
Eggue Gregoire Baadhio

PREFACE

This book constitutes a review volume on the relatively new subject of Quantum Topology. It is a review volume in the form of a snapshot of ongoing research activity. This activity goes across the board, among problems, techniques, and mysteries in topology and in mathematical physics.

The snapshot was taken at a special session of the American Mathematical Society on Topological Quantum Field Theory. This special session was held in Dayton, Ohio on October 30 – November 1, 1992 at the general meeting of the American Mathematical Society, in which one of us (Louis Kauffman) gave an invited address.

We have used ourselves and the participants in this special session as camera, film and darkroom for the development of the snapshot. Participants in the special session were asked to submit papers for this book that reflected the contents of their talk at the meeting, and their current research. The request was made at the special session, and the papers in this volume were all produced within a few months of that session. Thus, the film for our snapshot developed over this period.

In many cases, there were interesting surprises and new results. Some problems have not been resolved, but the sense of urgency they had at our meeting was not abated. Most of the work discussed in this session stems from two intertwined sources. The first source is the discovery in the mid 1980's of a number of new invariants of knots and links (Jones, Homfly and Kauffman polynomials), followed by a very wide extension of these invariants through the use of quantum groups and methods of statistical mechanics. The second source was Edward Witten's 1988 introduction of methods of quantum field theory into the subject, and the discovery (by Witten, Reshetikhin and Turaev) of new invariants of 3-manifolds directly related to the new knot invariants. This led Sir Michael Atiyah and Witten to formulate the concept of topological quantum field theories, and the subject of quantum topology was born!

In our conference many problems were discussed, but no single problem held more attention than the possibility of extending the combinatorial and quantum group methods from 3-manifolds to 4-manifolds and possibly higher dimensions. This was a key theme in the conference, and the work continues. This may be seen in the papers of Carter-Saito, Crane-Yetter,

Lawrence, Michielsen-Nijhoff, Baadhio, and Kauffman. It is in addition reflected in the paper by Crane-Kauffman-Yetter, where, their evaluation of the Crane-Yetter invariant of four-manifolds yields a formula for the signature of a closed four-manifold in terms of local combinatorial data. Their formula uses a topological quantum field theory state summation, and gives rise to many questions on the boundary between quantum topology and geometric topology.

In addition to the distributed discussion of problems, the paper in this volume by Przytycki gives specific treatment of problems in classical Knot Theory.

Difficulties occur in trying to extend, discretize or otherwise better understand Chern-Simons-Witten theories. This theme appears in the papers of Eliezer-Semenoff, Freed, King, McLaughlin-Brylinski, Phillips-Stone and Pullin. In these cases the methods are quite diverse, and the authors should not be lumped entirely under one theme. Eliezer-Semenoff are concerned with lattice gauge theory, Freed with special structures, King with expectations associated with Wilson loops, McLaughlin-Brylinski with the Pontrjagin class of a 4-manifold, and Pullin with a multifold relationship of these ideas with the Ashtekar theory of quantum gravity – centered around the role of Wilson loops and their expectations as states of quantum gravity.

Knot theory and related topological problems go back and forth in this arena, as exemplified by the papers of Baadhio (which associates physical pathologies known as global anomalies to some invariants of knots in dimension 3 and studies their extensions to higher dimensions), Carter-Saito, Gilmer, Kauffman, Mullins, Pullin, Mattes-Polyak-Reshetikhin, and Rong.

These are but a few of the main themes in the interaction. But what is quantum topology anyway? As a tentative definition, we suggest that quantum topology constitutes the use of fundamental ideas, methods and techniques of quantum mechanics in the study of both topology and mathematical physics. A sample of this interaction is presented in this book.

The book is organized as follows. It begins with an article by Louis H. Kauffman entitled *Introduction to Quantum Topology*. This article represents the development of the invited address at the Dayton meeting coupled with the contents of the meeting, and the other papers in this volume. No attempt at a complete synthesis has been made, but in those cases where relationships could be made directly, there has been an attempt to give some picture of these matters. The article is a self-contained account of quantum topology from the knot theoretic point of view.

The book then unfolds – by alphabetical order – into the research papers by the authors mentioned in this introduction. We thank them warmly for their participation in the conference and in the preparation of this book.

Louis H. Kauffman, *Chicago, Illinois*
Randy A. Baadhio, *Berkeley, California*
July 1993

CONTENTS

Quantum Topology

Introduction to Quantum Topology

by Louis H. Kauffman
Department of Mathematics, Statistics and Computer Science
University of Illinois at Chicago
Chicago, Illinois 60680
U10451@UICVM.BITNET

Abstract.
This paper is a general introduction to quantum topology and it is an exposition of a combinatorial approach to the Jones polynomial and the Witten-Reshetikhin-Turaev and Turaev-Viro invariants, via knot diagrams, tangles and the Temperley Lieb algebra.

I. Introduction

This paper is an introduction to quantum topology. It is an introduction to interactions between ideas and methods from quantum theory, and problems and techniques in the topology of low dimensional manifolds.

This paper is *not* an introduction to the quantization of point set topology or to the theory of quantum sets. These are subjects one might dream about. I am indebted to Dan Sandin for the impetus to put this disclaimer about quantum sets and quantum topologies right at the beginning. It would seem that the words quantum topology lead one into certain intellectual dreams. A geometric topologist who dreams in the light of these words will find a different world from a logician or set theorist. The dream that came to pass into our mathematical reality is an interlacing of quantum field theory with macroscopic topology problems. These are problems about combinatorial spaces, whose point-set properties are almost irrelevant. Nevertheless, as we proceed into matters involving four dimensions, I expect to see subtle interactions with point set topology. For this is the topologist's legacy in that dimension.

There are nine sections to this paper beyond the introduction. The second section is the longest, and it can be read independently of the rest. This section, Section 2, entitled Quantum Mechanics and Topology, begins with a review of the principles and ideas of quantum mechanics: De Broglie waves, wave packets, Schrodinger equation, operators and observables, rotation, unitary symmetry and the role of the Lie algebra. Then comes the Dirac string trick in a modern form that reconstructs SU(2) from the strings. There follows a discussion of Dirac bracket notation, amplitudes, knot amplitudes (quantum link invariants), and amplitudes for other manifolds (topological quantum field theory). This section then gets down to work with the formalism of Witten's path integral, and shows how the Homfly polynomial arises in SU(n) gauge and how the Lie algebra expressions for the top row of the corresponding Vassiliev invariants can be seen from this formalism.

This last part of section 2 (formalism of the functional integral) sets the stage both for the rest of this paper and for the other papers in this volume and for many problems in the subject. We illustrate how it is possible to see very broadly, by using physical formalism, results that require much work and reworking on other grounds. There are extraordinary depths to this predicament, and it is a motivating factor in the development of the subject. The use of the physical formalism is shaped by the topological results and the topological results are shaped by that formalism. Topologists desire simple combinatorial methods along with the analytic and the formal. These methods appear but at a cost in vision. In the case of four manifold invariants the appropriate combinatorial methods have not yet appeared, but there are hints. In the case of 3-manifolds the combinatorial methods exist but are difficult to understand. The simplest case, of link invariants such as the Jones polynomial are wonderfully tantalizing: Here one can construct the invariant in a flash by combinatorial reasoning, but only the formal functional integral comes close to a 3-dimensional conceptualization of this invariant.

It is recommended that the interested reader skip from section 2 of this paper to *both* the other sections of this paper *and* to the other papers in this volume on Quantum Topology!

Nevertheless, the present paper does have eight more sections. Section 3 discusses the Temperley Lieb Algebra in flat tangle form, and the inductive construction of the Jones-Wenzl projectors. Section 4 is a description of the recoupling theory that can be accomplished via these tangles and projectors. Section 5 constructs the bracket polynomial model for the Jones polynomial, and relates this construction to the Temperley Lieb algebra and to the projectors of sections 3 and 4. Section 6 constructs a purely combinatorial version of the Witten-Reshetikhin-Turaev (WRT) invariant in the case of SU(2) (aka the Temperley Lieb algebra). Section 7 reformulates section 6 in terms of the shadow world of Reshetikhin and Kirillov as adapted to our recoupling theory. Section 8 discusses a pair of manifolds (discovered by the author and Sostenes Lins) that are distinct but have the same WRT invariants (using "encirclement" properties of the recoupling theory). Section 9 is a sketch of recent work of Justin Roberts, and its relationship with the work of Crane and Yetter presented in this volume. This section raises a number of questions about the construction of invariants of 4-manifolds by combinatorial means.

Section 10 is about the Turaev-Viro invariant and its relationship with theories of quantum gravity. It is fitting that we end the essay in quantum gravity. Quantum gravity is necessarily about the integration over "all possible" metrics on a manifold. One of the residues of such a process is the underlying topology. Topology and quantum gravity must be intimately related. How this actually works out in a toy 2+1 model is the theme of this section. We also briefly mention the Rovelli-Smolin loop states for 3+1 quantum gravity in the Ashtekar formulation (See also the article by Pullin in this volume.). In these cases, we are confronted by situations that demand the use of all the tools and all the levels of rigor involved with these tools. This is the place of interdisciplinary work.

It gives the author pleasure to acknowledge the support of NSF Grant Number DMS 9205277 and the Program for Mathematics and Molecular Biology of the University of California at Berkeley, Berkeley, CA.

II. Quantum Mechanics and Topology

We recall principles of quantum mechanics, and point out how topology has come into relationship with ideas from physics.

Lets start with the fundamental equations of DeBroglie:

$E = \hbar w$ and $p = \hbar k$.

Here $\hbar = h/2\pi$ where h is Planck's constant.

In DeBroglie's equations E stands for the energy of an electron and p for its momentum, while w and k stand for the frequency and wave number of a plane wave associated with that particle.

DeBroglie had the idea that the discrete energy levels of the orbits of electrons in an atom of hydrogen could be explained by restrictions on the vibrational modes of waves associated with the motion of the electron. His choices for the energy and the momentum in relation to a wave are not arbitrary. They are designed to be consistent with *the notion that the wave or wave packet moves among with the electron.* To see how this works we discuss the notion of a wave packet.

Wave Packets

The equation for a plane wave in one dimension is

$$f(x,t) = \sin(kx-wt) = \sin((2\pi/\lambda)(x - ct))$$

where λ is the wavelength and c is the velocity of the wave. (This serves to define the frequency w and the wave number k.)

Linear interference of two or more waves of slightly differing frequency produces a wave packet that moves with its own velocity. To see the essence of this phenomenon, consider

$$g(x,t) = \sin(kx-wt) + \sin(k'x - w't).$$

Using $\sin(X+Y)+\sin(X-Y) = 2\sin(X)\sin(Y)$, we deduce that
$g(x,t) = [\cos((k-k')/2)x - ((w-w')/2)t)] \sin(((k+k')/2)x - ((w+w')/2)t)$.
Thus, if k is very close to k' and w is very close to w', then we can approximate $(k+k')/2$ by k and $(w+w')/2$ by w and write $\delta k = (k-k')$, $\delta w = (w-w')$, obtaining

$$g(x,t) = [\cos(\delta k/2)x - (\delta w/2)t)] \sin(kx - wt).$$

The pattern described by $[\cos(\delta k/2)x - (\delta w/2)t)]$ moves with its own velocity and frequency. This is the wave packet. The velocity of the wave packet (the so-called group velocity) is therefore given by the formula $v_g = dw/dk$.

Applying DeBroglie, we get $v_g = d(\hbar w)/d(\hbar k) = dE/dp$.

Now suppose that we are speaking of a classical particle of momentum p and velocity v. Then $p = mv$ and $E = (1/2)mv^2$ (where m is the mass of the particle). Thus $v_g = dE/dp = v$. The group velocity is equal to the classical velocity! This identification of wavepacket velocity and classical velolcity via DeBroglie's equations $E = \hbar w$, $p = \hbar k$ led to the idea that *material particles could be modelled by wave phenomena,* hence to the beginnings of *wave mechanics.*

Schrodinger's Equation

Writing our wave in complex form $\psi = \psi(x,t) = \exp(i(kx - wt))$, we see that we can extract DeBroglie's energy and momentum by differentiating: $i\hbar\partial\psi/\partial t = E\psi$ and $-i\hbar\partial\psi/\partial x = p\psi$. This led Schrodinger to postulate *the identification of dynamical variables with operators* so that the first equation , $i\hbar\partial\psi/\partial t = E\psi$, is promoted to the status of an equation of motion while the second equation becomes the definition of momentum as an operator: $p \longleftrightarrow -i\hbar\partial/\partial x$. In this formulation, the position operator is just x itself. In this way, energy becomes an operator via substitution of the momentum operator in the classical formula for the energy:

$$E = (1/2)mv^2 + V$$

$$E = p^2/2m + V \longleftrightarrow E = -(\hbar^2/2m)\partial^2/\partial x^2 + V.$$

With this operator identification for E, Schrodinger's equation is an equation in the first derivatives of time and in second derivatives of space. (V is the potential energy and its corresponding operator depends upon the details of the application.) In this form of the theory one considers general solutions to the differential equation and this in turn has led to excellent results in a myriad of applications.

Observation is modelled by the concept of *eigenvalues for corresponding operators.* The mathematical model of an observation is a projection of the wave function into an eigenstate. A spectrum of energy $\{E_k\}$ corresponds to the wave function ψ satisfying the

Schrodinger equation, and such that there are constants E_k with $E\psi = E_k\psi$. An *observable* E is a Hermitian operator on a Hilbert space H of wavefunctions. Since Hermitian operators have real eigenvalues, this provides the link with measurement for the quantum theory.

Note that the operators for position and momentum satisfy the equation $xp - px = \hbar i$. This corresponds directly to the equation obtained by Heisenberg, on other grounds, that dynamical variables can no longer necessarily commute with one another. In this way, the points of view of DeBroglie, Schrodinger and Heisenberg came together, and quantum mechanics was born. In the course of this development, interpretations varied widely. Eventually, physicists came to regard the wave function not as a generalized wave packet, but as a carrier of information about possible observations. In this way of thinking, ψ itself can be used mathematically, while $\psi^*\psi$ (ψ^* denotes the complex conjugate of ψ.) represents the probability of finding the "particle" (A particle is an observable with local spatial characteristics.) at a given point in spacetime.

Rotations and Symmetry

By going into three dimensions of space, one can see how the mathematics of quantum theory involves rotational symmetry. Consider angular momentum.

The classical formula for angular momentum is $L = r \times p$ where r and p are, respectively, the position vector of the particle, and the momentum vector of the particle (and x denotes vector cross product in three-space). Replacing p by its corresponding quantum operator, we have

$$p \longleftrightarrow -i\hbar(\partial/\partial x,\ \partial/\partial y,\ \partial/\partial z).$$

This gives rise to a vector (L_x, L_y, L_z) of angular momentum operators with

$$L_x = -i\hbar(y\partial/\partial z \ - z\partial/\partial y)$$
$$L_y = i\hbar(x\partial/\partial z \ - \ z\partial/\partial x)$$
$$L_z = -i\hbar(x\partial/\partial y \ - \ y\ \partial/\partial x).$$

It is then easy to see that $[L_x, L_y] = L_x L_y - L_y L_x = i\hbar L_z$ (and the same equation for cyclic permutations of x, y z.). This means that the angular momentum operators behave in exactly the same pattern as the generators for the Lie algebra of the group SU(2).

More precisely, consider the matrices J_1, J_2, J_3 illustrated below.

$$J_1 = (1/2)\begin{pmatrix} 0 & 1 \\ 1 & 0 \end{pmatrix}$$

$$J_2 = (1/2)\begin{pmatrix} 0 & -i \\ i & 0 \end{pmatrix}$$

$$J_3 = (1/2)\begin{pmatrix} 1 & 0 \\ 0 & -1 \end{pmatrix}$$

Then $[J_1, J_2] = iJ_3$ (and cyclic permutations thereof) so that we can write $[J_r, J_s] = i\varepsilon_{rst}J_t$ where ε_{rst} is the *epsilon symbol* whose value is 1 when **rst** is a cyclic permutation of 123, -1 when **rst** is a non-cyclic permutation of 123 and zero if **rst** has a repetition.. The matrices J_k are the linear generators of the set of Hermitian matrices of trace zero.

If $H = aJ_1 + bJ_2 + cJ_3$ with a,b,c real, then

$$\exp(iH)\exp(-iH^*) = \exp(i(H-H^*))=\exp(0) = 1$$

and $\det(\exp(iH)) = \exp(\text{trace}(iH))=\exp(0)=1$. Thus $\exp(iH)$ belongs to the special unitary group SU(2) of matrices U such that U^* equals U^{-1} and of determinant equal to 1. (U^* is the conjugate transpose of U.) The matrices H can be identified with the Lie algebra of SU(2).

Thus we see that the angular momentum operators give (up to

adjusting a factor of $\hbar$) a representation of the SU(2) Lie algebra. The direct relationship with rotations in three space comes from the fact that SU(2) double covers the rotation group SO(3) and that rotations must act on the wave function $\psi(x,y,z,t)$ by unitary transformations in order to preserve $\psi^*\psi$.

This relationship of Lie algebra with the elements of angular momentum illustrates the principle that the observables in quantum mechanics live in the Lie algebra of a unitary group. This is a restatement of the fact **exp(iH)** = U where H is Hermitian and U is unitary. As we have just mentioned, a unitary transformation is the form of a symmetry in quantum mechanics.
Observables live in the Lie algebras of symmetries.

Representations
Quantization of angular momentum depends upon the representations of the Lie algebra of SU(2). The simplest instance is the spin (1/2) representation of dimension two generated by $u=(1,0)^t$ and $d=(0,1)^t$, where t denotes transpose.
Then $J_3u = (1/2)u$ and $J_3d = (-1/2)d$. This representation generates eigenvalues for observing rotation about the z-axis. Higher representations are obtained from the two dimensional representation by taking symmetrized sums of tensor products of u's and d's. As we move into the topology related to these structures in section 5, this simple algebraic act of symmetrization will take on a special significance.

The Dirac String Trick
This brings us to one of the earliest relations of topology and quantum mechanics, the *Dirac string trick.* Consider an observer walking in a continuous circle around an electron. The vector valued wave function describing this situation must undergo a unitary transformation as the observer and associated experimental apparatus undergo a rotation in three space.

We are therefore describing a representation of SO(3) to SU(2). Such a representation is necessarily multiple valued. As the observer moves around a full circle, the corresponding element in SU(2) must change sign. The wave function is not returned to its original state after the full circle. The wave function has changed its sign. After two full turns the wave function is back to its original condition. We arrive at the strange conclusion that a physical system that is continuously turned through 360 degrees is not quite back in its original (un-turned) quantum state. Dirac devised a remarkable topological visualization of this fact.

Visualize a belt stretched between two concentric spheres. The belt is fixed to both the inner sphere and to the outer sphere, but it is free to move in between the spheres in the three dimensional annular space between them. In this situation a belt with a 360 degree twist in it can be deformed to a belt with a minus 360 degree twist in it. See the illustration below.

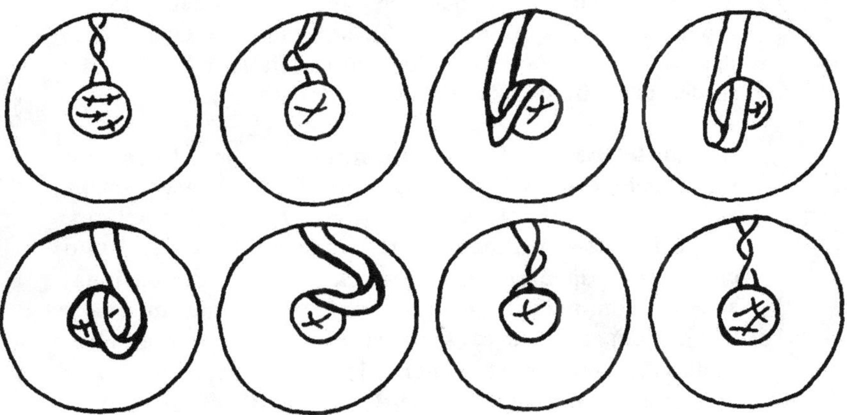

As a consequence of this effect, a belt with a 720 degree twist on it can be deformed to a belt with no twist at all. This is illustrated below.

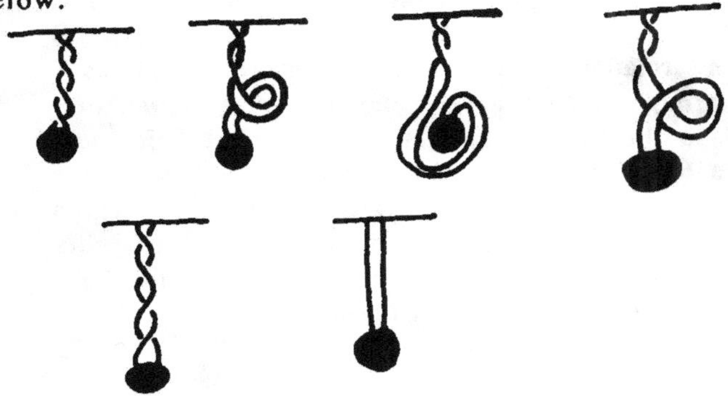

Note that that key to our illustrations is the basic deformation of a "curl" to a twist and back:

Spherical Digression
These motions of the belt are intimately related to the topology of SU(2) and SO(3), and that fact that SU(2) double covers SO(3). For more details we refer the reader to ([BL79],[Kau87],[KA91],[MTW73]).

It can be observed with no further technicalities that the topological deformations we have shown can be accomplished by an orchestration of spatial rotations. Regard the 3-dimensional region between two concentric spheres as a stack of two-dimensional spheres continuously parametrized from the interval [0,1]. Let the sphere at level t be called the *t-shell*. Thus the innermost sphere is the 0-shell and the outermost sphere is the 1-shell. Let there be a belt, **B**, embedded between the innermost and outermost spheres so that the belt is anchored on these spheres. Choose an axis v of rotation centered at the origin of the spheres. *Rotate the t-shell by $2\pi t$ about the axis v.* Let v(**B**) denote the image of the belt **B** under this rotation. Form the set of images v(**B**), as v turns through 180 degrees, starting from a position parallel to the belt. This is the simplest belt trick, taking a belt with a +360 degree twist to a belt with a -360 degree twist. The cancellation of the 720 degree twist is accomplished by a composite of this orchestration with a constant rotation about the axis parallel to the belt. **End of digression.**

Group theory is seen as a flowering of topology. SU(2) is the unit quaternions. That is, SU(2) can be described as the set of elements **a + bi + cj + dk** where the vector of real numbers **(a,b,c,d)** has unit length in Euclidean 4-space, and the algebra generators **i,j,k** satisfy **ii=jj=kk=ijk=-1**, the basic equations for Hamilton's quaternions.

The belt trick provides a topological/mechanical construction for the quaternion group itself ([Kau87],[KA91]). Attach a belt to a fixed wall and to a movable card. Let **i** , **j** and **k** denote turns of the card by 180 degrees about perpendicular axes in 3-space. After such turns, let the state of the belt be returned to normal by the use of the belt trick with the the ends fixed to the wall and the card. (The belt may be moved around the card without twisting the card.) The illustrations below show how the quaternion relations arise directly from this description.

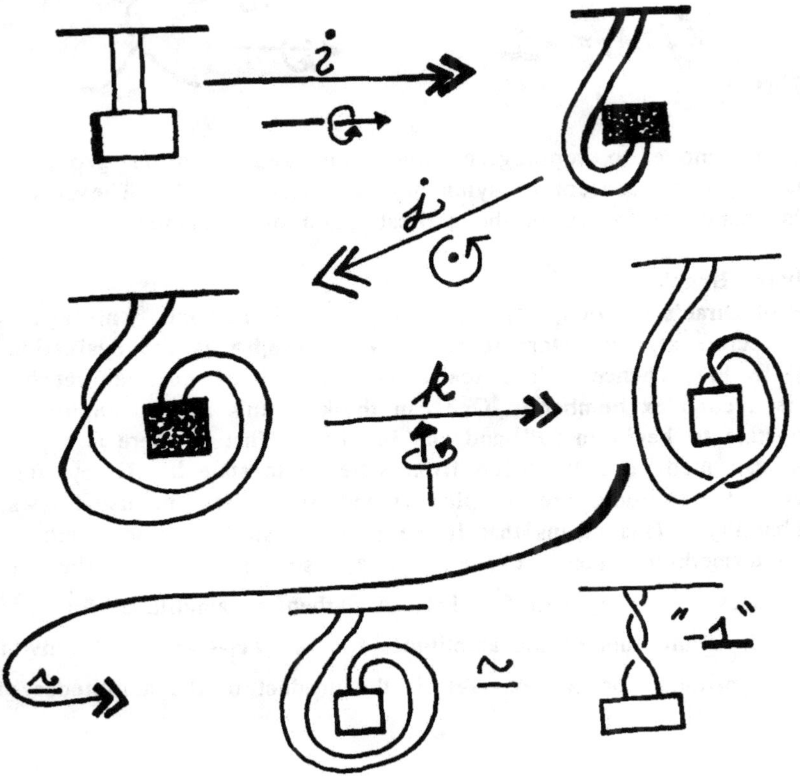

You can also use your own arm as shown below in cartoon form. This quaternionic arm is the discovery of the author and Eddie Oshins (See [KA91], p. 436).

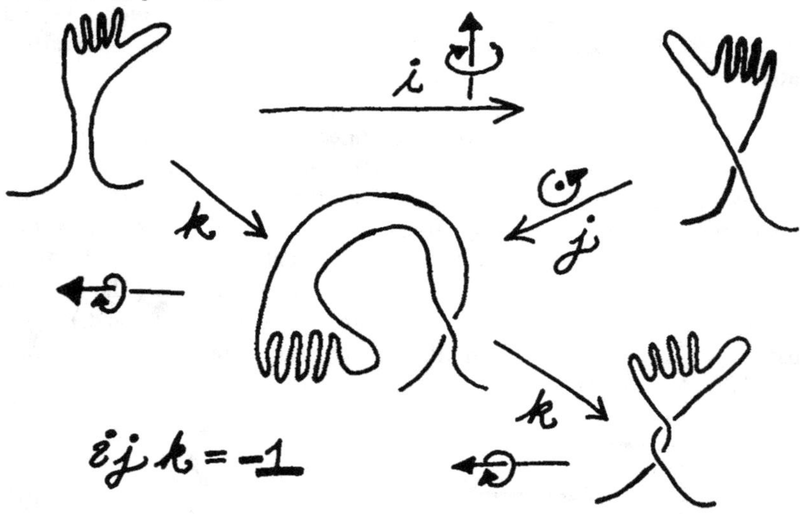

As we move to topological invariants and quantum groups, the change in the concept of symmetry is more radical. The seeds for this change are sown in the earliest quantum mechanics.

Dirac Brackets

Recall Dirac's notation, <a|b>, [D58]. In this notation <a| and |b> are vectors and covectors respectively. <a|b> is the evaluation of <a| by |b>, hence it is a scalar, and in ordinary quantum mechanics it is a complex number. One can think of this as the amplitude for the state to begin in "a" and end in "b". That is, there is a process that can mediate a transition from state **a** to state **b**. Except for the fact that amplitudes are complex valued, they obey the usual laws of probability. This means that if the process can be factored into a set of intermediate states $c_1, c_2, \ldots, c_n$ so that we have the set of processes a--->c_i--->b for i=1,...,n , then the amplitude for a--->b is the sum of the amplitudes for a--->c_i--->b. Meanwhile, the amplitude for a--->c_i--->b is the product of the amplitudes of the two subconfigurations $a \dashrightarrow c_i$ and $c_i \dashrightarrow b$. Formally we have

$$\langle a|b\rangle = \Sigma_i \langle a|c_i\rangle\langle c_i|b\rangle$$

where the summation is over all the intermediate states $i=1, ..., n$.

In general, the amplitude for mutually disjoint processes is the *sum* of the amplitudes of the individual processes. The amplitude for a configuration of disjoint processes is the *product* of their individual amplitudes.

Dirac's division of the amplitudes into *bras* $\langle a|$ and *kets* $|b\rangle$ is done mathematically by taking a vector space V (a Hilbert space, but it can be finite dimensional) for the bras: $\langle a|$ belongs to V. The dual space V* is the home of the kets. Thus $|b\rangle$ belongs to V* so that $|b\rangle$ is a linear mapping $|b\rangle:V \dashrightarrow C$ where C denotes the complex numbers. We restore symmetry to the definition by realizing that an element of a vector space V can be regarded as a mapping from the complex numbers to H. Given $\langle a|: C \dashrightarrow V$, the corresponding element of V is the image of 1 (in C) under this mapping. In other words, $\langle a|$ (1) is a member of V. Now we have $\langle a| :C \dashrightarrow V$ and $|b\rangle : V \dashrightarrow C$. The composition $\langle a|\,|b\rangle = \langle a|b\rangle : C \dashrightarrow C$ is regarded as an element of C by taking $\langle a|b\rangle$ (1). The complex numbers are regarded as the "vacuum", and the entire amplitude $\langle a|b\rangle$ is a "vacuum to vacuum" amplitude for a process that includes the creation of the state a, its transition to b, and the annihilation of b to the vacuum once more.

Knot Amplitudes

At this point a rich imagery arises that goes beyond quantum mechanics -- into modern knot theory. Consider first a circle in a spacetime plane with time represented vertically and space horizontally.

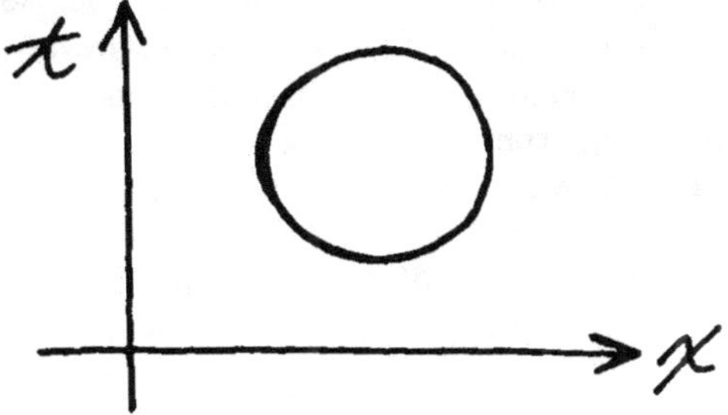

The circle represents a vacuum to vacuum process that includes the creation of two "particles"

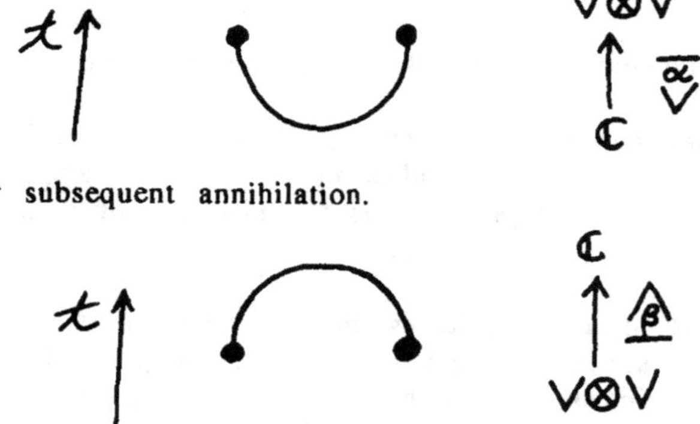

and their subsequent annihilation.

In accord with our previous description, we could divide the circle into these two parts (creation(a) and annihilation (b)) and consider the amplitude <a|b>. Since the diagram for the creation of the two particles ends in two separate points, it is natural to take a vector space of the form $V \otimes V$ as the target for the bra and as the domain of the ket. We imagine at least one particle property being catalogued by each factor of the tensor product. For example, a basis of V could enumerate the spins of the created particles. In this language the creation ket is a map alpha,
$\alpha = \langle a| : C \dashrightarrow V \otimes V$, and the annihilation bra is a mapping beta,
$\beta = |b\rangle : V \otimes V \dashrightarrow C$.

The first hint of topology comes when we realize that it is possible to draw a much more complicated simple closed curve in the plane that is nevertheless decomposed with respect to the vertical direction into many alphas and betas. In fact, any non-selfintersecting differentiable curve can be rigidly rotated until it is in general position with respect to the vertical. It will then be seen to be decomposed into these minima and maxima. Our prescriptions for amplitudes suggest that we regard any such curve as an amplitude via its description as a mapping from C to C.
The decomposition of the curve indicated below corresponds to the composition of maps

$$C \xrightarrow[\alpha_1]{} V \otimes V = V \otimes C \otimes V \xrightarrow[\alpha_2]{} V \otimes (V \otimes V) \otimes V$$

$$= (V \otimes V) \otimes (V \otimes V) \xrightarrow[\beta_1]{} C \otimes V \otimes V = V \otimes V \xrightarrow[\beta_2]{} C .$$

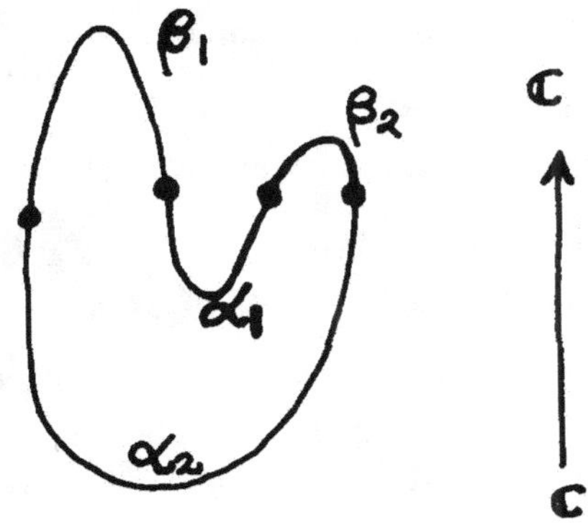

Each simple closed curve gives rise to an amplitude, but any simple closed curve in the plane is isotopic to a circle, by the Jordan Curve Theorem. If these are *topological amplitudes*, then they should all be equal to the original amplitude for the circle. Thus the question: What condition on creation and annihilation will insure topological amplitudes? The answer derives from the fact that all isotopies of the simple closed curves are generated by the cancellation of adjacent maxima and minima as illustrated below.

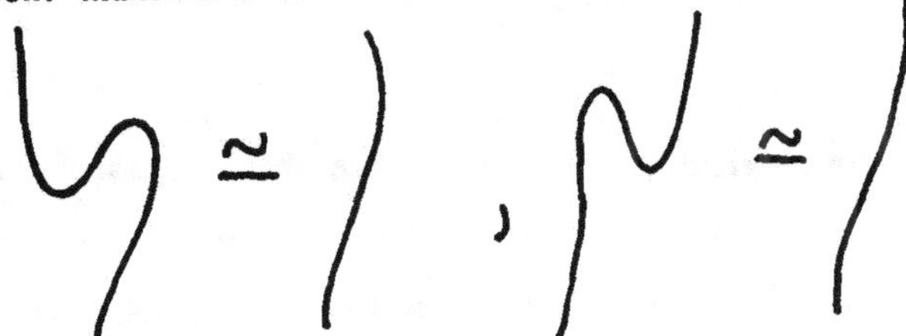

These diagrams show that we need that the compositions $(\beta \otimes 1)(1 \otimes \alpha)$ and $(\alpha \otimes 1)(1 \otimes \beta)$ are each the identity on V. More precisely, we require that the composition (and its dual obtained by reversing all the arrows) shown below is the identity.

$$V = V \otimes C \xrightarrow[\alpha]{} V \otimes (V \otimes V) = (V \otimes V) \otimes V \xrightarrow[\beta]{} C \otimes V = V.$$

This condition is said very simply by taking a matrix representation for the corresponding operators.

Specifically, let $\{e_1, e_2, \ldots, e_n\}$ be a basis for V. Let $e_{ab} = e_a \quad e_b$ denote the elements of the tensor basis for V V. Then there are matrices M_{ab} and M^{ab} such that $\alpha(1) = \Sigma\, M_{ab} e_{ab}$ with the summation taken over all values of a and b from 1 to n. Similarly, β is described by $\beta(e_{ab}) = M^{ab}$. Thus the amplitude for the circle is $\beta\alpha(1) = \beta\Sigma\, M_{ab} e_{ab}$ $= \Sigma\, M_{ab}\beta(e_{ab}) = \Sigma\, M_{ab} M^{ab}$. In general, the value of the amplitude on a simple closed curve is obtained by translating it into an "abstract tensor expression" in the M_{ab} and M^{ab}, and then summing over these products for all cases of repeated indices. For example,

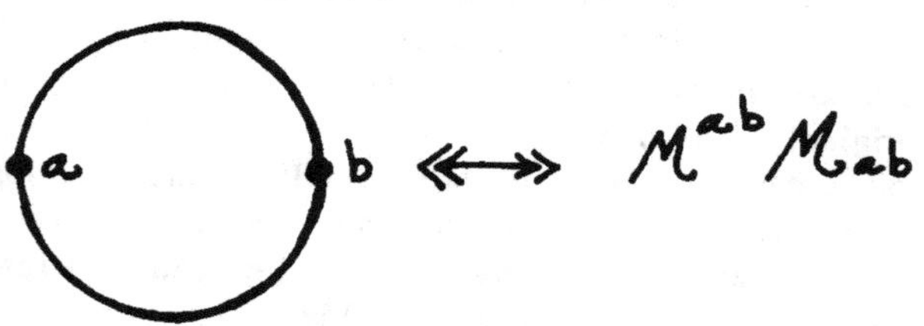

Returning to the topological conditions we see that they are just that the matrices M_{ab} and M^{cd} are inverses in the sense that $\Sigma M_{ai} M^{ib} = I_a{}^b$ and $\Sigma M^{ai} M_{ib} = I^a{}_b$ are the identity matrices:

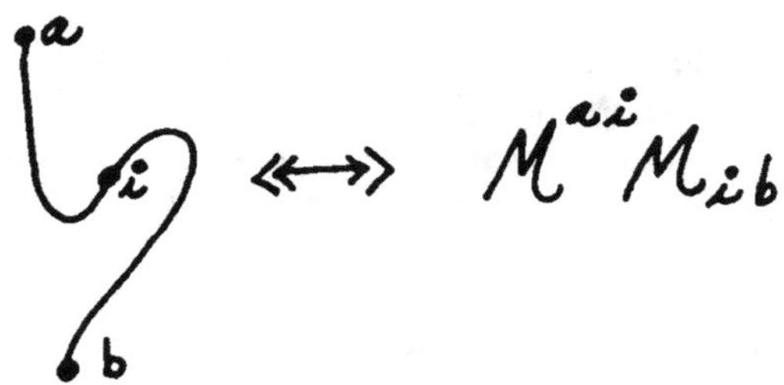

In the simplest case alpha and beta are represented by 2 x 2 matrices. The topological condition implies that these matrices are inverses of each other. Thus the problem of the existence of topological amplitudes is very easily solved for simple closed curves in the plane.

Now we go to knots and links. Any knot or link can be represented by a picture that is configured with respect to a vertical direction in the plane. The picture will decompose into minima (creations) maxima (annihilations) and crossings of the two types shown below. (Here I consider knots and links that are unoriented. They do not have an intrinsic preferred direction of travel.)

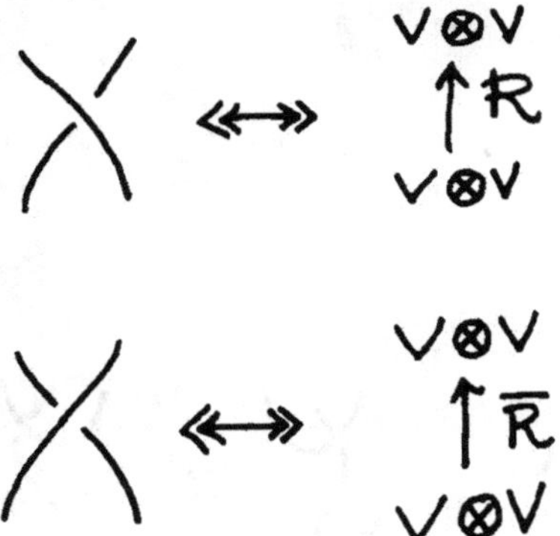

Next to each of the crossings we have indicated mappings of $V \otimes V$ to itself, called R and R^{-1} respectively. These mappings represent the transitions corresponding to these elementary configurations. That R and R^{-1} really are inverses follows from the isotopy shown below (This is the second Reidemeister move.)

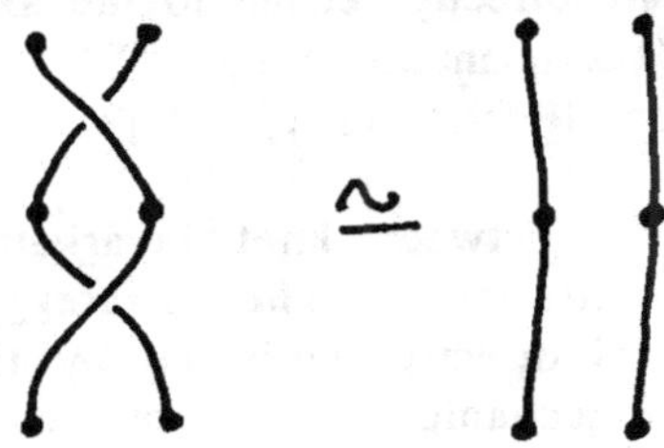

We now have the vocabulary of α, β, R and R^{-1}. Any knot or link can be written as a composition of these fragments, and consequently a choice of such mappings determines an amplitude for knots and links. In order for such an amplitude to be topological (i.e. an invariant of framed links) we want it to be invariant under the following list of local moves on the diagrams. These moves are an augmented list of the Reidemeister moves, adjusted to take care of the fact that the diagrams are arranged with respect to a given direction in the plane.

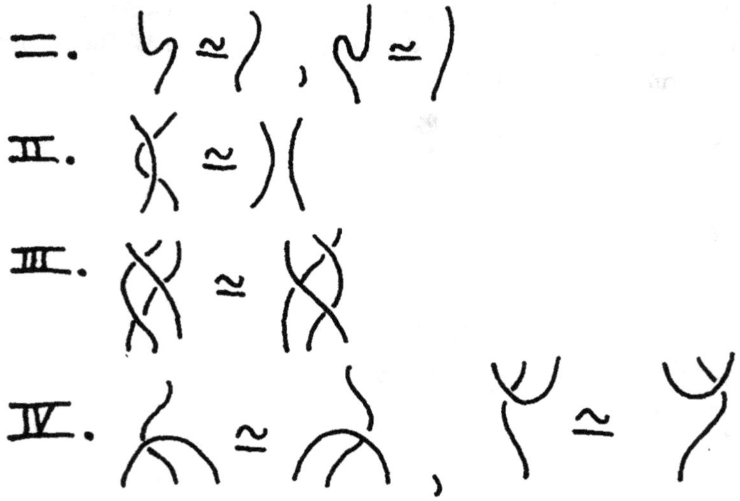

As the reader can see, we have already discussed the algebraic meaning of moves 0. and 2. The other moves translate into very interesting algebra. Move 3. is the famous Yang-Baxter equation (See [BA82],[TU87],[JO89]) that occurred for the first time in problems of exactly solved models in statistical mechanics.
All the moves taken together are directly related to the axioms for a quasi-triangular Hopf algebra (aka quantum group). We shall not go into this connection here (but see [RT90],[RT91],[K93]).

There is an intimate connection between knot invariants and the structure of generalized amplitudes. The strategy for the construction of such invariants is directly motivated by the concept of an amplitude in quantum mechanics. It turns out that the invariants that can actually be produced by this means are incredibly rich. They encompass, at present, all of the known invariants of polynomial type (Alexander polynomial, Jones polynomial and their generalizations.). We shall return to the specifics of the construction of the Jones polynomial via the bracket polynomial model in section 5, but it is now possible to indicate this model as an amplitude by specifying its matrices.

They are $(M_{ab}) = (M^{ab}) = M$ where M is the 2×2 matrix (with $ii=-1$)

$$M = \begin{pmatrix} 0 & iA \\ -iA^{-1} & 0 \end{pmatrix}$$

Note that $MM = I$ where I is the identity matrix. Note also that the amplitude for the circle is

$$\Sigma\, M_{ab}M^{ab} = \Sigma M_{ab}M_{ab} = \Sigma\, (M_{ab})^2$$

$$= (iA)^2 + (-iA^{-1})^2 = -A^2 - A^{-2}.$$

The matrix R is then defined by the equation

$$R^{ab}{}_{cd} = AM^{ab}M_{cd} + AI^{a}{}_{c}I^{b}{}_{d}\,,$$

or diagrammatically by

$$R = A \;\genfrac{}{}{0pt}{}{\smile}{\frown}\; + A^{-1}\;)($$

Since, diagrammatically, we identify R with the crossing this equation can be written symbolically as

$$\times = A \;\genfrac{}{}{0pt}{}{\smile}{\frown}\; + A^{-1}\;)($$

Taken together with the loop value of $-A^2 - A^{-2}$, this equation can be regarded as a recursive algorithm for computing the amplitude. In this form it is the bracket state model for the (unnormalized)

Jones polynomial [KA87]. This model can be studied on its own grounds, and we shall use it in later sections of this paper.

The Life of Notation and the Feynman Integral

Dirac notation has a life of its own.
Let $\mathbf{P} = |y\rangle\langle x|$.
Let $\langle x|\ |y\rangle = \langle x|y\rangle$.
Then $\mathbf{PP} = |y\rangle\langle x|\ |y\rangle\langle x| = |y\rangle\ \langle x|y\rangle\ \langle x| = \langle x|y\rangle\ \mathbf{P}$.
Up to a scalar multiple, **P** is a projection operator.
In this language, the completeness of intermediate states becomes the statement that a certain sum of projections is equal to the identity: Suppose that $\Sigma_i\ |ci\rangle\langle ci| = 1$ (summing over i). Then

$$\langle a|b\rangle = \langle a|\ |b\rangle = \langle a|\ \Sigma_i\ |ci\rangle\langle ci|\ |b\rangle = \Sigma_i\ \langle a|\ |ci\rangle\langle ci|\ |b\rangle$$
$$= \Sigma_i\ \langle a|ci\rangle\langle ci|b\rangle.$$

Iterating this principle of expansion over a complete set of states leads to the most primitive form of the Feynman integral [Fey65]. Imagine that the initial and final states a and b are points on the vertical lines x=0 and x=n+1 respectively in the x-y plane, and that (c(k)i(k) , k) is a given point on the line x=k for 0<i(k)<m. Suppose that the sum of projectors for each intermediate state is complete. That is, we assume that following sum is equal to one, for each k from 1 to n-1: $|c(k)1\rangle\langle c(k+1)1| + \dots + |c(k)m\rangle\langle c(k+1)m| = 1$.

Applying the completeness iteratively, we obtain the following expression for the amplitude $\langle a|b\rangle$:

$$\langle a|b\rangle = \Sigma\Sigma\dots\Sigma\ \langle a|c(1)i(1)\rangle\langle c(2)i(2)|c(3)i(3)\rangle\ \dots\ \langle c(n)i(n)|b\rangle$$

where the sum is taken over all i(k) ranging between 1 and m, and k ranging between 1 and n. Each term in this sum can be construed as a combinatorial path from a to b in the two dimensional space of the x-y plane. Thus the amplitude for going from a to b is seen as a summation of contributions from all the "paths" connecting a to b. Feynman used this description to produce his famous path integral expression for amplitudes in quantum mechanics. His path integral takes the form

$$\int dP \ \exp(iS)$$

where i is the square root of minus one, the integral is taken over all paths from point a to point b, and S is the *action* for a particle to travel from a to b along a given path. For the quantum mechanics associated with a classical (Newtonian) particle the action S is given by the integral along the given path from a to b of the difference T-V where T is the classical kinetic energy and V is the classical potential energy of the particle.

The beauty of Feynman's approach to quantum mechanics is that it shows the relationship between the classical and the quantum in a particularly transparent manner. Classical motion corresponds to those regions where all nearby paths contribute constructively to the summation. This classical path occurs when the variation of the action is null. To ask for those paths where the variation of the action is zero is a problem in the calculus of variations, and it leads directly to Newton's equations of motion. Thus with the appropriate choice of action, classical and quantum points of view are unified.

The drawback of this approach lies in the unavailability at the present time of an appropriate measure theory to support all cases of the Feynman integral.

To summarize, Dirac notation shows at once how the probabilistic interpretation for amplitudes is tied with the vector space structure of the space of states of the quantum mechanical system. Our strategy for bringing forth relations between quantum theory and topology is to pivot on the Dirac bracket. The Dirac bracket intermediates between notation and linear algebra. In a very real sense, *the connection of quantum mechanics with topology is an amplification of Dirac notation.*

Topological Quantum Field Theory - First Steps

In order to justify this idea of the amplification of notation, consider the following scenario. Let M be a 3-dimensional manifold. Suppose that F is a closed orientable surface inside M dividing M into two pieces M_1 and M_2. These pieces are 3-manifolds with boundary. They meet along the surface F. Now consider an amplitude $\langle M_1|M_2\rangle = Z(M)$. The form of this amplitude generalizes our previous considerations, with the surface F constituting the distinction between the "preparation" M_1 and the "detection" M_2. This generalization of the Dirac amplitude <a|b> amplifies the notational distinction consisting in the vertical line of the bracket to a topological distinction in a space M. The amplitude Z(M) will be said to be a *topological amplitude for M* if it is a toplogical invariant of the 3-manifold M. Note that a topological amplitude does not depend upon the choice of surface F that divides M.

From a physical point of view the independence of the topological amplitude on the particular surface that divides the 3-manifold is the most important property. An amplitude arises in the condition of one part of the distinction carved in the 3-manifold acting as "the observed" and the other part of the distinction acting as "the observer". If the amplitude is to reflect physical (read topological) information about the underlying manifold, then it should not depend upon this particular decomposition into observer and observed. The same remarks apply to 4-manifolds and interface with ideas in relativity. We mention 3-manifolds because it is possible to describe many examples of topological amplitudes in three dimensions. The matter of 4-dimensional amplitudes is a topic of current research. The notion that an amplitude be independent of the distinction producing it is prior to topology. Topological invariance of the amplitude is a convenient and fundamental way to produce such independence.

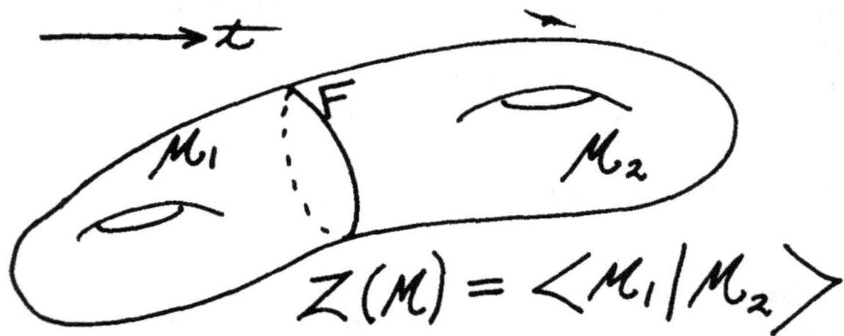

This sudden jump to topological amplitudes has its counterpart in mathematical physics. In [WIT89] Edward Witten proposed a formulation of a class of 3-manifold invariants as generalized Feynman integrals taking the form Z(M) where

$$Z(M) = \int dA\exp[(ik/4\pi)S(M,A)].$$

Here M denotes a 3-manifold without boundary and A is a gauge field (also called a qauge potential or gauge connection) defined on M. The gauge field is a one-form on M with values in a representation of a Lie algebra. The group corresponding to this Lie algebra is said to be the *gauge group* for this particular field. In this integral the "action" S(M,A) is taken to be the integral over M of the trace of the Chern-Simons three-form CS = AdA + (2/3)AAA. (The product is the wedge product of differential forms.)

Instead of integrating over paths, the integral Z(M) integrates over all gauge fields modulo gauge equivalence (See [AT79] for a discussion of the definition and meaning of gauge equivalence.) This generalization from paths to fields is characteristic of quantum field theory. Quantum field theory was designed in order to accomplish the quantization of electromagnetism. In quantum electrodynamics the classical entity is the electromagnetic field. The question posed in this domain is to find the value of an amplitude for starting with one field configuration and ending with another. What we call the photon is such a jump in the condition of the field. The analogue of all paths from point a to point b is "all fields from field A to field B".

Witten's integral Z(M) is, in its form, a typical integral in quantum field theory. In its content Z(M) is highly unusual. The formalism of the integral, and its internal logic supports the existence of a large class of topological invariants of 3-manifolds and associated invariants of knots and links in these manifolds.

The invariants associated with this integral have been given rigorous combinatorial descriptions (See [RT91],[TW91],[KM91],[LI91],[KL93].), but questions and conjectures arising from the integral formulation are still outstanding. (See [GA93].) Later in this paper we will describe one combinatorial approach to the invariants that corresponds to the gauge group SU(2). In this case it is possible to build the entire structure in an elementary fashion, using nothing but the combinatorics for diagrams of knots and links.

Links Again , and The Wilson Loop

We now look at the formalism of the Witten integral in more detail and see how it implicates invariants of knots and links corresponding to each classical Lie algebra. In order to accomplish this task, we need to introduce a gadget from gauge theory - the *Wilson loop.* The Wilson loop is an exponentiated version of integrating the gauge field along a loop K in three space that we take to be an embedding (knot) or a curve with transversal self-intersections. For the purpose of this discussion, the Wilson loop will be denoted by the notation $\langle K|A\rangle$ to denote the dependence on the loop K and the field A. It is usually indicated by the symbolism $tr(Pexp(\int_K A))$. Thus $\langle K|A\rangle = tr(Pexp(\int_K A))$. Here the P denotes path ordered integration - that is we are integrating and exponentiating matrix valued functions, and so one must keep track of the order of the operations. The symbol tr denotes the trace of the resulting matrix.

With the help of the Wilson loop functional on knots and links, Witten [WIT89] writes down a functional integral for link invariants in a 3-manifold M:

$$Z(M,K) = \int dA \exp[(ik/4\pi)S(M,A)] \; tr(Pexp(\int_K A))$$
$$= \int dA \exp[(ik/4\pi)S] \; \langle K|A\rangle.$$

Here S(M,A) is the Chern-Simons Lagrangian, as in the previous discussion.
We abbreviate S(M,A) as S and write $\langle K|A\rangle$ for the Wilson loop. Unless otherwise mentioned, the manifold M will be the three-dimensional sphere S^3.

We shall now give an analysis the formalism of this functional integral that reveals quite a bit about its role in knot theory. This analysis depends upon some key facts relating the curvature of the gauge field to both the Wilson loop and the Chern-Simons Lagrangian. To this end, let us recall the local coordinate structure of the gauge field A(x), where x is a point in three-space. We can write $A(x) = A^a{}_k(x)T_a dx^k$ where the index a ranges from 1 to m with the Lie algebra basis $\{T_1, T_2, T_3, ..., T_m\}$. The index k goes from 1 to 3. For each choice of a and k, $A^a{}_k(x)$ is a smooth function defined on three-space. In A(x) we sum over the values of repeated indices. The Lie algebra generators T_a are actually matrices corresponding to a given representation of an abstract Lie algebra. We assume some properties of these matrices as follows:

1. $[T_a , T_b] = i f_{abc} T_c$ where $[x ,y] = xy - yx$, and f_{abc} (the matrix of structure constants) is totally antisymmetric. There is summation over repeated indices.

2. $tr(T_a T_b) = \delta_{ab}/2$ where δ_{ab} is the Kronecker delta ($\delta_{ab} = 1$ if a=b and zero otherwise).

We also assume some facts about curvature. (The reader may enjoy comparing with the exposition in [KA91]. But note the difference in conventions on the use of i in the Wilson loops and curvature definitions.) The first fact is the relation of Wilson loops and curvature for small loops:

Fact1. The result of evaluating a Wilson loop about a very small planar circle around a point x is proportional to the area enclosed by this circle times the corresponding value of the curvature tensor of the gauge field evaluated at x. The curvature tensor is written $F^a{}_{rs}(x)T_a dx^r dy^s$. It is the local coordinate expression of $dA + AA$.

Application of Fact 1. Consider a given Wilson line <K|S>. Ask how its value will change if it is deformed infinitesimally in the neighborhood of a point x on the line. Approximate the change according to Fact 1, and regard the point x as the place of curvature evaluation. Let δ<K|A> denote the change in the value of the line. δ<K|A> is given by the formula

$\delta\langle K|A\rangle = dx^r dx^s F^a{}_{rs}(x) T_a \langle K|A\rangle$. This is the first order approximation to the change in the Wilson line.

In this formula it is understood that the Lie algebra matrices T_a are to be inserted into the Wilson line at the point x, and that we are summing over repeated indices. This means that each $T_a\langle K|A\rangle$ is a new Wilson line obtained from the original line <K|A> by leaving the form of the loop unchanged, but inserting the matrix T_a into that loop at the point x. See the figure below.

Remark. In thinking about the Wilson line $\langle K|A\rangle = tr(P\exp(\int_K A))$, it is helpful to recall Euler's formula for the exponential: $e^x = \lim_{n\to\infty}(1+x/n)^n$.

The Wilson line is the limit, over partitions of the loop K, of products of the matrices $(1 + A(x))$ where x runs over the partition. Thus we can write symbolically,

$$\langle K|A\rangle = \prod_{x\in K}(1 + A(x)) = \prod_{x\in K}(1 + A^a{}_k(x)T_a dx^k).$$

It is understood that a product of matrices around a closed loop connotes the trace of the product. The ordering is forced by the one dimensional nature of the loop. Insertion of a given matrix into this product at a point on the loop is then a well-defined concept. If T is a given matrix then *it is understood that* $T\langle K|A\rangle$ *denotes the insertion of T into some point of the loop.* In the case above, it is understood from context of the formula $dx^r dx^s F^a{}_{rs}(x)T_a\langle K|A\rangle$ that the insertion is to be performed at the point x indicated in the argument of the curvature.

Remark. The previous remark implies following formula for the variation of the Wilson loop with respect to the gauge field:

$$\delta\langle K|A\rangle/\delta(A^a{}_k(x)) = dx^k T_a\langle K|A\rangle.$$

Varying the Wilson loop with respect to the gauge field results in the insertion of an infinitesimal Lie algebra element into the loop.

Proof. $\delta\langle K|A\rangle/\delta(A^a{}_k(x))$

$= \delta \prod_{y \varepsilon K}(1 + A^a{}_k(y)T_a dy^k)/\delta(A^a{}_k(x))$

$= [\prod_{y<x}(1 + A^a{}_k(y)T_a dy^k)] [T_a dx^k][\prod_{y>x}(1 + A^a{}_k(y)T_a dy^k)]$

$= dx^k T_a \langle K|A\rangle$. QED.

Fact2. The variation of the Chern-Simons Lagrangian S with respect to the gauge potential at a given point in three-space is related to the values of curvature tensor at that point by the following formula (where ε_{abc} is the epsilon symbol for three indices):

$$F^a{}_{rs}(x) = \varepsilon_{rst}\ \delta S/\delta(A^a{}_t(x)).$$

With these facts at hand we are prepared to determine how the Witten integral behaves under a small deformation of the loop K.

Proposition 1. (Compare [KA91].)
All statements of equality in this proposition are up to order $(1/k)^2$.
1. Let $Z(K) = Z(K, S^3)$ and let $\delta Z(K)$ denote the change of $Z(K)$ under an infinitesimal change in the loop K. Then
$\delta Z(K) = (4\pi i/k) \int dA \ \exp[(ik/4\pi)S][\varepsilon_{rst}dx^r dy^s dz^t]T_a T_a \langle K|A\rangle$.
The sum is taken over repeated indices, and the insertion is taken of the matrix products $T_a T_a$ at the chosen point x on the loop K that is regarded as the "center" of the deformation. The volume element $[\varepsilon_{rst}dx^r dy^s dz^t]$ is taken with regard to the infinitesimal directions of the loop deformation from this point on the original loop.
2. The same formula applies, with a different interpretation, to the case where x is a double point of transversal self intersection of a loop K, and the deformation consists in shifting one of the crossing segments perpendicularly to the plane of intersection so that the self-intersection point disappears. In this case, one T_a is inserted into each of the transversal crossing segments so that $T_a T_a \langle K|A\rangle$ denotes a Wilson loop with a self intersection at x and insertions of

T_a at $x+\varepsilon_1$ and $x+\varepsilon_2$ where ε_1 and ε_2 denote small displacements along the two arcs of K that intersect at x.. In this case, the volume form is nonzero, with two directions coming from the plane of movement of one arc, and the perpendicular direction is the direction of the other arc.

Proof.

$\delta Z(K) = \int dA \ \exp[(ik/4\pi)S] \ \delta<K|A>$

$= \int dA \ \exp[(ik/4\pi)S] dx^r dy^s \ F^a{}_{rs}(x) \ T_a<K|A>$ (Fact 1)

$= \int dA \ \exp[(ik/4\pi)S] \ dx^r dy^s \ \varepsilon_{rst} \ \delta S/\delta(A^a{}_t(x)) \ \ T_a<K|A>$
(Fact 2)

$= \int dA \ \{\exp[(ik/4\pi)S] \ \delta S/\delta(A^a{}_t(x))\} \ \ \varepsilon_{rst} \ dx^r dy^s \ T_a<K|A>$

$= (-4\pi i/k)\int d\,A \ \delta\{\exp[(ik/4\pi)S]\}/\delta(A^a{}_t(x)) \ \ \varepsilon_{rst} \ dx^r dy^s T_a<K|A>$

$= (4\pi i/k)\int d\,A \ \exp[(ik/4\pi)S] \ \ \varepsilon_{rst} \ dx^r dy^s \ \delta\{T_a<K|A>\}/\delta(A^a{}_t(x))$

(integration by parts)

$= (4\pi i/k)\int d\,A \ \exp[(ik/4\pi)S] \ [\ \varepsilon_{rst} \ dx^r dy^s dz^t] \ T_a T_a<K|A>.$

(differentiating the Wilson line).

This completes the formalism of the proof. In the case of part 2., the change of interpretation occurs at the point in the argument when the Wilson line is differentiated. Differentiating a self intersecting Wilson line at a point of self intersection is equivalent to differentiating the corresponding product of matrices at a variable that occurs at two points in the product (corresponding to the two places where the loop passes through the point). One of these derivatives gives rise to a term with volume form equal to zero, the other term is the one that is described in part 2. This completes the proof of the proposition. //

Applying Proposition 1.

As the formula of Proposition 1 shows, the integral Z(K) is unchanged if the movement of the loop does not involve three independent space directions (since $\varepsilon_{rst}dx^r dy^s dz^t$ computes a volume.). This means that $Z(K) = Z(S^3, K)$ is invariant under moves that slide the knot along a plane. In particular, this means that if the knot K is given in the nearly planar representation of a knot diagram. Then Z(K) is invariant under regular isotopy of this diagram. That is , it is invariant under the Reidemeister moves II and III. We expect more complicated behaviour under move I since this deformation does involve three spatial directions. This will be discussed momentarily.

We first determine the difference between Z(K+) and Z(K-) where K+ and K- denote knots that differ only by switching a single crossing. We take the given crossing in K+ to be of positive type as indicated below, and the crossing in K- to be of negative type.

The strategy for computing this difference is to use K# as an intermediate, where K# *is the link with a transversal self-crossing replacing the given crossing in* K+ *or* K-. Thus we must consider $\Delta+ = Z(K+) - Z(K\#)$ and $\Delta- = Z(K-) - Z(K\#)$. The second part of Proposition 1 applies to each of these differences and gives

$$\Delta+ = (4\pi i/k)\int dA \exp[(ik/4\pi)S] \; [\varepsilon_{rst} \; dx^r dy^s \; dz^t] \, T_a T_a \langle K\#|A\rangle$$

where, by the description in Proposition 1, this evaluation is taken along the loop K# with the singularity and the T_aT_a insertion occurs along the two transversal arcs at the singular point. The sign of the volume element will be opposite for Δ- and consequently we have that

$$(\Delta+) + (\Delta-) = 0.$$

[The volume element $[\epsilon_{rst}\, dx^r dy^s dz^t]$ *must be given a conventional value in our calculations. There is no reason to assign it different absolute values for the cases of* Δ+ *and* Δ- *since they are symmetric except for the sign.]*

Therefore $Z(K+) - Z(K\#) + (Z(K-) - Z(K\#)) = 0$. Hence

$$Z(K\#) = (1/2)(Z(K+) + Z(K-)).$$

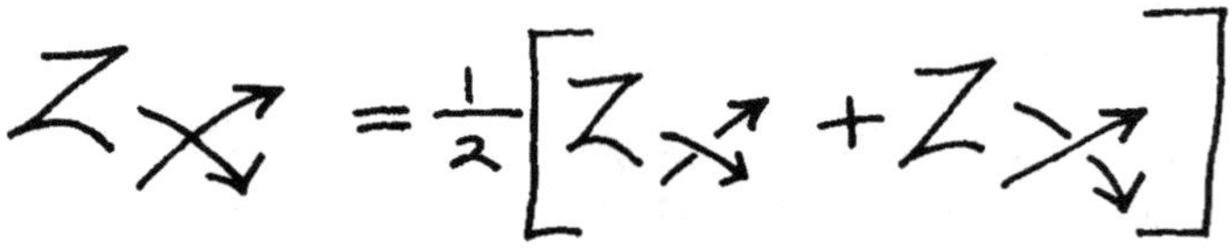

This result is central to our further calculations. It tells us that the evaluation of a singular Wilson line can be replaced with the average of the results of resolving the singularity in the two possible ways.

Now we are interested in the difference $Z(K+) - Z(K-)$:

$$Z(K+) - Z(K-) = \Delta+ - \Delta- = 2\Delta+$$
$$= (8\pi i/k)\int dA\, \exp[(ik/4\pi)S]\ [\epsilon_{rst}\ dx^r dy^s dz^t]\, T_aT_a \langle K\#|A\rangle$$

Volume Convention. It is useful to make a specific convention about the volume form.

We take $[\epsilon_{rst}\ dx^r dy^s\ dz^t] = 1/2$ for D+ and -1/2 for D-.

Thus $Z(K+) - Z(K-) = (4\pi i/k)\int dA\, \exp[(ik/4\pi)S]\, T_aT_a \langle K\#|A\rangle.$

Integral Notation. Let $Z(T_aT_aK\#)$ denote the integral

$$Z(T_aT_aK\#) = \int dA \exp[(ik/4\pi)S]\, T_aT_a\langle K\#|A\rangle.$$

Difference Formula. Write the difference formula in abbreviated form

$$Z(K+) - Z(K-) = (4\pi i/k)\, Z(T_aT_aK\#).$$

This formula is the key to unwrapping many properties of the knot invariants. The rest of this section will be devoted to discussion of these matters. For diagrammatic work it is convenient to rewrite the difference equation in the form shown below. The crossings denote small parts of otherwise identical larger diagrams, and the Casimir insertion $T_aT_aK\#$ is indicated with crossed lines entering a disk labelled C.

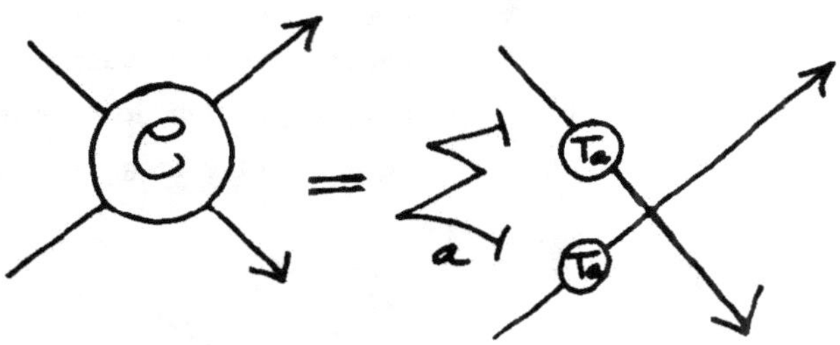

The Casimir

The Lie algebra element $\Sigma_a T_aT_a$ is called the Casimir. Its key property is that it is in the center of the algebra and so can have common eigenvalues with other elements. Note that by our conventions

$tr(\Sigma_a T_aT_a) = \Sigma_a \delta_{aa}/2 = d/2$ where d *is the dimension of the Lie algebra.* This implies that an unknotted loop with one singularity and a Casimir insertion will have Z-value $d/2$.

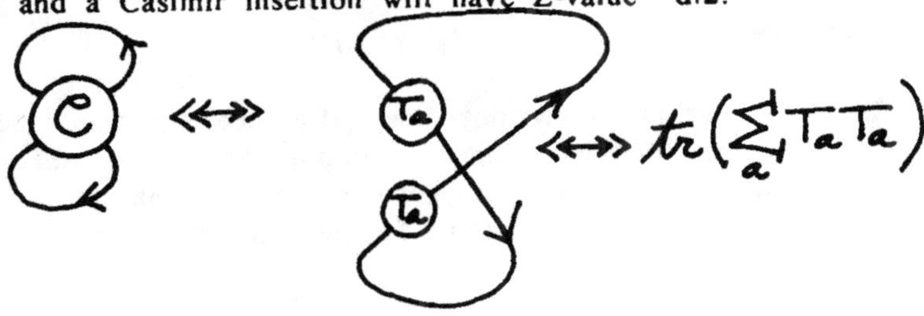

In fact, for the classical semi-simple Lie algebras *one can choose a basis so that the Casimir is a diagonal matrix with identical values* (d/2D) *on its diagonal.* D is the dimension of the representation. We then have the general formula: $Z(T_aT_aK\#^{loc}) = (d/2D)\ Z(K)$ for any knot K. Here $K\#^{loc}$ denotes the singular knot obtained by placing a local self-crossing loop on K as shown below.

Note that $Z(K\#^{loc}) = Z(K)$. (Let the flat loop shrink to nothing. Wilson line is still defined on a loop with an isolated cusp and it is equal to the Wilson loop obtained by smoothing that cusp.)

Let K_+^{loc} denote the result of adding a positive local curl to the knot K, and K_-^{loc} the result of adding a negative local curl to K.

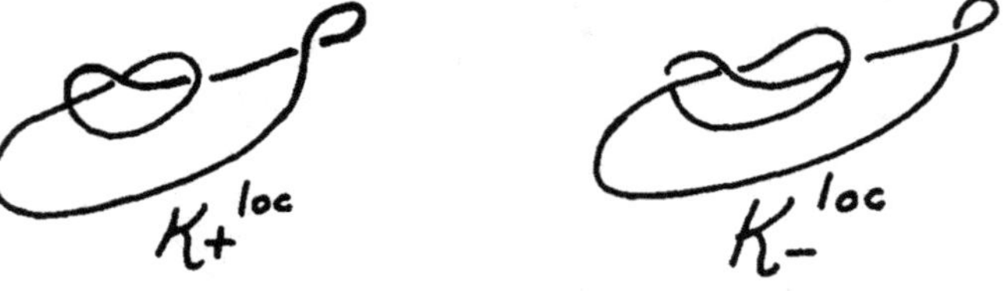

Then by the above discussion and the difference formula, we have

$Z(K_+^{loc}) = Z(K\#^{loc}) + (2\pi i/k)\ Z(T_aT_aK\#^{loc})$

$= Z(K) + (2\pi i/k)(d/2D)\ Z(K).$

Thus, $Z(K_+^{loc}) = (1 + (\pi i/k)(d/D))\ Z(K).$

Similarly, $Z(K_-^{loc}) = (1 - (\pi i/k)(d/D))\ Z(K).$

These calculations show how the difference equation, the Casimir, and properties of Wilson lines determine the framing factors for the knot invariants. In some cases we can use special properties of the Casimir to obtain skein relations for the knot invariant.

For example, in the fundamental representation of the Lie algebra for SU(N) the Casimir obeys the following equation (See [KA91],[BN92]):

$$\Sigma_a (T_a)_{ij}(T_a)_{kl} = (1/2)\delta_{il}\delta_{jk} - (1/2N)\delta_{ij}\delta_{kl}$$

Hence $Z(T_aT_aK\#) = (1/2)Z(Ko) - (1/2N)\ Z(K\#)$ where Ko denotes the result of smoothing a crossing as shown below:

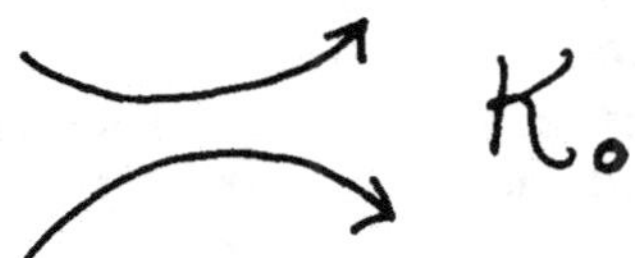

Using $Z(K\#) = (Z(K+) + Z(K-))/2$ and the difference identity, we obtain

$$Z(K+) - Z(K-)$$
$$= (4\pi i/k)\{(1/2)Z(Ko) - (1/2N)[(Z(K+) + Z(K-))/2]\}.$$

Hence
$$(1 + \pi i/Nk)Z(K+) - (1 - \pi i/Nk)Z(K-) = (2\pi i/k)Z(Ko)$$

or $e(1/N)Z(K+) - e(-1/N)Z(K-) = \{e(1) - e(-1)\}Z(Ko)$

where $e(x) = \exp((\pi i/k)x)$ taken up to $O(1/k^2)$.

Here $d = N^2 -1$ and $D = N$, so the framing factor is
$\alpha = (1 + (\pi i/k)((N^2 -1)/N)) = e((N -(1/N))$.

Therefore, if $P(K) = \alpha^{-w(K)}Z(K)$ denotes the normalized invariant of ambient isotopy associated with $Z(K)$ (with $w(K)$ the sum of the crossing signs of K), then

$$\alpha e(1/N)P(K+) - \alpha^{-1} e(-1/N)P(K-) = \{e(1) - e(-1)\}P(Ko).$$

Hence $e(N)P(K+) - e(-N)P(K-) = \{e(1) - e(-1)\}P(Ko)$.

This last equation shows that **P(K)** is a specialization of the Homfly polynomial for arbitrary N, and that for N=2 (SU(2)) it is a specialization of the Jones polynomial.

Graph Invariants and Vassiliev Invariants

Our last application of this integral formalism is to the structure of rigid vertex graph invariants that arise naturally in this context of knot polynomials. If V(K) is a (Laurent polynomial valued, or more generally - commutative ring valued) invariant of knots, then it can be naturally extended to an invariant of rigid vertex graphs by defining the invariant of graphs in terms of the knot invariant via an "unfolding" of the vertex as indicated below ([KV92]):

$$V(K\$) = aV(K+) + bV(K-) + cV(Ko)$$

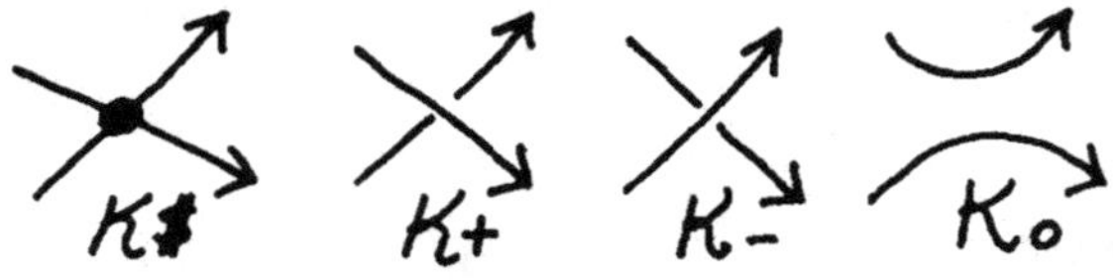

Here **K$** indicates an embedding with a transversal 4-valent vertex ($). We use the symbol $ to distinguish this choice of vertex designation from the previous one we used involving a self-crossing Wilson line.

Formally, this means that we define **V(G)** for an embedded 4-valent graph G by taking the sum over $a^{i+(S)}b^{i-(S)}c^{io(S)}V(S)$ for all knots S obtained from G by replacing a node of G with either a crossing of positive or negative type, or with a smoothing (denoted o). It is not hard to see that if V(K) is an ambient isotopy invariant of knots, then, this extension is an rigid vertex isotopy invariant of graphs. In rigid vertex isotopy the cyclic order at the vertex is preserved, so that the vertex behaves like a rigid disk with flexible strings attached to it at specific points.

There is a rich class of graph invariants that can be studied in this manner. The *Vassiliev Invariants* ([V90],[BL91],[BN92]) constitute the important special case of these graph invariants where a=+1, b=-1 and c=0. Thus V(G) is a Vassiliev invariant if

$$V(K\$) = V(K+) - V(K-).$$

V(G) is said to be of *finite type k if V(G) = 0 whenever #(G) >k* where #(G) denotes the number of 4-valent nodes in the graph G.

With this definition in hand, lets return to the invariants derived from the functional integral Z(K). We have shown that

$Z(K+^{loc}) = \alpha Z(K)$ with $\alpha = e(d/D)$.

Hence $P(K) = \alpha^{-w(K)}Z(K)$ is an ambient isotopy invariant. The equation $Z(K+) - Z(K-) = (4\pi i/k)Z(T_a T_a K\#)$ implies that if $w(K+) = w +1$, then we have the *ambient isotopy difference formula:*

$$P(K+) - P(K-) = \alpha^{-w}(4\pi i/k)\{Z(T_a T_a K\#) - (d/2D)Z(K\#)\}.$$

We leave the proof of this formula as an exercise for the reader.

This formula tells us that for the Vassiliev invariant associated with P we have $P(K\$) = \alpha^{-w}(4\pi i/k)\{Z(T_a T_a K\#) - (d/2D)Z(K\#)\}$.

Furthermore, if $V_j(K)$ denotes the coefficient of $(4\pi i/k)^j$ in the expansion of P(K) in powers of (1/k), then the ambient difference formula implies that $(1/k)^j$ divides P(G) when G has j or more nodes. Hence $V_j(G) = 0$ if G has more than j nodes. Therefore $V_j(K)$ is a Vassiliev invariant of finite type. (This result was proved by Birman and Lin [BL91] by different methods and by Bar-Natan [BN92] by methods equivalent to ours.)

The fascinating thing is that the ambient difference formula , appropriately interpreted, actually tells us how to compute $V_k(G)$ when G has k nodes. Under these circumstances each node undergoes a Casimir insertion, and because the Wilson line is being evaluated abstractly, independent of the embedding, this suggests that we insert nothing else into the line. Thus we take the pairing structure associated with the graph (the so-called chord diagram) and use it as a prescription for obtaining a trace of a sum of products of Lie algebra elements with T_a and T_a inserted for each pair or a simple crossover for the pair multiplied by (d/D). This yields the graphical evaluation implied by the recursion

$$V(G\$) = \{V(T_a T_a G\#) - (d/2D)V(G\#)\}.$$

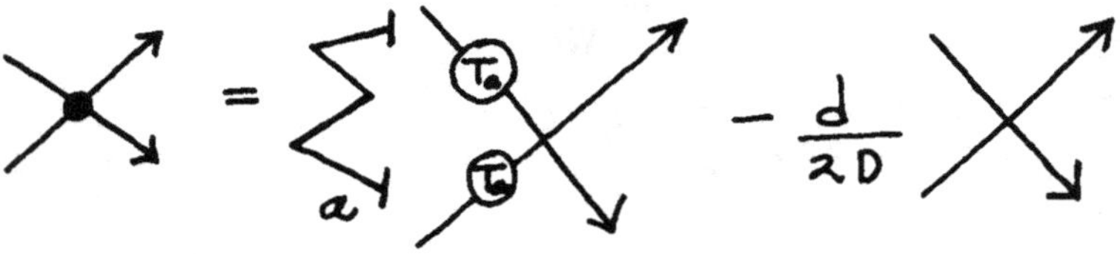

At each stage in the process one node of G disappears or it is replaced by these insertions. After k steps we have a fully inserted sum of abstract Wilson lines, each of which can be evaluated by taking the indicated trace. This result is equivalent to Bar-Natan's result, but it is very interesting to see how it follows from a minimal approach to the Witten integral.

In particular, it follows from Bar-Natan[BN92] and Kontsevich [KO92] that the condition of topological invariance is translated into the fact that the Lie bracket is represented as a commutator *and* that it is closed with respect to the Lie algebra. Diagrammatically we have

$$T_a T_b - T_b T_a = i f_{abc} T_c$$

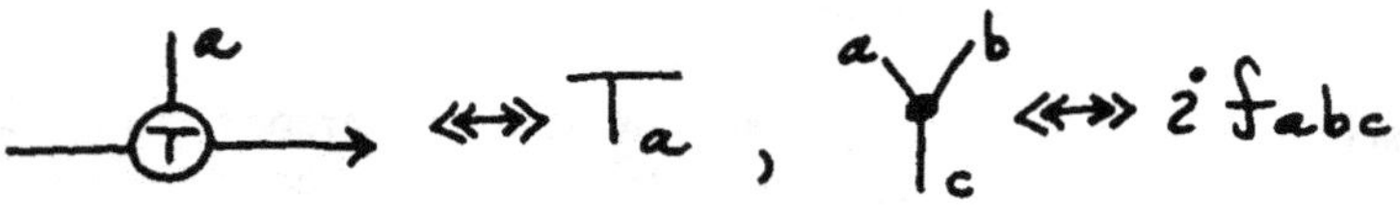

Hence

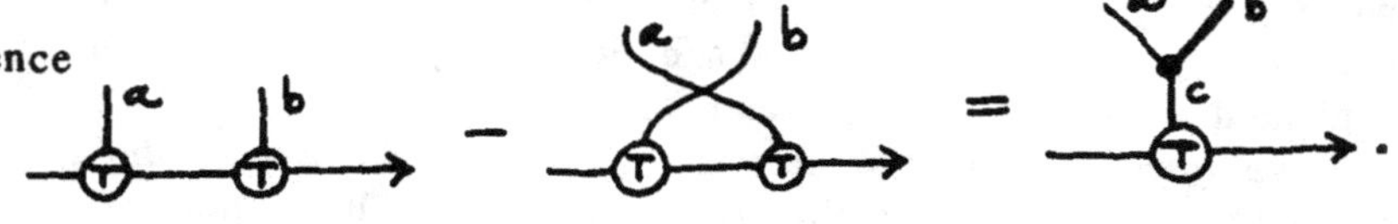

This relationship on chord diagrams is the seed of all the topology. In particular, it implies the basic 4-term relation:

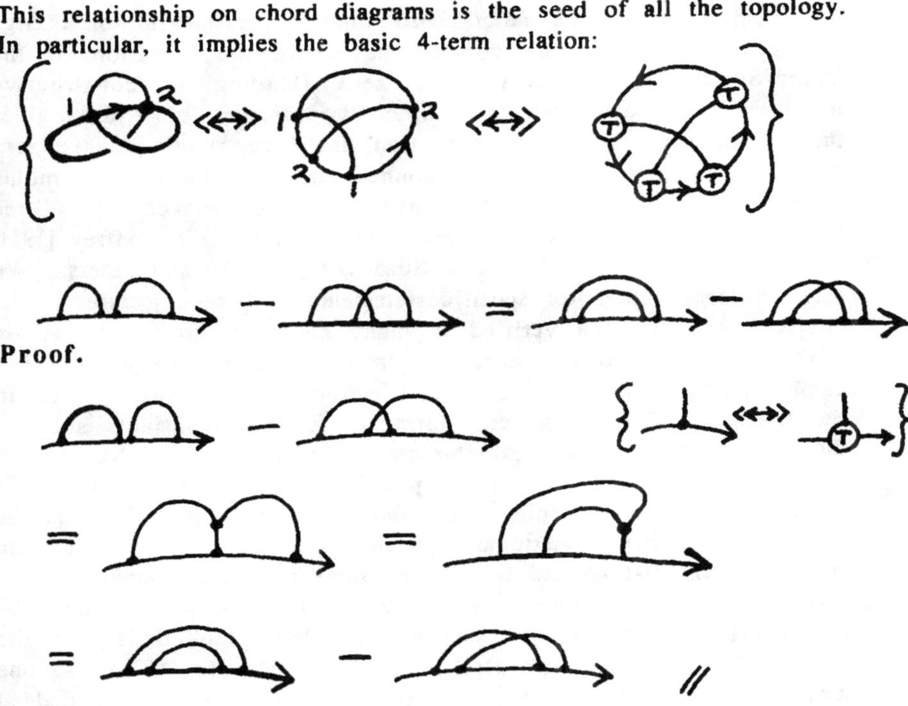

The presence of this relation on chord diagrams for $V_i(G)$ with $\#(G) = i$ is the basis for the existence of a corresponding Vassiliev invariant. There is not room here to go into more detail about this matter, and so we shall bring this discussion to a close. Nevertheless, it must be mentioned that this brings us to the core of the main question about Vassiliev invariants: Are there non-trivial Vassiliev invariants of knots and links that cannot be constructed through combinations of Lie algebraic invariants? There are many other open questions in this arena, all circling this basic problem.

Conjectures Related to the Functional Integral

It should be mentioned that there are a number of conjectures about link invariants and 3-manifold invariants that arise from the formalism of the functional integral that is seen on taking k very large. This is the so-called stationary phase approximation to the functional integral. The major contribution to the integral for k large comes from those gauge connections where the variation of the Chern-Simons action S(M,A) is zero (leading to constructive interference in the summation). Since the variation of S(M,A) gives the curvature of the gauge connection, these contributions have zero curvature. They are the flat connections. Asymptotic formulas involving summing over flat connections are derived by Witten [WIT89] and studied by Freed and Gompf [FG91], Lisa Jeffrey [J91], Stavros Garoufalidis [GA93], Lev Rozansky [RO93] and others. We refer to these papers for specific statements of the conjectures. The conjectures have been verified in many cases by direct comparison with exact formulas derived from specific examples via combinatorial techniques. Full verification remains elusive even in the presence of exact general formulas for the invariants such as those given in section 7 of this paper in the case of SU(2). A complete resolution of this problem would have fantastic consequences. For example, Garoufalidis shows in [GA93] that the strong form of the stationary phase conjecture implies that the Homfly polynomial applied to all the cables of a given embedding of a loop in Euclidean three-space detects whether that loop is knotted! (A stronger conjecture that has not yet been faulted is that the original Jones polynomial itself detects knottedness. That is, one may conjecture that if the Jones polynomial applied to an embedded loop in 3-space is equal to 1, then the loop is unknotted.)

Caboose

We have, in this first section, given an almost unbroken line of argument from the beginnings of quantum mechanics to the construction of topological quantum field theories, link invariants associated with the classical Lie algebras, and the structure of the Vassiliev invariants. While this approach using the formalism of the functional integral gives a direct route into the heart of the subject, it involves a significant number of leaps of faith. These leaps are slowly being filled by rigorous mathematics. In the remaining sections of this essay we will look at one brand of this rigor in the form of combinatorial knot theory and the use of surgery on manifolds.

III. The Temperley Lieb Algebra

We begin with the combinatorial underpinnings of the Temperley Lieb algebra.

First recall the tangle-theoretic interpretation of the Temperley-Lieb Algebra [KA87]. In this interpretation, the additive generators of the algebra are *flat tangles* with equal numbers of inputs and outputs. We denote by T_n the (Temperley Lieb) algebra generated by flat tangles with n inputs and n outputs. A flat n-tangle is a an *embedding* of disjoint curves and line segments into the plane so that the free ends of the segments are in one-to-one correspondence with the input and output lines of a rectangle in the plane that is denoted the *tangle box.* Except for these inputs and outputs, the disjoint curves and line segments are in the interior of the rectangle.

Two such tangles are *equivalent* if there is a regular isotopy carrying one to the other occurring within the rectangle and keeping the endpoints fixed. Regular isotopy is generated by the Reidemeister Moves of type II and type III for link diagrams. (See [KA87].) The reason we adhere to regular isotopy at this point is that it is necessary to be able to freely move closed curves in such a tangle.

The Temperley Lieb algebra T_n is freely additively generated by the flat n-tangles, over the ring $C[A,A^{-1}]$ where C denotes the complex numbers. A closed loop in a tangle is identified with a central element d in this algebra to be specified later. The familiar multiplicative generators of the Temperley Lieb algebra then appear as the following special flat tangles $U_1, U_2, \ldots, U_{n-1}$ in T_n:

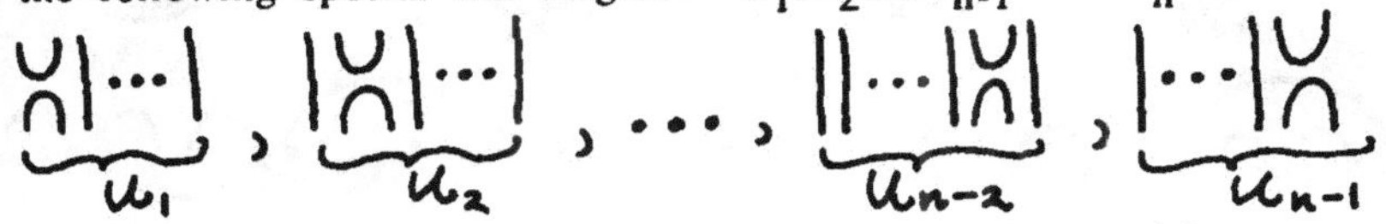

U_1, U_2, ..., U_{n-2}, U_{n-1}

These generators enjoy the relations shown above in diagrams and below in algebra.

$$U_iU_{i+1}U_i = U_i$$
$$U_iU_{i-1}U_i = U_i$$
$$U_iU_i=dU_i$$
$$U_iU_j=U_jU_i \text{ for } |i-j|>1$$

These relations generate equivalence of flat tangles [KA90].

Note that we can convert flat n-tangles to forms of parenthesization by bending the upper ends downward and to the left as illustrated below.

Call such a structure a *capform.*
In this way we convert the tangles to capforms with 2n strands restricted to the bottom of the form. These are capforms with n caps. The tangle multiplication then takes the form of tying the rightmost n strands of the left capform to the leftmost n strands of the right capform.

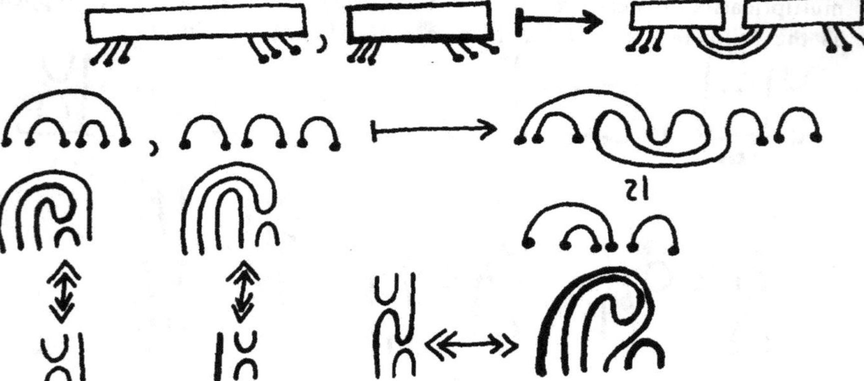

Certain structural features of the Temperley Lieb algebra are easy to see from the point of view of capforms. For example, consider the following natural map $T_n \times T_n$ -----> T_{n+1} given by the formula A,B -----> A*B = A $|_{n-1}$ B where A $|_{n-1}$ B denotes the result of joining only (n-1) strands at the base.

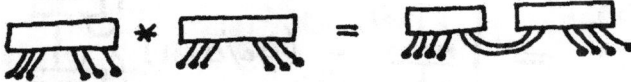

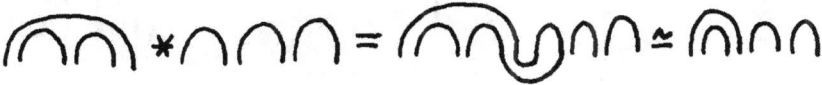

Theorem. Every element in T_{n+1} other than the identity element is of the form A*B = A $|_{n-1}$ B for some elements A and B in T_n. This decomposition is not unique.

Proof. Exercise.//

Another operation for going from T_n to T_{n+1} is x -----> x' where x' is obtained by adding an innermost cap as in

The operation x -----> x' takes the identity to the identity, and so together with x*y encompasses all of T_{n+1} from T_n. This suggests combining these operations to produce inductive constructions in the Temperley Lieb algebra.

An example that fits this idea is the well known ([JO83],[KA91], [LI91]) inductive construction of the Jones-Wenzl projectors. In tangle language these projectors, g_n in T_n, are constructed by the recursion $g_{n+2} = g_{n+1} - (\Delta_n/\Delta_{n+1})\, g_{n+1} U_1 g_{n+1}$,

$g_1 = 1$. Note that T_n is included in T_{n+1} by adding an extra line to the tangle.

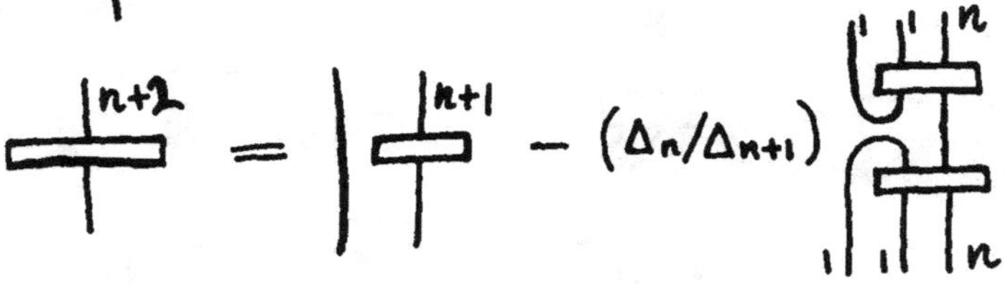

$$= g_n \in T_n$$

$$= \quad - (\Delta_n/\Delta_{n+1})$$

Here Δ_n is a Chebyschev polynomial defined recursively as shown below.

$$\Delta_{n+1} = d\Delta_n - \Delta_{n-1}$$
$$\Delta_0 = 1$$
$$\Delta_1 = d$$

These projectors are nontrivial idempotents in the Temperley Lieb algebra T_n, and they give zero when multiplied by the generators U_i for $i=1,\ldots,n-1$.

Example.

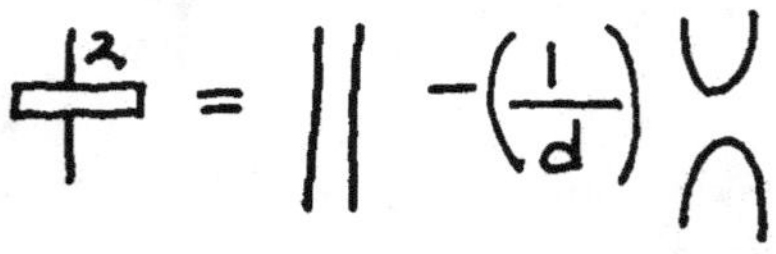

$$= \; || \; - \left(\frac{1}{d}\right)$$

Now note the capform interpretation of the projector recursion.

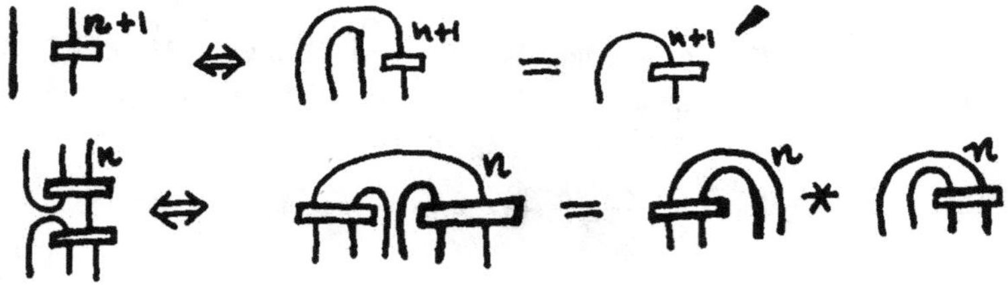

In the capform algebra the projectors are constructed via the recursion

$$g_{n+2} = g_{n+1}' \quad - \quad (\Delta_n/\Delta_{n+1})\, g_{n+1} \;*\; g_{n+1}.$$

We shall return to these projectors in the next two sections.

IV. Temperley Lieb Recoupling Theory

By using the Jones-Wenzl projectors, one builds a recoupling theory for the Temperley Lieb algebra that is essentially a version of the recoupling theory for the SL(2)q quantum group. (See [KR88],[KAU92],[KL90], [KL92]). From the vantage of this theory it is easy to construct the Witten-Reshetikhin-Turaev invariants of 3-manifolds.

We begin by recalling the basics of the recoupling theory. The 3-vertex in this theory is built from three interconnected projectors (see section 3) in the pattern indicated below.

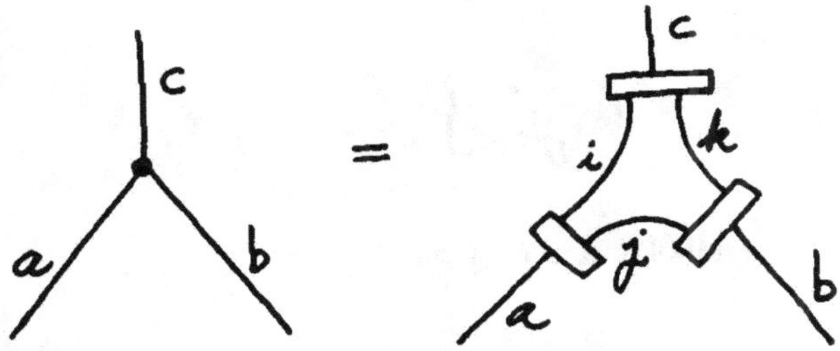

The internal lines must add up correctly and this forces the sum of the external lines to be even and it also forces the sum of any two external line numbers to be greater than or equal to the third. This condition on the lines is called *admissibility*. In the case where the loop value in the Temperley-Lieb algebra is $d = -A^2 - A^{-2}$ and A is a 4r-th root of unity $(A=\exp(\pi i/2r))$, we need the extra condition at each 3-vertex that a+b+c <= 2r-4 where a,b and c denote the multiplicity of the lines incident to the vertex.. See [KL93]. The term *admissible* will refer to either the generic case described above, or to the case with a+b+c < 2r-3 when A is a 4r-th root of unity.

With these 3-vertices, we have a recoupling formula, valid generically for sums over admissibles. In the case of the root of unity, the sums are again over admissibles, but the multiplicities of the lines are restricted to the set {0,1,2,..., r-2}. The recoupling formula is indicated below.

Here the symbol

$$\begin{Bmatrix} a & b & i \\ c & d & j \end{Bmatrix}$$

is a generalized 6j symbol (a *q-6j symbol*).

A specific formula for the evaluation of this q-6j symbol arises as the consequence of the following identity (See [KA91], [Kau92],[KL93]):

From this identity it is easy to deduce that the q-6j symbol is given by the network evaluation shown below:

$$\begin{Bmatrix} a & b & i \\ c & d & j \end{Bmatrix} = \frac{\mathrm{Tet}\begin{bmatrix} b & c & \\ & j & i \\ a & d & \end{bmatrix} \Delta_i}{\theta(a,d,i)\,\theta(b,c,i)}$$

The key ingredients are the tetrahedral and theta nets. They, in turn, can be evaluated quite specifically. (See [KL93],[LI91],[MV92]) There are a number of methods for obtaining these specific evaluations. For the general case one can induct using the recursion formula for the Jones-Wenzl projectors. In the special case where d=-2 there is a method to obtain the results via counting loops and colorings of loops in the networks. See [PEN79], [MOU79], [KA91], [KL93]. It should be mentioned that the case d= -2 corresponds to the classical theory of SU(2) recoupling.

V. The Bracket Polynomial Model for the Jones Polynomial

There is a different path to this recoupling theory. That path begins with topology. Start with an identity in the form

$$\langle \times \rangle = A \langle \asymp \rangle + B \langle \supset\,\subset \rangle$$

$$\langle O K \rangle = d \langle K \rangle$$

and ask that it be topologically invariant in the sense of regular isotopy generated by the knot diagrammatic moves II and III.

We see at once that

$$\langle \; \rangle = AB \langle \; \rangle + (ABd + A^2 + B^2) \langle \; \rangle$$

Proof.

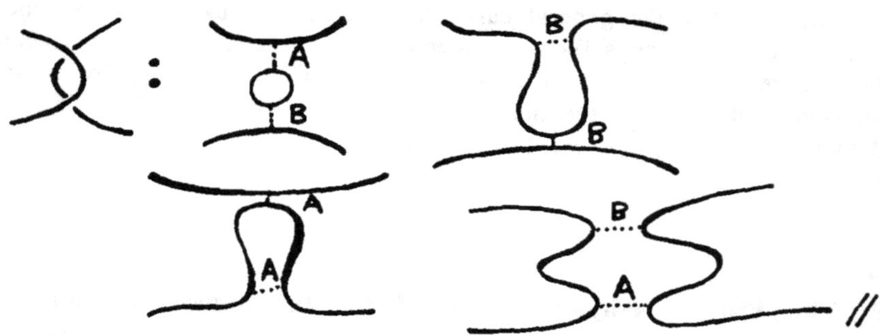

Hence we can achieve the invariance

$$\langle \; \rangle = \langle \; \rangle$$

by taking $B = A^{-1}$ and $d = -A^2 - A^{-2}$. Then, a miracle happens, and we are granted invariance under the triangle move with no extra restrictions:

$$\langle \; \rangle = A \langle \; \rangle + A^{-1} \langle \; \rangle$$

$$= A \langle \; \rangle + A^{-1} \langle \; \rangle = \langle \; \rangle .$$

Call this invariant the *bracket polynomial* [KA87].
Note that the bracket polynomial is not invariant under the first Reidemeister move. It should be regarded as an invariant of framed links, whose framing is expressed in the plane. We have the formulas

$$\langle \cdots \rangle = (-A^{3}) \langle \cdots \rangle$$
$$\langle \cdots \rangle = (-A^{-3}) \langle \cdots \rangle .$$

This allows normalization of the bracket by multiplication by a power of $(-A^3)$. Up to this normalization, the bracket gives a model for the original Jones polynomial [JO86]. The Jones polynomial is denoted $V_K(t)$. The precise relationship with the bracket is [KA87] that $V_K(t) = f_K(t^{-1/4})$ where $f_K(A) = (-A^3)^{-w(K)}\langle K\rangle(A)$ where $w(K)$ is the sum of the crossing signs of the oriented link K, and $\langle K\rangle$ is the bracket polynomial obtained by ignoring the orientation of K.

For braids, we see that the bracket provides a representation, ρ, of the Artin braid group into the Temperley Lieb algebra with loop value $d = -A^2 - A^{-2}$.

$$\rho(\sigma_i) = AU_i + A^{-1}$$
$$\rho(\sigma_i^{-1}) = A^{-1}U_i + A$$

The diagrams below should make the source of this representation transparent.

$$\sigma_1, \quad \sigma_2, \quad \sigma_3, \quad \ldots$$

$$\langle \sigma_i \rangle = \langle \cdots \rangle = A\langle \cdots \rangle + A^{-1}\langle \cdots \rangle = A\langle U_i \rangle + A^{-1}\langle \mathbb{1} \rangle .$$

The projectors in the Temperley Lieb algebra can then be construed as generalizations of the algebraic process of symmetrization. In algebra, symmetrization is accomplished by summing over all permutations of a set. Permutations are represented by tangle diagrams where the tangle lines connect points to permuted points. Thus the permutations on three letters are indicated by the tangles

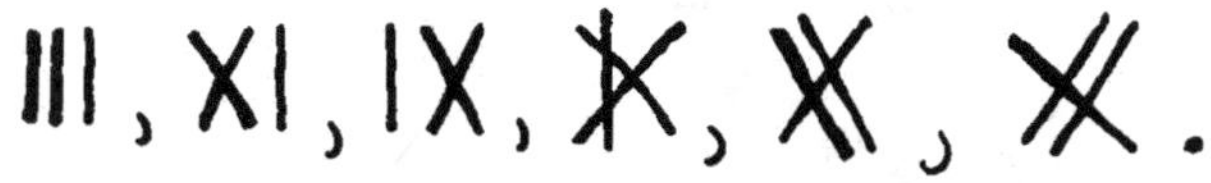

The reader will find it a pleasant exercise to check that in the case A=1 (loop value -2), the n-th Temperley Lieb projector is equal to the sum over all permutations for n lines, divided by n!, given that we apply the bracket identity to represent permutations into the Temperley Lieb algebra. For example,

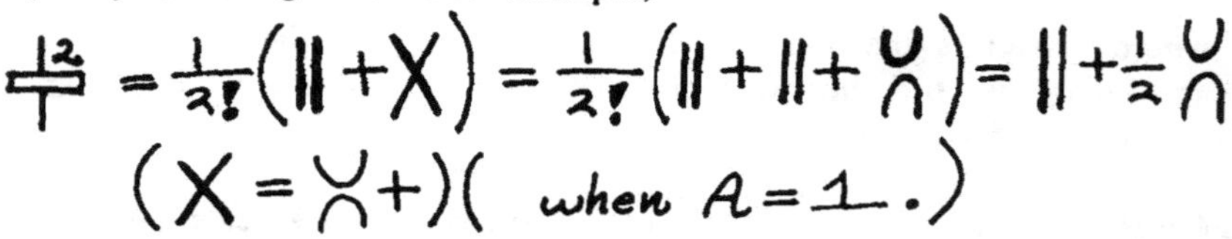

(when A = 1.)

For arbitrary A, a construction of projectors through symmetrization still exists by summing over all permutations σ, but replacing each permutation by a braid σ^ such that all the crossings of the braid are of the same type. More precisely, let

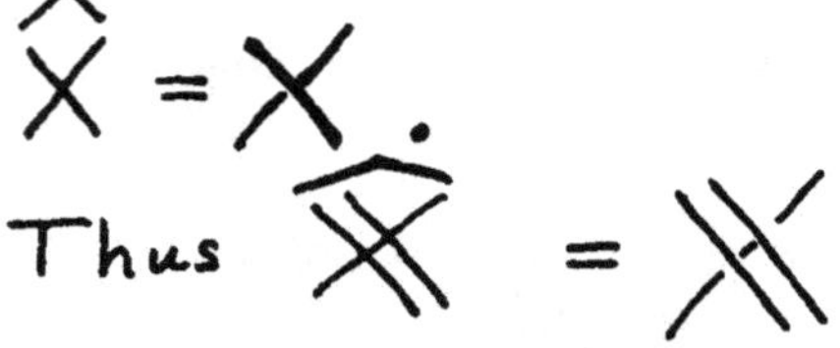

Thus

Let t(σ) denote the least number of transpositions needed to return the permutation σ to the identity. Let {n}! denote the following generalization of n! :

$$\{n\}! = \Sigma_{\sigma} (A^{-4})^{t(\sigma)}$$

The summation is over all permutations of n letters.

Then the projectors in the Temperley Lieb algebra are given by the formula

$$\boxed{n} = \frac{1}{\{n\}!} \sum_{\sigma \in S_n} (A^{-3})^{t(\sigma)} \boxed{\hat{\sigma}}$$

with the caveat that the braids are represented into the Temperley Lieb algebra via the bracket formula

$$\times = A \asymp + A^{-1})($$

This alternate construction of the projectors is useful because it gives a global picture of their structure. In particular, in the special cases A=1 and A=-1, one can use these formulas to calculate network evaluations, and thereby find the patterns that work in the general case. This construction also provides the link with SU(2) and the quantum group SU(2)q and shows that we are doing q-deformed angular momentum calculations in the recoupling theory of the Temperley Lieb algebra (KAU92]).

VI. Links and Invariants of 3-Manifolds

A link diagram is a code for a specific three dimensional manifold. This is accomplished by regarding the diagram as instructions for doing surgery to build the manifold. Each component of the diagram is seen as an embedding in three space of a solid torus, and the curling of the diagram in the plane gives instructions for cutting out this torus and repasting a twisted version to produce the 3-manifold. (See [KI78], [KA91] for a more detailed description of the process.) Extra moves on the diagrams give a set of equivalence classes that are in one-to-one correspondence with the topological types of three-manifolds. These moves consist in the addition or deletion of components with one curl as shown below

plus "handle sliding" as shown below

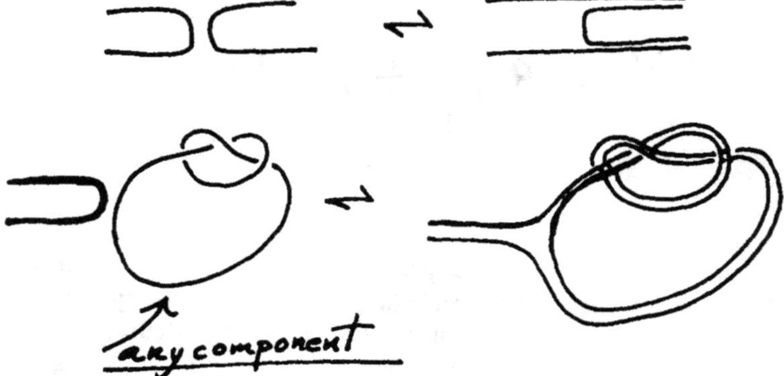

We also need the "ribbon" equivalence of curls shown below.

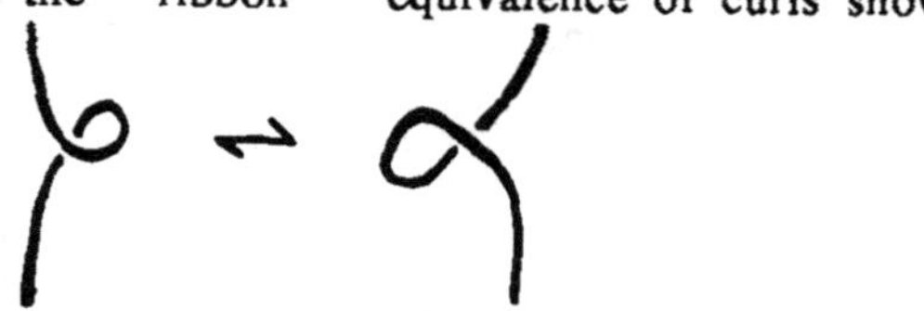

With these extra moves the links codify the topology of three dimensional manifolds.

It is at the roots of unity, $A=\exp(i\pi/2r)$, that one can define invariants of 3-manifolds. There are many ways to make this definition.

A particularly neat version using the Jones-Wenzl projectors is given by Lickorish [LIC92]. We reproduce his definition here:

Let a link component ,K, labelled with ω (as in ω*K) denote the sum of i-cablings of this component over i belonging to the set {0,1,2,...,r-2}. A projector is applied to each cabling. Thus

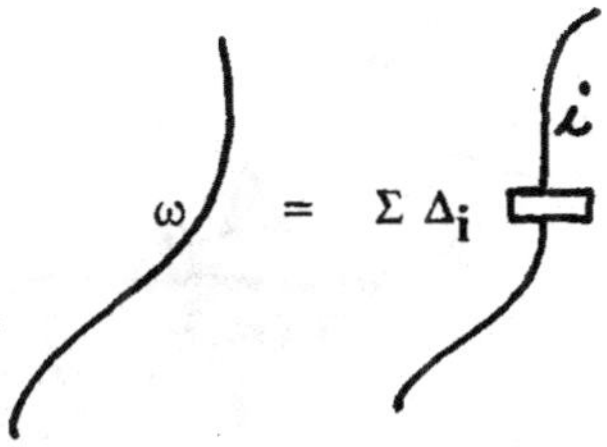

$$\omega \;=\; \Sigma\, \Delta_i$$

and

$$= \sum_{i,j} \Delta_i \Delta_j$$

$$\omega * [\;]$$

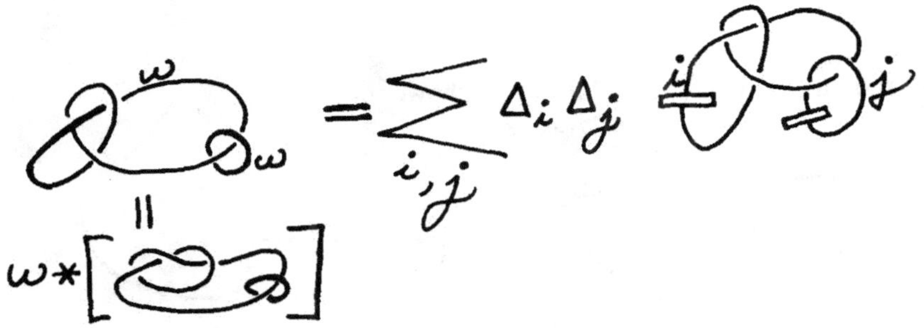

Then <ω*K> denotes the sum of evaluations of the corresponding bracket polynomials. If K has more than one component, then ω*K denotes the result of labelling each component, and taking the corresponding formal sum of products for the different cableings of individual components.

It is then quite an easy matter to prove that <ω*K> is invariant under handle sliding.

This is the basic ingredient in producing an invariant of 3-manifolds as represented by links in the blackboard framing. (A normalization is needed for handling the fact the 3-sphere is returned after surgery on an unknot with framing plus or minus one.) A quick proof of handle sliding invariance using this definition and recoupling theory has been discovered by Oleg Viro, and independently by Justin Roberts (See [LIC92].). The proof is based on the formula (a special case of the general recoupling formula)

$$a\,\big)\big(\,b \;=\; \sum_i \frac{\Delta_i}{\theta(a,b,i)} \;\; \underset{a\quad b}{\overset{a\quad b}{\succ\!\!\!\!\prec}}\, i$$

and the sequence of "events" shown below.

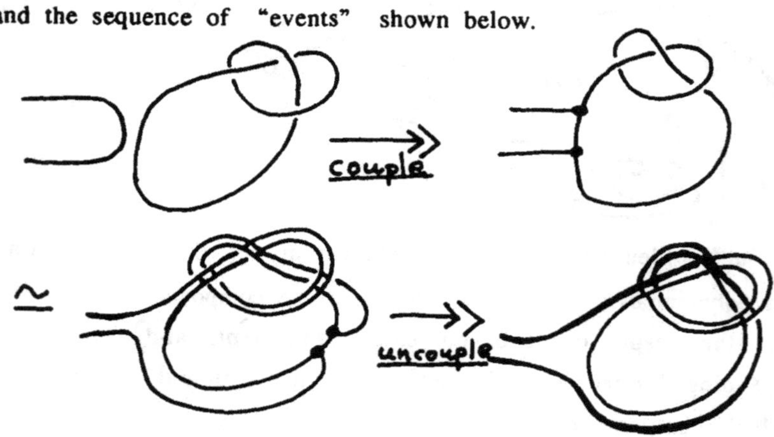

We turn these events into a proof of invariance under handle sliding by adding the algebra.

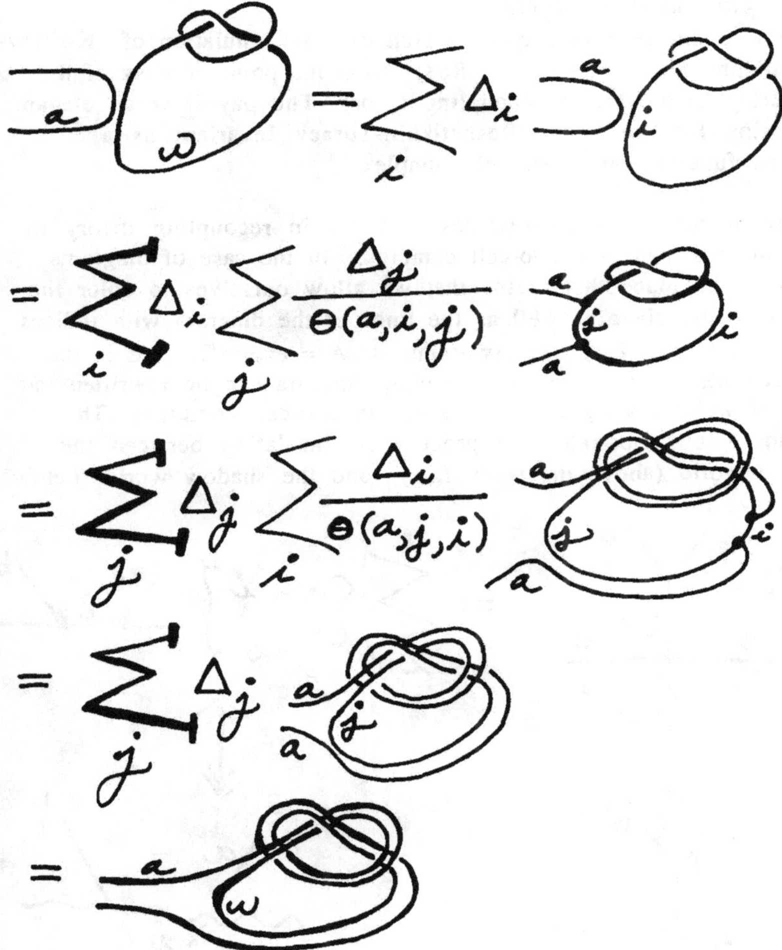

The logic of this construction of an invariant of 3-manifolds depends directly on formal properties of the recoupling theory, plus subtle properties of that theory at the roots of unity. In order to completely define the invariant it is necessary to introduce a normalization factor. This normalization is a constant(depending only on the root of unity) raised to the signature of the linking matrix (consisting of linking numbers of different components of the surgery data and framing numbers on its diagonal).
See [LI90],[KI91],[KL93] for discussions of this normalization.

VII. The Shadow World

In this section we give a quick sketch of a reformulation of Kirillov-Reshetikhin Shadow World [KR88] from the point of view of the Temperley Lieb Algebra recoupling theory. The payoff is an elegant expression for the Witten-Reshetikhin-Turaev Invariant as a partition function on a two-cell complex.

Shadow world formalism rewrites formulas in recoupling theory in terms of colorings of a two-cell complex. In the case of diagrams drawn in the plane this means that we allow ourselves to color the regions of the plane as well as the lines of the diagram with indices from the set $\{0,1,2,\ldots, r-2\}$ (working at $A = \exp(i\pi/2r)$, as in the last section). In this way a recoupling formula can be rewritten in terms of weights assigned to parts of the two-cell complex. The diagram below illustrates the process of translating between the daylight world (above the wavy line) and the shadow world (below the wavy line).

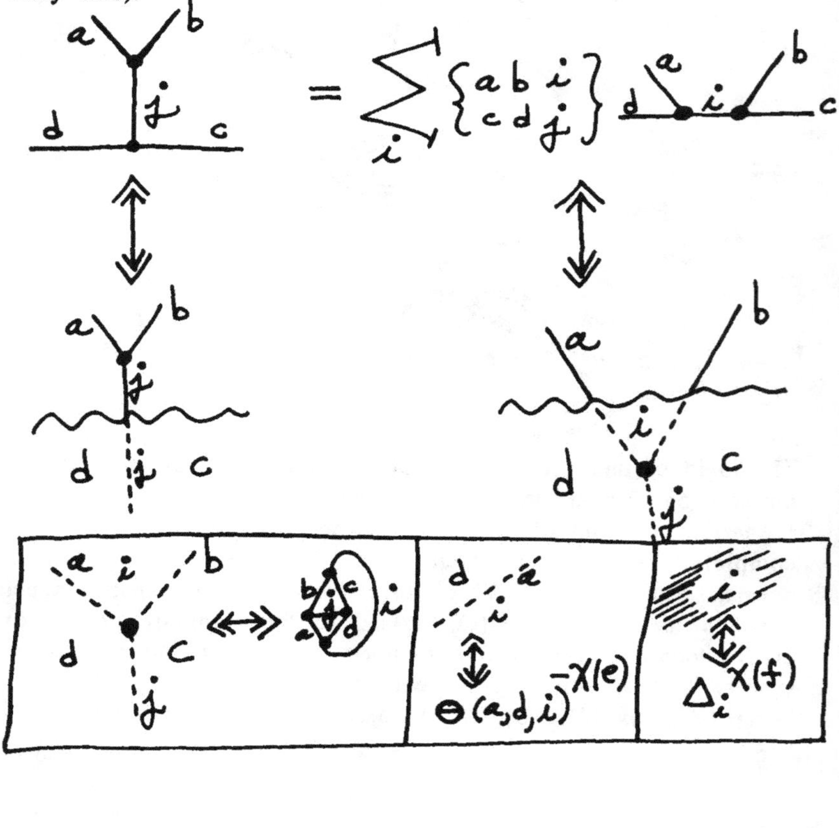

$$= \sum_i \begin{Bmatrix} a & b & i \\ c & d & j \end{Bmatrix}$$

$$\Theta(a,d,i)^{-\chi(e)} \qquad \Delta_i^{\chi(f)}$$

In this diagram we have indicated how if the tetrahedral evaluation is assigned to the six colors around a shadow world vertex (either 4-valent or 3-valent) , the theta symbol is assigned to the three colors corresponding to an edge (the edge itself is colored as are the regions to either side of the edge) then we can take the shadow world picture as holding the information about 6-j symbols that are in the recoupling formula.

To complete this picture we assign Δ_i to a face labelled i (when this face is a disk, otherwise we assign Δ_i raised to the Euler characteristic of the face). There are phase factors corresponding to the crossings. The shadow world diagram is then interpreted as a sum of products of these weights over all colorings of its regions edges and faces.

The result is an expression for the handle sliding invariant $\langle \omega * K \rangle$ (of section 6) as a partition function on a two-cell complex. The extra Δ_i's coming from the assignments of ω to the components of the link can be indexed by attaching a 2-cell to each component. The result is a 2-cell complex that is a "shadow" of the 4-dimensional handlebody whose boundary is the 3-manifold constructed by surgery on the link.

We shall spend the rest of this section giving more of the details of this shadow world translation. A complete description can be found in [KL93]. First, here are some elementary identities. The first is the one we have already mentioned above.

1.

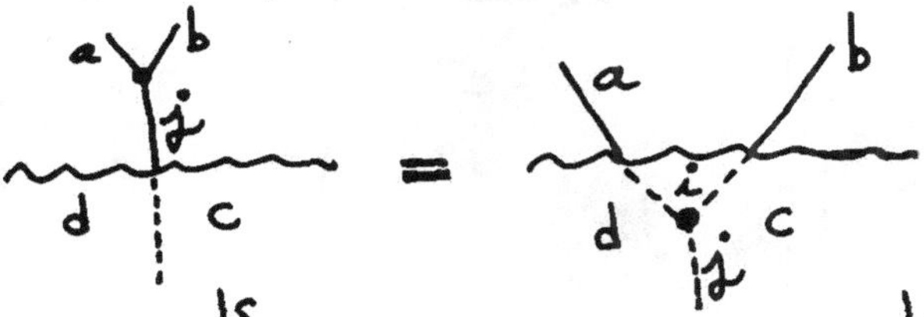

2.

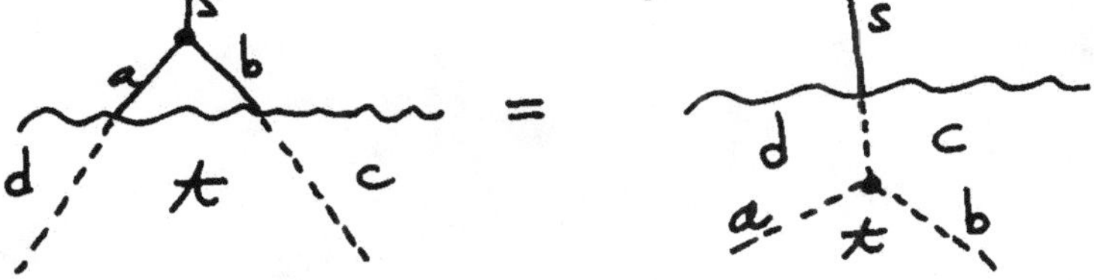

3.

4.

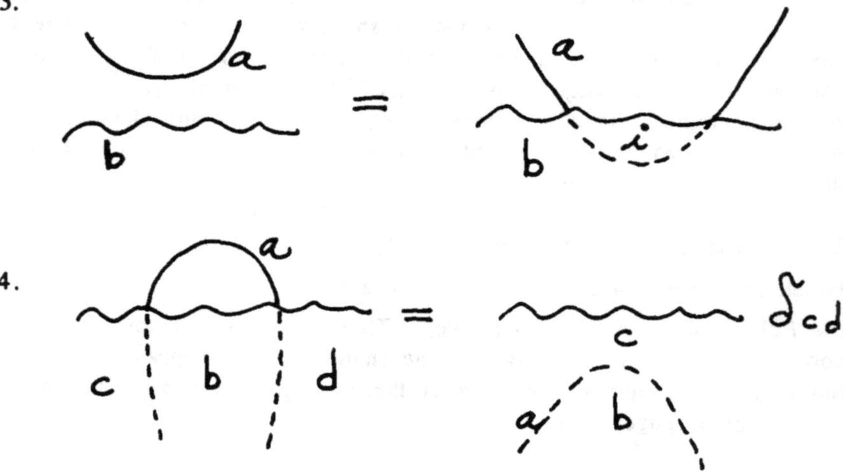

To illustrate this technique, we prove the next identity. This one shows how to translate a crossing across the shadow line. In order to explicate this identity, we need to recall a formula from the recoupling theory - namely that a 3-vertex with a twist in two of the legs is equal to an untwisted vertex multiplied by a factor of λ^{ab}_{c}. We refer to [KL93] for the precise value of lambda as a function of a, b and c.

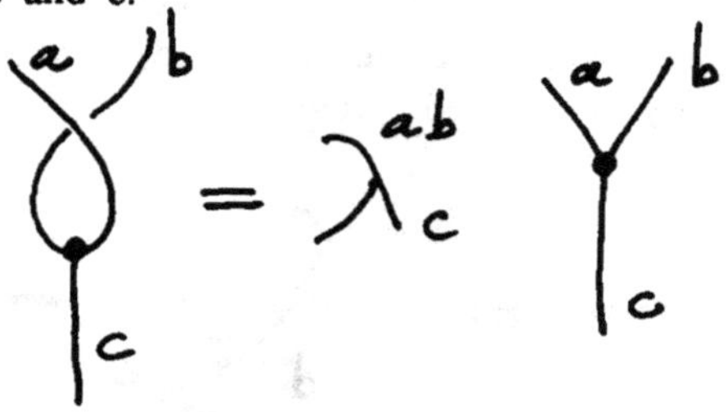

We define two shadow vertices to correspond to the two types of unoriented crossings that occur with respect to a vertical direction in the plane. Each shadow crossing corresponds to a tetrahedral evaluation multiplied by a lambda factor. This lambda factor is the phase referred to above. This correspondence is illustrated below.

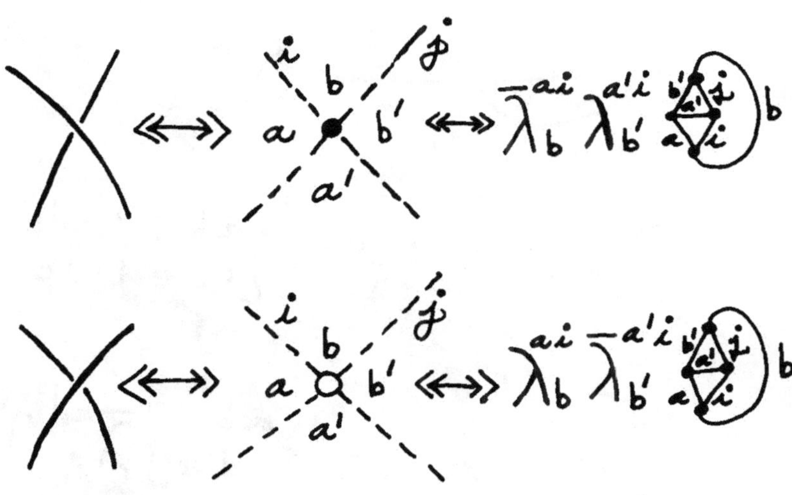

The crossing identity then becomes the following shadow equation.

5.

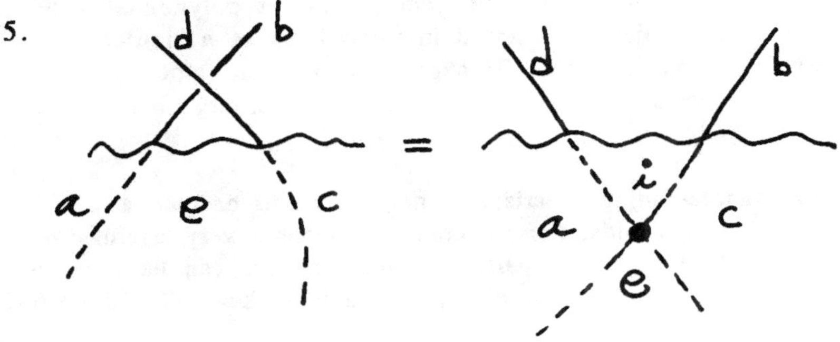

Proof.

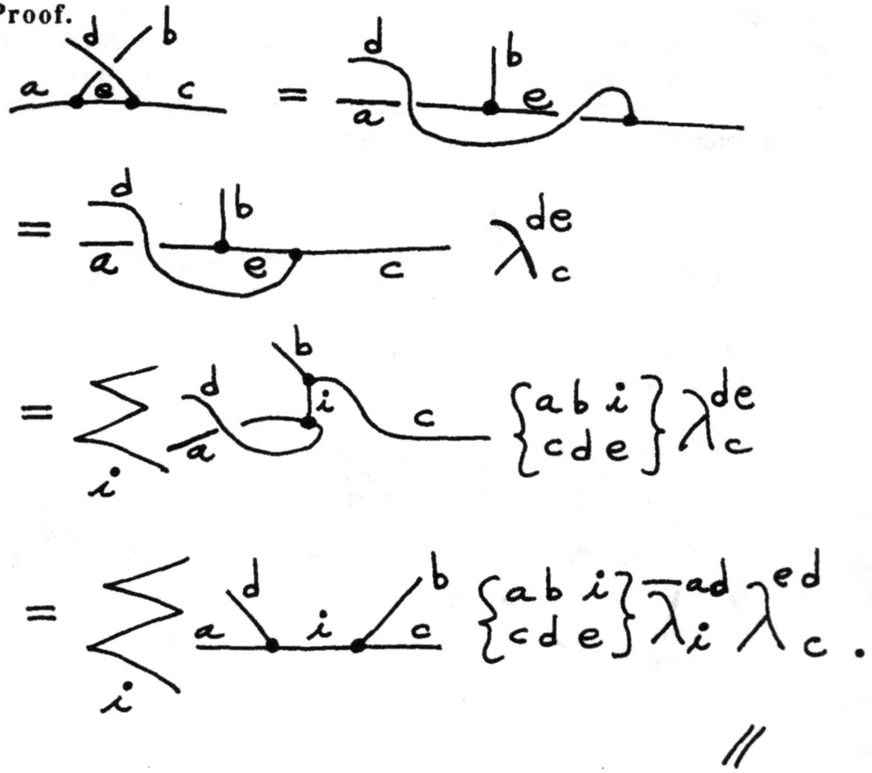

These formulas provide the means to rewrite the handle sliding invariant as a partition function on a two-cell complex. They also provide the means to rewrite the general bracket polynomial with Temperley Lieb projectors inserted into the lines as a similarly structured partition function. This gives an explicit (albeit complicated) formula for these invariants for arbitrary link diagrams.

Once the handle-sliding invariant is normalized to become an invariant of 3-manifolds, these techniques become very useful for studying the invariants. In particular this structure can be used to give a proof of the theorem of Turaev and Walker [TU92],[WA91] relating the Witten-Reshetikhin-Turaev invariant with the Turaev-Viro invariant. It remains to be seen what further applications and relationships will arise from this point of view.

VIII. On Manifolds With The Same Invariants

In this section we give an example of two manifolds that are easily seen to be non-homeomorphic (They have different first homology groups.) yet they share the same WRT invariants. We give the proof that they share the same invariants by using a very nice property of the SU(2) theory in its Temperley-Lieb recoupling mode. We shall refer to this property as the *Encirclement Lemma* . This property (See [LIC92],[KL93]) is summarized by the equation below.

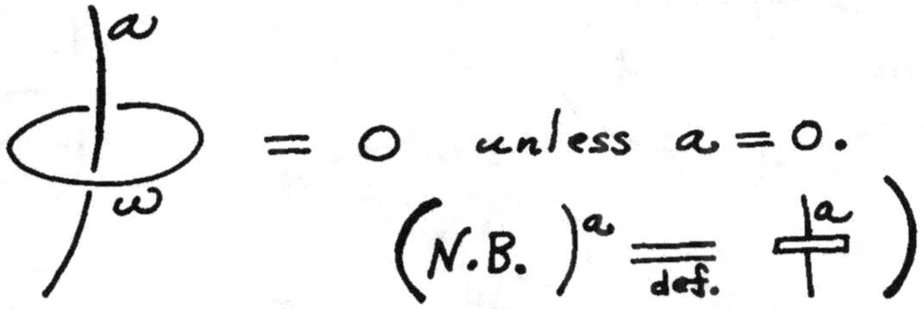

= 0 unless a = 0.

(N.B. $)^a \underset{\text{def.}}{=}$)

In other words, if an unknotted curve of framing number one encircles a line colored by label a with a in the set {0,1,...r-2} and A=exp(2πi/4r), then the resulting bracket evaluation is zero unless the cable multiplicity a is equal to zero. The proof follows from the handle-sliding equivalence shown below, and the fact that

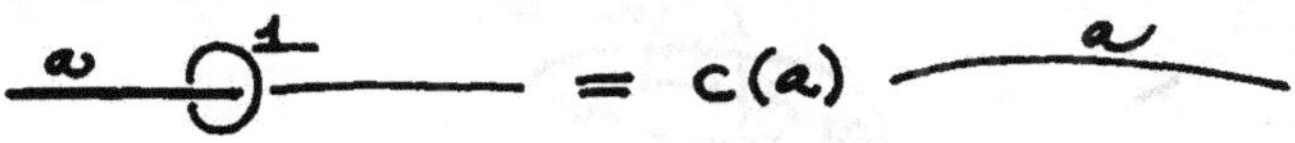

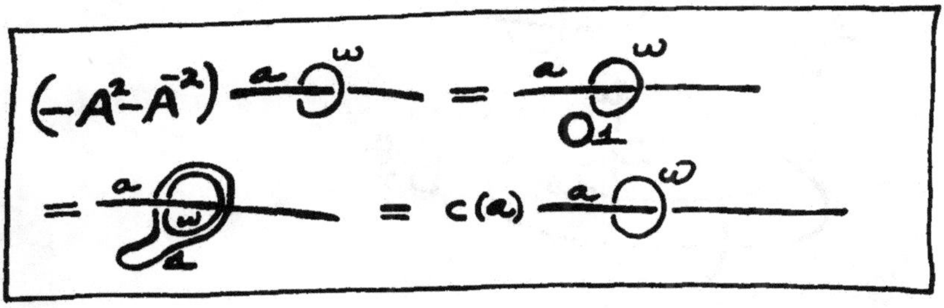

= c(a)

$(-A^2-A^{-2})$ = = = c(a)

for a constant c that is not equal to $-A^2-A^{-2}$ for a not equal to zero.

Note the consequence of this encirclement for two labelled lines passing through a circle:

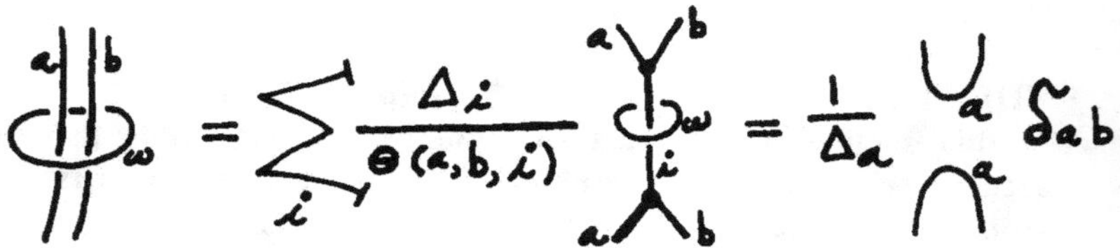

$$\text{[diagram: lines } a, b \text{ through circle } \omega] = \sum_i \frac{\Delta_i}{\Theta(a,b,i)} \text{[diagram]} = \frac{1}{\Delta_a} \text{[diagram]}\, \delta_{ab}$$

Now apply this to the Borommean rings:

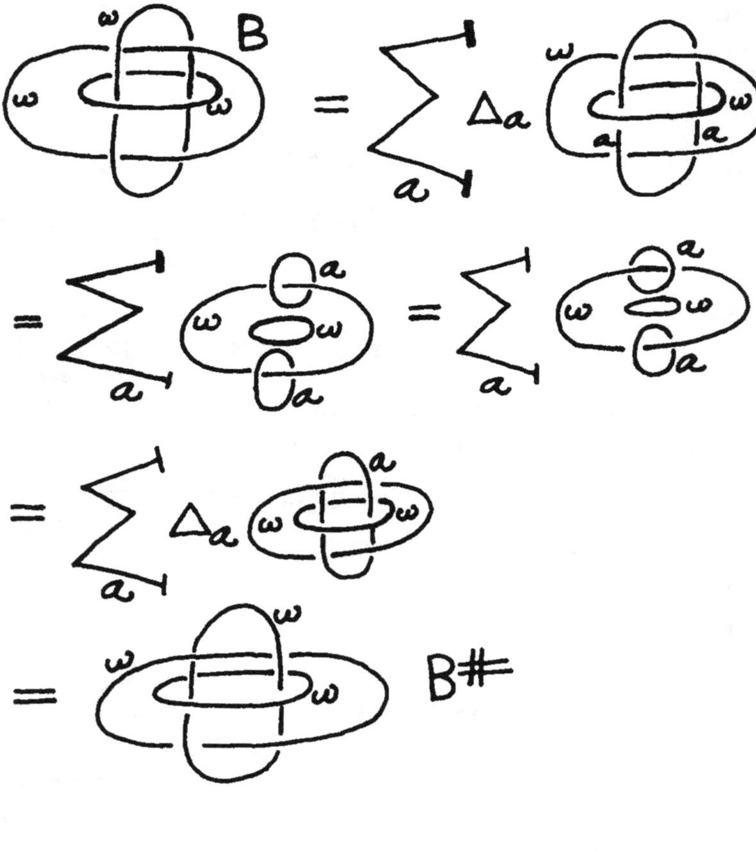

$$B = \sum_a \Delta_a \text{[diagram]}$$

$$= \sum_a \text{[diagram]} = \sum_a \text{[diagram]}$$

$$= \sum_a \Delta_a \text{[diagram]}$$

$$= \text{[diagram]}\quad B\#$$

We conclude that the Borommean rings, B, and the link, B#, obtained from them by switching two crossing as shown above must have the same unnormalized invariants. That is <w*B> = <w*B#> for any r. It is easy to see that B and B# have the same normalizing factors as well. Hence B and B# share the same WRT invariants for all r. The 3-manifold M(B) is the cartesian product of three circles, $S^1 \times S^1 \times S^1$. The 3-manifold M(B#) is a non-trivial torus bundle over the circle, and it has different homology from M(B). This example appears in [KL93]. It is an example of different character from the examples in [BAR91] and [LIC91]. Heretofore examples of this phenomenon of coincident invariants depended upon the properties of mutant tangles. In this case the coincidence depends upon recoupling properties and the Encirclement Lemma.

It remains an open problem to characterize when manifolds receive the same WRT invariants. The most intriguing related problem is the possibility that there may be manifolds with trivial fundamental group and non-trivial WRT invariant. An affirmative example of this type would be a counterexample to the classical Poincare Conjecture (which states that a 3-manifold with trivial fundamental group is homeomorphic to the the 3-sphere.)

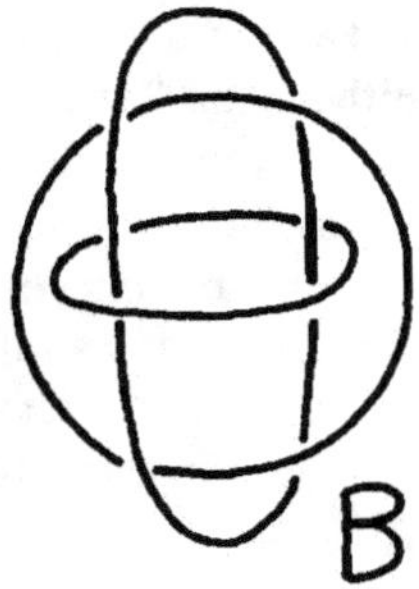

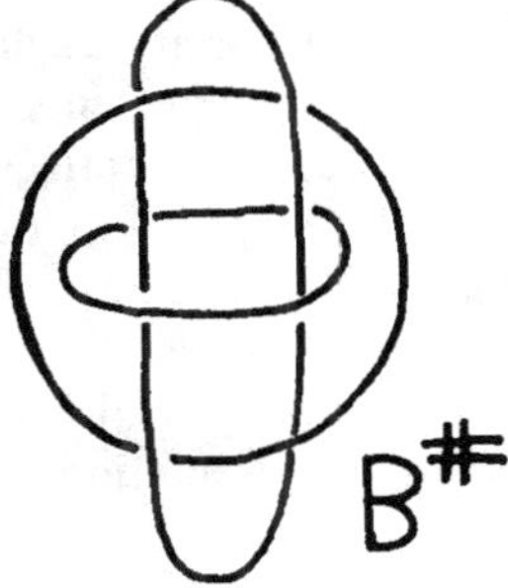

IX. Encirclement and Invariants of 4-Manifolds

The encirclement lemma in combination with recoupling yields formulas for several colored strands with an Omega belt about them:

$$= \sum_{i,j} \frac{\Delta_i \Delta_j}{\theta(a,b,i)\,\theta(i,c,j)}$$

$$= \frac{\Delta_c}{\theta(a,b,c)}$$

In [R93], this trick is used to analyse the structure of some invariants of 3-manifolds and 4-manifolds. In the case of 3-manifolds, one can present a 3-manifold by pasting together two handlebodies. (A handlebody is a 3-ball with 1-handles $(D^2 \times [0,1])$ attached along embeddings of $D^2 \times 0$ and $D^2 \times 1$ in the boundary of the 3-ball.) It is sufficient, to specify such a pasting, to give two sets of curves on a surface S that represents the boundary of either handlebody. The first set of curves is a collection of the basic curves that bound disks in the first handlebody (so that this handlebody is returned to a ball upon being cut along these disks). The second set of curves is in correspondence with the bounding curves for the second handlebody. It is customary to draw the first set of curves as the meridian curves on a standard representation of the surface. See the diagram below.

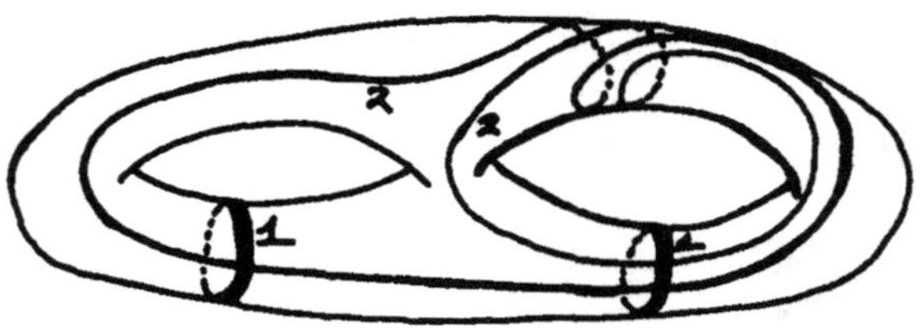

The second set of curves will then have the complexity that corresponds to the structure of the pasting map between the two handle-bodies. The resulting picture of two sets of curves on a given surface is called a *Heegard Diagram* for the 3-manifold.

For example, the pictures below show Heegard diagrams for S^3 and for $S^2 \times S^1$.

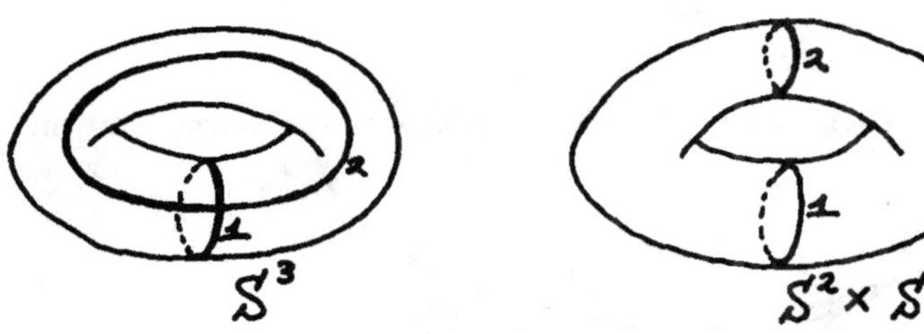

Now take such a diagram, embed it in 3-space (any way you please). Push each of the meridians in the first set of curves along the outward normal to the surface so that they do not intersect the curves in the second set. Call this link L(H) where H is the original Heegard decomposition. Let M(H) denote the 3-manifold that corresponds to H. The usual normalization of $\langle\omega^* L(H)\rangle$ (See section 6) is an invariant of the 3-manifold M(H). In fact, Roberts shows that it is the Kauffman-Lins [KL93] version of the Turaev-Viro invariant.

The proof is based on the encirclement recoupling formulas and the geometry of the handlebody decomposition that is obtained from the dual 1-skeleton of a triangulation of M = M(H). In the dual 1-skeleton we have four edges incident to each 0-simplex. The handle-bodies consist in a tubular neighborhood of this structure, and its complement.

Thus L(H) looks locally as shown below.

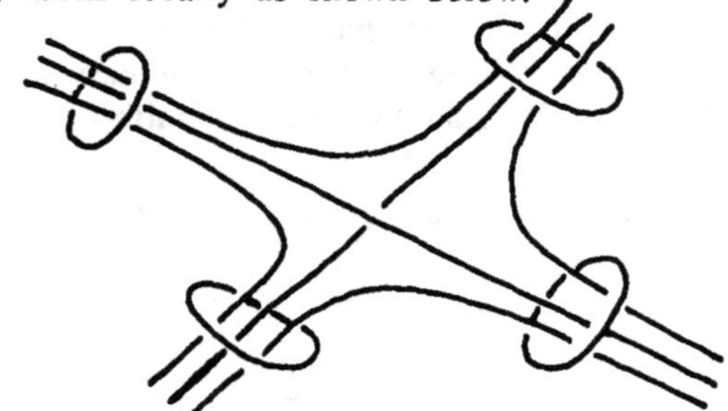

Applying the encirclement lemma yields isolated tetrahedral nets at each vertex.

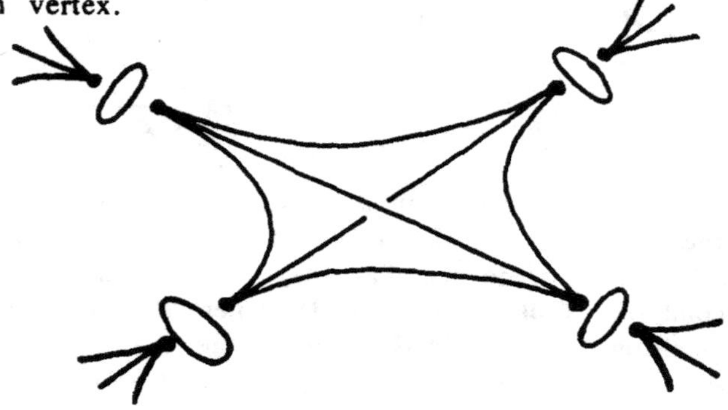

The invariant <w*L(H)> is therefore seen as a sum of products of evaluations of tetrahedral nets and other terms corresponding to the details of the recoupling formulas. It is easy to see that this is exactly the KL version of the Turaev-Viro invariant. Roberts uses this method to give a simple proof that the Turaev-Viro invariant is the product of the WRT invariant and its complex conjugate. See the next section for a further note on this theorem, first proved by Turaev [TU92] and Walker [WA91].

Roberts then goes on to discuss a four dimensional version of the construction M ---> M(H) ----->L(H)-----> <ω*L(H)>. In the four dimensional version, one again takes the dual 1-skeleton of a triangulation of the 4-manifold. A tubular neighborhood N of this dual 1-skeleton in the four manifold is regarded as the result of building 0-handles and 1-handles. The remainder of the four manifold - obtained from the complementary handlebody- is btained relative to N by adding 2-handles to the boundary of N along ie bounding curves from the second handlebody. The result will be a onnected sum of $S^1 \times S^2$'s and can be filled in to form the four-manifold. y possibly adding more 1-handles at the beginning of the process, one can egard the dual 1-skeleton as embedded in a four-ball with all 0-handles nd dimensionally reduced neighborhoods of the 1-handles in the 3-phere boundary. The curves corresponding to the 2-handle attachments re then in the 3-sphere on the surface of the dimensionally reduced dual -skeleton.

he upshot of this description is that one can again compute $<\omega^*L>$ for the nk obtained from curves for the 2-handles together with pushoffs for the ieridians of the thickened and dimensionally reduced dual 1-skeleton. he story about computing $<\omega^*L>$ is then the same as before except that iere are five tubes meeting at a node in this dual thickened skeleton since we are working with the four simplices of the triangulation of M). herefore the encirclement and recoupling works as shown below.

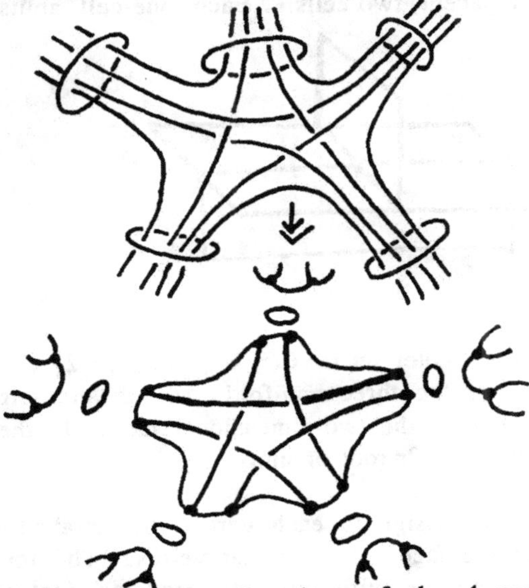

his means that the resulting invariant of the 4-manifold is a sum of roducts of "15-j" symbols and some other terms. Roberts proves that fter normalization, this invariant is a constant raised to the signature of he 4-manifold. As the reader can see by comparing this description with hat of the invariant of Crane and Yetter in this volume [CR93], there is a emarkable similarity in the definitions involving 15-j symbols. This raises an interesting problem of translation and articulation. It appears to be the case that the Crane-Yetter invariant is, up to form factors, the same as Robert's invariant.

One hopes for non-trivial 4-manifold invariants constructed as combinatorial topological quantum field theories. The present examples are food for thought.

XI. The Turaev-Viro Invariant and Quantum Gravity

We use the Matveev-Piergallini representation (see [KL93], and [TV92]) of three-manifolds in terms of special spines. In such a spine , a typical vertex appears as shown below with four adjacent one-cells , and six adjacent two-cells. Each one-cell abuts to three two cells.

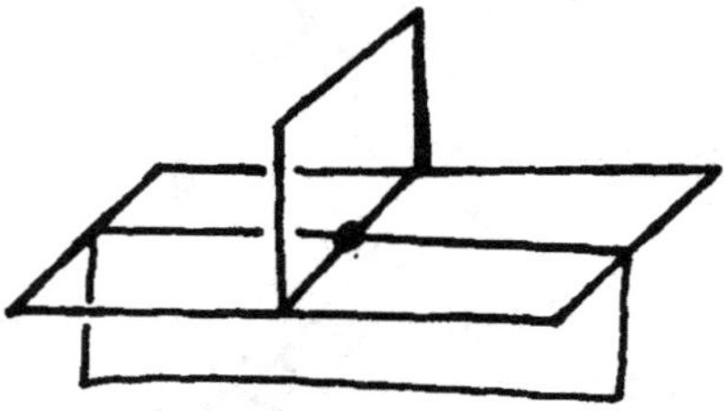

For an integer r>=3 the color set is C(r) = {0,1,2,..., r-2}.
A *state* at level r of the three-manifold M is an assignment of colors from C(r) to each of the two-dimensional faces of the spine of M. Let q be a primitive 2r-root of unity.

Given a state S of M, assign to each vertex the tetrahedral symbol whose edge colors are the face colors at that vertex. The form of this assignment is shown below with the standard orientation at the vertex .

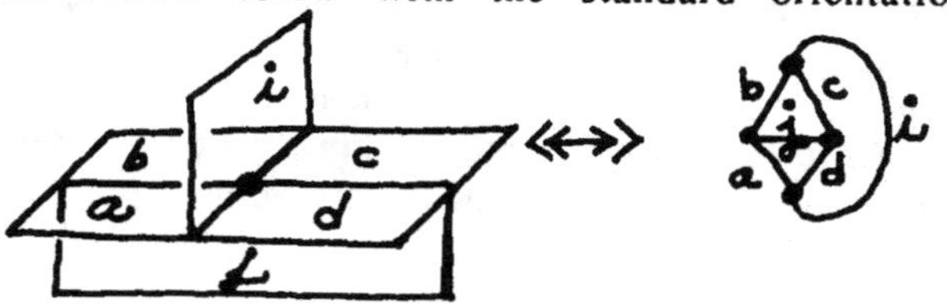

Assign to each edge the theta - symbol associated with its triple of colors.

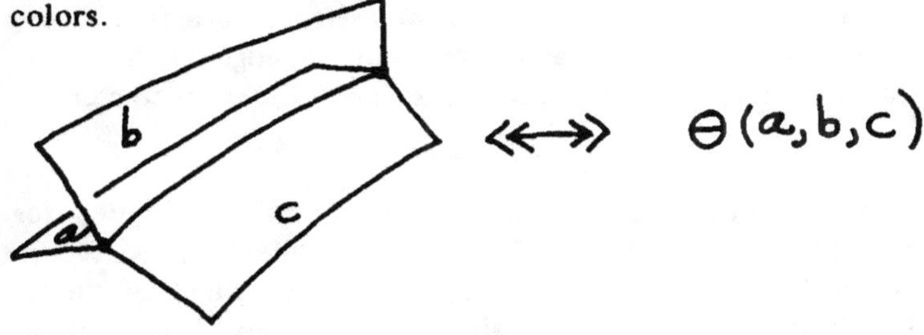

Assign to each face the Chebyshev polynomial

whose index is the color of that face.

Let I(M,r) denote the sum over all states for level r, of the products of the vertex evaluations and the face evaluations divided by the edge evaluations for those evaluations that are admissible (See section 4).

$$I(M,r) = \sum_{S}{}' \frac{\prod_{v} Tet(v,S) \prod_{f} \Delta_{S(f)}}{\prod_{e} \theta(S_a(e), S_b(e), S_c(e))}.$$

Here TET(v,S) denotes the tetrahedral evaluation associated with a vertex v , for the state S. S(f) is the color assigned to a face f, and Sa,Sb,Sc are the triplet of colors associated with an edge in the spine. It is assumed, for simplicity, that all cells have Euler characteristic zero.

It follows via the orthogonality and Elliot-Biedenharn identities for the q-6j symbols, that I(M,r) is invariant under the Matveev-Piergallini moves. Hence, I(M,r) is a topological invariant of the 3-manifold M. This is our [KL93] version of the Turaev-Viro invariant [TV92].

I(M,r) does not depend upon the orientation of the 3-manifold M. This follows easily from the symmetries in the evaluation of the tetrahedron. It is an open question whether a state sum of this sort can be constructed to produce an invariant of 3- manifolds that detects orientation. Perhaps a subtle combinatorial insight will produce the full WRT invariant as a state summation on the 3-manifold.

Quantum Gravity

As mentioned in the previous section, the Turaev-Viro invariant is equal to the product of the Witten-Reshetikhin-Turaev invariant with its complex conjugate: $TV(M) = WRT(M)WRT(M)^* = |WRT(M)|^2$ To this end it is interesting to examine the functional integral formalism for $|WRT(M)|^2$. We find [WITT89]:

$$|WRT(M)|^2 = |\int dA \ \exp((ik/4\pi)\int tr(AdA \ +(2/3)A^3)|^2$$

$$= \int dAdB \ \exp((ik/4\pi)\int tr(AdA \ +(2/3)A^3 - BdB \ - \ (2/3)B^3)$$

$$= \int ded\omega \ \exp(i\int tr(e\wedge R \ +(\Lambda/3)e\wedge e\wedge e))$$

where $e = (k/8\pi)(A - B)$, $\omega = (1/2)(A + B)$,

$\Lambda = (4\pi/k)^2$ and $R = d\omega + \omega\wedge\omega$.

The upshot is that $|WRT(M)|^2$ has the formal structure of a functional integral for $2 + 1$ quantum gravity. (Two spatial dimensions and one time dimension.) Here e is interpreted as a metric, while w is interpreted as a connection so that R is the curvature. For quantum gravity one expects to integrate over all metrics (and connections) with $\int tr(e \wedge R + (\Lambda/3) e \wedge e \wedge e))$ as the action with a "cosmological constant" Λ. The i in this action makes the theory non-Euclidean, but it is formally a quantum gravity theory. This is quite remarkable since we know that this functional integral is in fact a topological invariant of the the 3-manifold M, and furthermore it can be computed by spin-network methods from any triangulation of M via the Turaev-Viro state summation. The first hint that such a state summation might be related to gravity appeared in the work of Regge-Ponzano [PR68] and Hasslacher-Perry [HP81]. These authors investigated analogous state sums using approximations to classical recoupling coefficients. They found that their state sums gave an approximation to the Regge calculus simplicial quantum gravity. The Chern-Simons interpretation of this situation is very appealing, and it leads to many interesting questions about the correspondence of the continuum and discrete models.

More Quantum Gravity

We cannot resist mentioning in this context the relationship of Wilson loops and quantum gravity that is forged in the theory of Ashtekar, Rovelli and Smolin [ASH92]. In this theory the metric is expressed in terms of a spin connection A, and quantization involves considering wavefunctions $\Psi(A)$. Smolin and Rovelli analyze the loop transform $\psi^{\wedge}(K) = \int dA\ \Psi(A) \langle A|K\rangle$ where $\langle A|K\rangle$ denotes the Wilson loop for the knot or singular embedding K. (See section 2 for a definition of the Wilson loop.) Then differential operators on the wavefunction can be referred, via integration by parts to corresponding statements about the Wilson loop. It turns out that the condition that $\psi^{\wedge}(K)$ be a knot invariant is equivalent to the so-called diffeomorphism constraint for these wave functions. In this way, knots and weaves and their topological invariants become a language for representing the states of quantum gravity.

Once again, one wants a detailed understanding of invariants expressed in the integral formalism. The search for a well-founded calculus in this domain is a major problem in topology and in quantum gravity.

References

[ASH92] Abhay Ashtekar, Carlo Rovelli, Lee Smolin. Weaving a classical geometry with quantum threads. (preprint 1992).

[AT79] M.F. Atiyah. **Geometry of Yang-Mills Fields.** Accademia Nazionale dei Lincei Scuola Superiore Lezioni Fermiare, Pisa (1979).

[AT90] M.F. Atiyah. **The Geometry and Physics of Knots.** Cambridge University Press (1990).

[BA82] R.J. Baxter. **Exactly Solved Models in Statistical Mechanics.** Acad. Press (1982).

[BAR91] J. Kania-Bartoszynska. Examples of different 3-manifolds with the same invariants of Witten-Reshetikhin-Turaev. (preprint 1991).

[BL79] L.C. Biedenharn and J.D. Louck. **Angular Momentum in Quantum Physics- Theory and Application.** (Encyclopedia of Mathematics and its Applications). Cambridge University Press (1979).

[BN92] D. Bar-Natan. On the Vassiliev knot invariants. (preprint 1992).

[BL92] J. Birman and X.S.Lin. Knot polynomials and Vassiliev's invariants. Invent. Math. (to appear).

[BR92] Bernd Brugmann, Rodolfo Gambini, Jorge Pullin. Knot invariants as nondegenerate quantum geometries. Phys. Rev. Lett. Vol. 68, No. 4, 27 Jan. 1992, pp. 431-434.

[CR91] L. Crane. Conformal Field Theory, Spin Geometry and Quantum Gravity. Physics Letters B. Vol. 259. No. 3. pp. 243-248. (1991).

[CR93] L. Crane and D. Yetter. A Categorical Construction of 4D Topological Quantum Field Theories. (Preprint 1993).

[D58] P.A.M. Dirac. **Principles of Quantum Mechanics.** Oxford University Press (1958).

[FR79] R.A.Fenn and C.P.Rourke. On Kirby's Calculus of Links. Topology, 18 (1979), pp. 1-15.

[FEY65] R. Feynman and A.R. Hibbs. **Quantum Mechanics and Path Integrals.** McGraw Hill (1965).

[FG91] D. Freed and R. Gompf. Computer calculations of Witten's 3-manifold invariants. Comm. Math. Phys. 41, pp. 79-117 (1991).

[GA93] S. Garoufalidis. Applications of TQFT to invariants in low dimensional topology. (preprint 1993).

[HP81] B. Hasslacher and M. J. Perry. Spin networks are simplicial quantum gravity. Physics Letters, Vol. 103B, No. 1, July 1981.

[J91] L. C. Jeffrey. On Some Aspects of Chern-Simons Gauge Theory. (Thesis - Oxford (1991)).

[JO83] V.F.R. Jones. Index for subfactors. Invent. Math. 72 (1983). pp 1-25.

[JO85] V.F.R.Jones.A polynomial invariant for links via von Neumann algebras. Bull.Amer.Math.Soc. 129 (1985) 103-112.

[JO86] V.F.R.Jones. A new knot polynomial and von Neumann algebras. Notices of AMS 33 (1986) 219-225.

[JO87] V.F.R.Jones. Hecke algebra representations of braid groups and link polynomials. Ann. of Math. 126 (1987), pp. 335-338.

[JO89] V.F.R.Jones. On knot invariants realted to some statistical mechanics models. Pacific J. Math., vol. 137, no. 2 (1989), pp. 311-334.

[KA87] L.H.Kauffman. State Models and the Jones Polynomial. Topology 26 (1987) 395-407.

[KAU87] L.H. Kauffman. **On Knots.** Annals of Mathematics Studies Number 115, Princeton University Press (1987).

[KA88] L.H.Kauffman. New invariants in the theory of knots. Amer. Math. Monthly Vol.95,No.3,March 1988. pp 195-242.

[KA89] L.H.Kauffman. Spin Networks and the Jones Polynomial. Twistor Newsletter No. 29. (1989) pp. 25-29.

[KAU89] L.H.Kauffman. Statistical mechanics and the Jones polynomial. AMS Contemp. Math. Series (1989), Vol. 78. pp. 263-297.

[KA90] L.H.Kauffman. An invariant of regular isotopy. Trans. Amer. Math. Soc. Vol. 318. No. 2 (1990). pp. 417-471.

[Kauff90] L.H.Kauffman. Spin networks and knot polynomials. Intl. J. Mod. Phys. A. Vol. 5. No. 1. (1990). pp. 93-115.

[Kauffm90] L.H. Kauffman. Map Coloring and the vector cross product. J. Comb. Theo. Ser.B, Vol. 48, No. 2, April 1990. pp.145-154.

[Kauff91] L.H.Kauffman. SU(2)q - Spin Networks. Twistor Newsletter No. 32. (1991),pp. 10-14.

[KA91] L.H. Kauffman. **Knots and Physics** , World Scientific Pub. (1991).

[KA92] L. H. Kauffman. Knots, Spin Networks and 3-Manifold Invariants. **Knots 90** (edited by A. Kawauchi) pp. 271-287 (1992).

[KAU92] L.H. Kauffman. Map Coloring. q-Deformed Spin Networks, and Turaev-Viro Invariants for 3-Manifolds. In the Proceedings of the Conference on Quantum Groups - Como, Italy June 1991 - edited by M. Rasetti - World Sci. Pub., Intl. J. Mod. Phys. B, Vol. 6, Nos. 11 & 12 (1992), pp. 1765 - 1794.

[KL90] L. H. Kauffman and S. Lins. A 3-manifold invariant by state summation. (announcement 1990).

[KL91] L.H. Kauffman and S. Lins. Computing Turaev-Viro Invariants for 3-Manifolds. Manuscripta Math. 72, pp. 81-94,(1991).

[KL93] L.H.Kauffman and S. Lins. Temperley Lieb Recoupling Theory and Invariants of 3-Manifolds. (to appear)

[KS92] L.H. Kauffman and H. Saleur. Map coloring and the Temperley Lieb algebra. (to appear in Comm. Math. Phys.)

[K93] L.H. Kauffman. Gauss codes, quantum groups and ribbon Hopf algebras. (to appear in Reviews of Modern Physics).

[KV92] L.H.Kauffman and P.Vogel. Link polynomials and a graphical calculus. Journal of Knot Theory and Its Ramifications. Vol. 1, No. 1 (March 1992).

[KI78] R. Kirby. A calculus for framed links in S^3. Invent. Math. 45 (1978), pp. 35-56.

[KM91] R. Kirby and P. Melvin. On the 3-manifold invariants of Reshetikhin- Turaev for sl(2,C). Invent. Math. 105, 473-545 (1991).

[KO92] M. Kontsevich. Graphs, homotopical algebra and low dimensional topology. (preprint 1992).

[KR88] A.N. Kirillov. and N.Y. Reshetikhin. Representations of the algebra $U_q(sl_2)$, q-orthogonal polynomials and invariants of links. In **Infinite Dimensional Lie Algebras and Groups**. ed. by V.G. Kac. Adv. Ser. in Math. Phys. Vol. 7. (1988). pp. 285-338.

[LI62] W.B.R. Lickorish. A representation of orientable combinatorial 3-manifolds. Ann. of Math. 76 (1962), pp. 531-540.

[LI90] W.B.R. Lickorish. Calculations with the Temperley-Lieb algebra. (preprint 1990).

[LI91] W. B. R. Lickorish. 3-Manifolds and the Temperley Lieb Algebra. Math. Ann. 290, 657-670 (1991).

[LIC91] W.B.R.Lickorish. Distinct 3-manifolds with all SU(2)q invariants the same. (preprint 1991).

[LI92] W.B.R. Lickorish. Skeins and handlebodies. (preprint 1992).

[LIC92] W.B.R.Lickorish. The Temperley Lieb Algebra and 3-manifold invariants. (preprint 1992). (and private communication)

[LD91] S. Lins and C. Durand. Topological classification of small graph-encoded orientable 3-manifolds. Notas Com. Mat. UFPE, 177, (1991).

[MTW73] C.W.Misner, K.S.Thorne and J.W.Wheeler. **Gravitation**. W.H. Freeman (1973).

[MV92] G. Masbaum and P. Vogel. 3-valent graphs and the Kauffman bracket. (preprint 1992).

[MOU79] J.P. Moussouris. The chromatic evaluation of strand networks. In **Advances in Twistor Theory**. ed. by Huston and Ward. Research Notes in Mathematics. Pitman Pub. (1979). pp. 308-312.

[MOU83] J. P. Moussouris. Quantum models of space-time based on recoupling theory. (Mathematics Thesis, Oxford University - 1983).

[OG91] H. Ooguri. Discrete and continuum approaches to three-dimensional quantum gravity. (preprint 1991).

[OG92] H. Ooguri. Topological Lattice Models in Four Dimensions. Mod.Phys.Lett.A, Vol. 7, No. 30 (1992) 2799-2810.

[PEN69] R. Penrose. Angular momentum: an approach to Combinatorial Space-Time. In **Quantum Theory and Beyond.** ed. T.A. Bastin. Cambridge Univ. Press (1969).

[PEN71] R.Penrose.Applications of negative dimensional tensors. **Combinatorial Mathematics and its Applications.** Edited by D.J.A.Welsh.Academic Press (1971).

[PEN79] R. Penrose. Combinatorial quantum theory and quantized directions. **Advances in Twistor Theory.** ed by L.P. Hughston and R.S. Ward. Pitman (1979). pp. 301-307.

[PR68] G. Ponzano and T. Regge. Semiclassical limit of Racah coefficients. In **Spectroscopic and Group Theoretic Methods in Physics.** (North Holland, Amsterdam (1968)).

[PIU92] Sergey Piunikhin. Turaev-Viro and Kauffman-Lins Invariants for 3-Manifolds Coincide. Journal of Knot Theory and its Ramifications, Vol. 1, No. 2, (1992), pp. 105 - 135.

[SM88] Lee Smolin. Quantum gravity in the self-dual representation. Contemp. Math. Vol. 71 (1988), pp. 55-97.

[R93] J. Roberts. Skein theory and Turaev-Viro invariants. (preprint 1993).

[RE87] N.Y.Reshetikhin. Quantized universal enveloping algebras, the Yang-Baxter equation and invariants of links, I and II. LOMI reprints E-4-87 and E-17-87, Steklov Institute, Leningrad, USSR.

[RT90] N.Y. Reshetikhin and V. Turaev. Ribbon graphs and their invariants derived from quantum groups. Comm. Math. Phys. 127 (1990). pp. 1-26.

[RT91] N.Y. Reshetikhin and V. Turaev. Invariants of Three Manifolds via link polynomials and quantum groups. Invent. Math. 103, 547-597 (1991).

[RO93] L. Rozansky. A large k asymptotics of Witten's invariant of Seifert manifolds. (preprint 1993).

[ST92] T. Stanford. Finite-type invariants of knots, links and graphs. (preprint 1992).

[TU87] V.G.Turaev. The Yang-Baxter equations and invariants of links. LOMI preprint E-3-87, Steklov Institute, Leningrad, USSR. Inventiones Math. 92 Fasc.3,527-553.

[TV92] V.G. Turaev and O. Viro. State sum invariants of 3-manifolds and quantum 6j symbols. Topology, Vol. 31, No. 4, pp. 865-902 (1992).

[TU90] V. G. Turaev. Quantum invariants of links and 3-valent graphs in 3-manifolds. (preprint 1990).

[TW91] V.G. Turaev and H. Wenzl. Quantum invariants of 3-manifolds associated with classical simple Lie algebras.

[TUR90] V.G. Turaev. Shadow links and face models of statistical mechanics. Publ. Inst. Recherche Math. Avance. Strasbourg (1990).

[TURA90] V.G. Turaev. Quantum invariants of 3-manifolds and a glimpse of shadow topology (preprint 1990).

[TU92] V.G. Turaev. Topology of Shadows. (preprint 1992).

[V90] V. Vassiliev. Cohomology of knot spaces. In **Theory of Singularities and Its Applications.** (V.I.Arnold, ed.), Amer. Math. Soc. (1990), pp. 23-69.

[WA91] K. Walker. On Witten's 3-Manifold Invariants. (preprint 1991).

[WA91] R. Williams and F. Archer. The Turaev-Viro state sum model and 3-dimensional quantum gravity. (preprint 1991).

[WIT89] E.Witten. Quantum field theory and the Jones polynomial. Commun.Math.Phys. 121, 351-399 (1989).

[WITT89] E. Witten. Nuc. Phys. B 311 (1989),p. 46.

[YA92] S. Yamada. A topological invariant of spatial regular graphs. **Knots 90** (edited by Akio Kawauchi), de Gruyter (1992), pp. 447-454.

Knot Theory, Exotic Spheres And Global Gravitational Anomalies*

RANDY A. BAADHIO†

Joseph Henry Laboratories, Department of Physics
Princeton University, Princeton, NJ 08544;
§Theoretical Physics Group, Physics Division,
Lawrence Berkeley Laboratory, and Department of Physics
University of California at Berkeley, Berkeley, CA 94720

Abstract

Global gravitational anomalies, which may ruin the physical consistency of a quantum field theory, are shown to be absent in $D = 2 + 1$ Topological Quantum Field Theories.

This comes about by establishing the absence of disconnected general coordinate transformations in the Hamiltonian version of the theories. Specifically, we prove the invariance of the Chern–Simons–Witten effective action under the mapping class group, that is, the group of equivalence classes of diffeomorphisms that cannot be smoothly deformed to the identity.

A newly found relationship between knot theory, exotic spheres and global anomalies, as exhibited in a joint work with L. H. Kauffman and the author, is described.

---

* Paper based on a talk given in the Special Session on Knots and Topological Quantum Field Theory as part of the American Mathematical Society October 1992 Annual Meeting in Dayton, Ohio

† Research Supported by the Director, Office of Energy Research, Office of High Energy and Nuclear Physics, Division of High Energy of The United States Department of Energy under Contracts DE–AC03–76–SF00098 and in part by a Grant from the Eppley Foundation for Research, Inc.

§ Permanent Address

I. Introduction

Quantum Field theories, the quantized version of Classical Field theories, are nowadays the preferred physical framework by which physics attempts to describe fundamental interactions in nature. Yet, despite their successes, they suffer from formal pathologies which manifest themselves as divergent expressions, and therefore they need to be regularized. The regularization procedure is a delicate one, as it can break the (original) classical symmetry of the theories. A celebrated example among physicists of the loss of classical symmetry is the breaking of the scale invariance by renormalization effects.

Renormalization often imposes the choices of an arbitrary energy scale, which spoils dimensional analysis; scale changes are then absorbed into a modification of the coupling constant during the renormalization procedure.

There are other processes that result in the loss of the classical scale invariance. The running of the coupling constant, for instance, is a delicate mechanism that compensates its loss of scale invariance with the advantage of relating anomalous dimensions to the dynamics of the field theory.

Another celebrated example of an anomaly is the breaking of chiral $U(1)$ invariance of a fermionic theory. By introducing a nonabelian gauge field in the theory, the original left-right chiral symmetry of a fermion field is no longer valid because of instanton effects. Thus, the axial charge is no longer conserved in the quantum version, and this gives a net production of chirality.

The two examples just mentioned are in fact harmless as long as they do not interfere with the quantization or renormalization procedure: one simply ends up with a consistent, but less symmetrical quantum version than we would have originally thought of.

Anomalies become a much more serious threat when the lost invariance is actually *a necessary ingredient* in the quantization procedure. This observation has mostly applied to gauge symmetries, since gauge degrees of freedom are known to compensate, in a sense, for unphysical modes of a given field; gauge symmetries are crucial in proving unitarity and renormalization in physics.

The quantum mechanical breaking of gauge symmetries was initially discovered by Bardeen in [1] using a perturbative calculation. Since its discovery in 1969, this pathology has been shown to have its roots deep in topological structures inherent to gauge theories [2]. This fact is certainly reinforced when it comes to the study of global anomalies, as we shall soon see.

Anomalies, in general, are symptoms of an ill-defined theory. A theory is known to be free of perturbative anomalies when its effective action is invariant under gauge and coordinate transformations that *can* be reached continuously from the identity.

Global Anomalies

Witten, in reference [3], introduced global anomalies through this rather interesting approach: even though a theory is free of perturbative anomalies, one may wonder whether its effective action is invariant under gauge or coordinate transformations that are *not* continuously connected to the identity. Gauge or coordinate transformations that *cannot* be reached continuously from the identity are referred to as *global gauge* or *coordinate transformations*, and lack of invariance of the effective action under such transformations is called a *global gauge* or *gravitational anomaly*. Global anomalies occur when *large* gauge transformations of a classical field theory fail to be symmetries of the corresponding quantum theory. Global gravitational anomalies, in particular [3, 9], reflect non–invariance of a theory under the mapping class group or homeotopy group [4], that is, the group of equivalence classes of diffeomorphisms that cannot be smoothly deformed to the identity.

In quantum field theories, diffeomorphism invariance is referred to as the perfect symmetry [5]. Consequently, the occurrence of global gravitational anomalies is taken as a sign that the theory is sick, and it becomes necessary to arrange to cancel them, meaning that one should establish invariance of the effective action under global coordinate transformations. According to the authors whose work is cited in references [2] and [3], situations in which one may encounter gravitational anomalies associated with disconnected general coordinate transformations are theories in which:

- the matter field in dimension $4k + 2$ cannot have bare masses – since they transform in a complex representation of the group $SO(4k + 2)$ [6];
- theories that are generally defined in dimensions $8k$ or $8k + 1$, with a single Majorana field or any odd number of them [3]; and
- theories in which bare masses are generally possible [3, 6].

II. Global Gravitational Anomalies in Topological Field Theory

The Manifestations

Global gravitational anomalies are difficult to evaluate in $D = 2+1$ Topological Quantum Field Theory (TQFT). The situation owes much to a somewhat vague relationship between TQFT and Quantum Gravity: in dimension three, the Chern–Simons–Witten (CSW) theory gives the first order formalism of the gravity; the gravity is then described by spin connections and dreibeins, provided that a gauge group of the CSW theory is chosen to give a structure group of the local frame bundle over a 3–manifold.

Yet, much remains to be known about the relationship between CSW theories and $2+1$ Quantum Gravity theories. In particular, the CSW gravitational anomaly counterterm is obtained by indirect derivation, proceeding via conformal field theory. A more straightforward derivation would certainly strengthen our understanding, and a better analysis of this term's behavior under the homeotopy group could be quite valuable.

There are, so far, two possible forms for the gravitational CSW term: it can be written in terms of a spin connection or an affine connection [11]. In the first form, the anomalous transformations are $SO(2,1)$ gauge transformations (i.e. local Lorentz transformations), while in the second form they are diffeomorphisms. The relationship between the two forms is very poorly understood.

The same can be said about the mapping class group of 3–manifolds: a complete set of generators and relations has yet to be discovered and classified.

Global gravitational anomalies in Topological Quantum Field Theory can be approached in various ways. The first symptom manifests itself in the CSW effective action. In reference [7], Witten showed through explicit path integral computation, that to the lowest order of perturbation, the action for a CSW theory requires a counterterm proportional to the gravitational CSW action. Since this action is not invariant under large diffeomorphism transformations, this term represents a global gravitational anomaly.

The second and perhaps the most difficult symptom that needs to be worked out in detail is the Wilson–line anomaly. This anomaly appears as a framing dependence because the choice of the framing is *not* mapping class group invariant. The coefficient of this anomaly can be calculated to the lowest order of perturbation–with the exact value corresponding to the central charge associated with the CSW theory.

Finally, the canonical quantization of CSW theories makes succinct the relationship between anomalies and mapping class groups. The geometric quantization procedure of [8] exhibits an anomaly that shows up as a dependence of the wave function on the moduli space of the the 2–dimensional Riemann surface Σ. Explicitly, canonical quantization gives a projectively flat–but not necessarily flat–bundle over the moduli space with a curvature given by the anomaly; the mapping class group of Σ is then known to have a non–trivial and computable action on this bundle.

The Anomaly Cancellation

I will now proceed to demonstrate the absence of global gravitational anomalies for $2+1$ Topological Quantum theories, particularly the ones that are defined on manifolds with simple topologies of the type $\Sigma \times [0,1]$. The interested reader can find a detailed account of this work in reference [9]. The proof will rely on the absence of disconnected coordinate transformations via establishing the invariance of the CSW effective action under the group of equivalence classes of diffeomorphisms that *cannot* be smoothly deformed to the identity.

We associate to a 3–manifold $M = \Sigma \times [0,1]$ the CSW topological invariant:

$$\begin{aligned} \mathcal{L} = \frac{1}{4\pi} \int_M tr\,(A \wedge dA + \frac{2}{3} A \wedge A \wedge A) = \\ \frac{k}{8\pi} \int_M tr\,(A_i\,(\partial_j\, A_k \;-\; \partial_k\, A_j) + \frac{2}{3}\, A_i[A_j, A_k]) \end{aligned} \qquad (1).$$

A is the connection, a one–form taking values in a given Lie Algebra $\mathcal{G}$; $[0,1]$ is the associated one–dimensional time component of Σ. The connection and gauge transformation can be understood as follows: if V is a C^∞ vector bundle over Σ, then a connection is a differential operator

$$d_A : \Omega^0\,(\Sigma; V) \rightarrow \Omega^1\,(\Sigma; V),$$

such that

$$d_A\,(f_s) = df \otimes s + f\, d_A\, s$$

for any C^∞ function f and section $s \in \Omega^0\,(\Sigma; V)$. The $\Omega^p\,(\Sigma; V)$ are p–forms on Σ with values in V.

Now, a gauge transformation is an automorphism of the bundle V. Locally, a gauge transformation g is a C^∞ function with values in $GL(n,C)$. Gauge transformations act on connections by conjugating the differential operator, $g^{-1}\, d_A\, g$.

We denote the group of orientation preserving diffeomorphism of Σ by $Diff\,(\Sigma)$, and furthermore, we define the group of isotopy classes of orientation preserving self–diffeomorphism of Σ as the mapping class group, which we write as Γ_Σ.

According to results of [6], one can exhibit the presence of global gravitational anomalies if the term $e^{-\Gamma_\Sigma[A,g]}$ provides a trivial, one–dimensional representation of the group of diffeomorphisms of the 3–manifold M. Working with this procedure however, presents the possibility of losing the topological invariance of equation (1). Instead, we choose the method of working out the invariance of the CSW action under the mapping class group Γ_Σ.

A given element λ of $Diff(\Sigma)$ yields the relation:

$$\mathcal{L} = \mathcal{L}(\lambda) = \mathcal{L}(1) \qquad (2),$$

for all λ, with 1 as the identity. We now take h_t, a one–form of λ to be:

$$h_t = \lambda_t; \qquad 0 \leq t \leq 1$$

with $\lambda \in \Gamma_\Sigma$. Using $\phi \in Diff(\Sigma)$ enables us to write down the mapping:

$$\phi \times 1 : \Sigma \times [0,1] \rightarrow \Sigma \times [0,1] \qquad (3).$$

$\phi \times 1$ is thus a diffeomorphic transformation, and can be shown to induce the following map:

$$\Sigma \times [0,1] \xrightarrow{\phi\times 1 \sim \phi^*(A)} \Sigma \times [0,1] \xrightarrow{A} \mathcal{G} \qquad (4).$$

Replacing A by the under diffeomorphism transformation $\phi^*(A)$ yields:

$$\mathcal{L}\,(\phi^*) = \frac{1}{4\pi}\int_M tr(\phi^* A \wedge d(\phi^* A) + \frac{2}{3}(\phi^* A) \wedge (\phi^* A) \wedge (\phi^* A)) \qquad (5).$$

Note that $\phi^* A$ is the one–form obtained by pulling back along equation (4).

Introducing at this point the mapping

$$\pi : Diff(\Sigma) \rightarrow \Gamma_\Sigma,$$

and identifying $\pi(\phi)$ with $\pi(\theta)$ allows us to verify that

$$\mathcal{L}\,(\phi^*(A)) = \mathcal{L}\,(\theta(A)) \qquad (6)$$

holds.

With these descriptions in hand, we can now consider a set of diffeomorphism transformations $\alpha, \beta : \Sigma \to \Sigma \times [0,1]$ for which one associates $h_0 = \alpha; h_1 = \beta$. Equation (6) then becomes:

$$\mathcal{L}(\alpha^*(A)) = \mathcal{L}(\beta^*(A)) \qquad (7);$$

where $\alpha^*(A)$ is the one–form generated by the mapping

$$\Sigma \times [0,1] \xrightarrow{\alpha \times 1} \Sigma \times [0,1] \xrightarrow{A} \mathcal{G}.$$

For $\mathcal{L}(\phi^*(A)) = \mathcal{L}(\theta(A))$ to be diffeomorphism invariant means that equations (6–7) should be invariant under the diffeomorphism transformations:

$$(e^{i\theta}, e^{i\phi}) \mapsto (e^{i(\theta+\epsilon)}, e^{i\phi}), \quad 0 < \epsilon < 2\pi.$$

Consequently, the homeotopy h_t reads:

$$h_t(e^{i\theta}, e^{i\phi}) = (e^{i(\theta+t)}, e^{i\phi})),$$

and we can see that

$$\mathcal{L}(h_t(A)) = \mathcal{L}(A) = \mathcal{L}_{eff} \qquad (8)$$

is indeed mapping class group invariant. This concludes the proof that Topological Quantum Field Theory is inherently a theory free of global gravitational anomalies, at least as far as the Hamiltonian version of the theory is concerned.

III. Knot Theory, Exotic Spheres and Global Anomalies

I would now like to introduce a relationship between knot theory, exotic spheres and global anomalies. These connections have been recently discovered in a joint work with L. H. Kauffman [10].

The initial idea of using exotic spheres as a solution for the cancellation of global anomalies is due to Witten [3]. His arguments stem from deep topological observations: in detecting global anomalies, a result by Milnor makes the use of exotic spheres rather straightforward, given that the group of diffeomorphisms that enters in their construction *may not* be continuously connected to the identity. The following short presentation gives a clearer picture of it.

We start by considering the $N = 2$, $D = 10$ Supergravity theory, or for that matter any 10–dimensional physical theory which is known to be free of perturbative anomalies. (Certain restrictions may apply that will make some choices trivial. For instance, the 10-dimensional Heterotic Superstring theory does not have global gauge anomalies because of the fact that $\pi_{10}(E_8 \times E_8) \cong 0$). The manifold in which the theory is defined is referred to as M^{10}; and g is the space of gauge transformations, with $g \times Diff(M^{10})$ its corresponding group of gauge transformations and diffeomorphisms. We write $\pi(g \times Diff(M^{10}))$ as the number of components of $g \times Diff(M^{10})$ – incidentally, it is a discrete group that gives all possible global gauge transformations.

Since the $N = 2$, $D = 10$ Supergravity theory is known to be free of perturbative anomalies [2], the next level of interest is to ask whether the exponentiation of the effective action provides a one–dimensional representation of $\pi(g \times Diff(\Sigma^{10}))$. In reference [3], it is argued that, when this is the case, the theory suffers from global gravitational anomalies.

How one evaluates these anomalies largely depends on how many components $Diff(\Sigma^{10})$ possesses. The answer to this question is unknown. We are therefore left with applying a somewhat indirect derivation. Consider a special class of diffeomorphisms which are different from the identity only in some small ball B^{10}, and become the identity outside. These diffeomorphisms are actually similar to diffeomorphisms on S^{10}, and consequently, one may wonder about the number of components of $(Diff(S^{10}))$, and whether $N = 2$, $D = 10$ Supergravity is invariant under the action of these disconnected diffeomorphisms.

This very question is related to the existence of Milnor and Kervaire's exotic eleven–spheres [12, 13]; Milnor in particular, has shown that exotic $(k+1)$ spheres and topological classes of diffeomorphisms of the k–sphere are in one–to–one correspondence.

For a twelve–dimensional manifold without boundary, the integrals of products of characteristic classes that are relevant as topological invariants are:

$$I_1 = \int_{M^{12}} (tr\, R^2)^3;$$

$$I_2 = \int_{M^{12}} tr\, R^2 \cdot tr\, R^4;$$

$$I_3 = \int_{M^{12}} tr\, R^6.$$

These quantities are independent of the metric and the connection. When M^{12} has a non-empty boundary, the expressions written above are topological invariants. In reference [3], Witten relates the generalization of these quantities (to 12-manifolds admitting a boundary) to solving $tr\,R^2 = dH$ in the boundary of $M^{12} = \Sigma^{11}$. Solving $tr\,R^2$ on Σ^{11}, according to Witten's claim, amounts to removing the obstruction in generalizing the topological invariant I_i. The generalization itself gives:

$$\bar{I}_1 = \int_{M^{12}} (tr\,R^2)^3 - \int_{\partial M^{12}} H\,(tr\,R^2)^2;$$

$$\bar{I}_2 = \int_{M^{12}} tr\,R^2 \cdot tr\,R^4 - \int_{\partial M^{12}} trR^4.$$

From now on, we take $\partial M^{12} = \Sigma^{11}$ to be an eleven-dimensional exotic sphere. What we have shown so far is simply that the exotic 11-sphere provides us with a powerful means to solve the geometric intricacies of 12-dimensional manifolds. But this relates, in turn, to the property of a twelve-manifold to bound an exotic 11-sphere. Therefore, the question of interest at this point is: when can one accurately state that such bounding exists? Using integral expressions for Pontrjagin classes, we can see that the resulting characteristic form:

$$\lambda(\Sigma^{11}) = \alpha p_1^3 + \beta p_1 p_2 - \frac{\sigma(M^{12})}{8 \cdot 992},$$

is modulo 1, an invariant of Σ^{11} (otherwise, $\lambda(\Sigma^{11})$ is an integer if Σ^{11} is a standard sphere; note that $\sigma(M^{12})$ is the signature of the twelve-manifold M^{12}.

The example is due to Milnor of an exotic sphere Σ^{11} bounding M^{12} with the property that

$$\lambda(\Sigma^{11}) = -\frac{1}{992};$$

$\sigma(M^{12}) = 8$ in this case. This gives rise to a family of 991 exotic spheres via connected sums of the Milnor sphere. In evaluating global gravitational anomalies, Witten's insight was to look at these connected components and relate them to large diffeomorphism transformations (that is, to small variations of the effective action under disconnected coordinate transformations). This is made explicit in what follows: First, we recall a result by Milnor [12], that every exotic 11-sphere bounds a 12-manifold whose signature is divisible by eight. Second, as noted in [3] and [12], exotic spheres are natural candidates for the cancellation of global gravitational anomalies because the group of diffeomorphisms that enters in their construction *need not* be continuously connected to the identity.

Under large diffeomorphism transformations, the variation of an effective action which is not continuously connected to the identity reads:

$$\Delta\mathcal{L} = 2\pi i\, \frac{\sigma(M^{12})}{8}.$$

Since Σ^{11} and the topological classes of diffeomorphisms of M^{10} are in one–to–one correspondence (see [12]), we can generalize this case to 10–dimensional $N = 2$ Supergravity: the theory is then known to be free of disconnected general coordinate transformations, or equivalently, global gravitational anomalies when

$$\Delta\mathcal{L} = 0 \ mod\ 2\pi i.$$

Recently, in a joint work with L. H. Kauffman [10], a novel method aimed at cancelling global gravitational anomalies was exhibited. A striking feature of the construction is the finding that the 10–dimensional global anomaly is in fact encoded in the 3–dimensional classical invariants of knots and links. To arrive at this result, we used several topological properties of a categorie of surgery–generated manifolds, known as Link Manifolds. These are smooth, closed manifolds of dimension $2n+1$ that admit a smooth action of the orthogonal group $O(n)$, such that all isotopy subgroups are conjugate to $O(n)$, $O(n-1)$, or $O(n-2)$, and their orbit space is diffeomorphic to the 4–ball, D^4. The fixed point set in the manifold M of the $O(n)$ action corresponds to a link $L \subset S^3 = \partial D^4$.

In reference [10], it is noted that certain classical invariants of links in the 3–sphere suffice to completely determine the differentiable (i.e. exotic) structure of a corresponding higher–dimensional link manifold. The signature and quadratic form invariants in high dimensions, for instance, can be calculated from invariants of the link in dimension three. What specific benefits, with repect to global anomalies, the present relationship between low and higher dimensional topology, as outlined in [10], can achieve is discussed at the end of this presentation. However, before getting to that point, I would like to discuss more explicitly the construction of [10], particularly the basic relationship between links in the 3–sphere and the link manifolds, and subsequently, emphasize their connection with exotic structures and global gravitational anomalies.

To an oriented link $L \subset S^3$, we associate a Seifert matrix V, computed from an oriented surface F spanning L. This yields the relations : $F \subset S^3$; $\partial F = L$. The Seifert matrix is taken to be the matrix

$$V_{ij} = \theta(a_i, a_j),$$

of the Seifert pairing

$$\theta : H_1(F) \times H_1(F) \longrightarrow Z$$

with respect to a basis for the first homology group of the surface F. We can actually write the Seifert pairing as

$$\theta(a, b) = lk(a^*, b),$$

where a^* denotes the result of pushing the cycle a on F into the complement $S^3 - F$, along a positive normal to the surface F; and lk is the linking number of curves in S^3.

The signature of the link $\sigma(L)$ is, by definition the signature of the matrix $V + V^{\top}$; here $V^{\top}$ stands for the transpose of V. Turning now to the quadratic form, we write a symmetric bilinear pairing

$$f : H_1(F) \times H_1(F) \longrightarrow Z$$

in virtue of the relation:

$$f(x, y) = \theta(x, y) + \theta(y, x).$$

When V is taken to correspond to $H_1(F)$, this pairing gives an exact sequence

$$V \longrightarrow Hom(V, Z) \xrightarrow{\pi} \mathcal{G} \longrightarrow 0,$$

with a torsion linking pairing on the torsion subgroup of $\mathcal{G}$. This linking form has an associated quadratic form $q(f) : \tau\mathcal{G} \rightarrow Q/Z$.

Assuming that M^{4k+1} is an $O(2k-1)$ link manifold corresponding to the link $L \subset S^3$, and taking $M^{4k-1} = \partial N^{4k}$ to be a smooth manifold, then [10]:

$$\sigma(L) = \sigma(N^{4k}).$$

The quadratic form $q(f) : \tau\mathcal{G} \rightarrow Q/Z$ is identical with the quadratic form on

$$\tau H_{2k-1}(M^{4k-1}) \rightarrow Q/Z$$

that is needed for the classification of M^{4k-1}.

By doing a sequence of surgeries –corresponding to the bands on the spanning surface F– to a $(2k+1)$–sphere, we obtain a link manifold, M^{2k+1}! The surgery obtained Link manifold is equivalent to the Brieskorn variety

$$\Sigma(3,2,...,2) = V(z_0^3 + z_1^2 + ... + z_{k+1}^2) \cap S^{2k+3}$$

, that is:

$$M^{2k+1}(\tau) \simeq \Sigma(3,2,2,...,2).$$

Note that $O(n)$ acts on $z_2, ..., z_{k+1}$; furthermore, M^{2k+1} is the link manifold for the right–handed trefoil knot, and the Brieskorn variety $\Sigma(3,2,...,2)$ is an exotic sphere. Perhaps the most well known example in this class is $M^7(3,5)$, that is, the Milnor exotic 7–sphere, which corresponds to the $(3,5)$ torus knot of signature 8.

What is the restriction for M^{2k+1} to truly become an exotic sphere? This depends (in the case where $k+1$ is even) on the signature of the manifold W^{2k+2} bounded by M, and also on the relationship

$$\sigma(W^{2k+2}) = \sigma(K);$$

where $\sigma(K) = \sigma(\theta + \theta^{\top})$; θ is the Seifert pairing of the spanning surface

$$F(\partial F = K).$$

It is then just a matter of procedure to cancel the higher dimensional global gravitational anomalies using the resulting 11-dimensional exotic sphere obtained by equivariant surgery.

IV. Acknowledgments

I wish to express my gratitude to L. H. Kauffman for many years of stimulating discussions. It is also a pleasure to acknowledge useful exchanges with E. Witten, F. Cohen, A. Casson and S. N'Djimbi–Tchikaya, and to thank R. Cahn for his support at Berkeley.

V. References

[1] Bardeen, W. A. : Phys. Rev. 184 (1969), 1848.

[2] Alvarez–Gaumé, L. and Ginsparg, P. : Nucl. Phys. B234 (1984), 449.
– Annals Phys. (N Y) 101 (1985), 423.

[3] Witten, E. : *Global Gravitational Anomalies*, Commun. Math. Phys. 100 (1985), 197.

[4] Birman, J. : *Braids and Mapping Class Groups*, Annals of Maths. Studies (Princeton) 82 (1975).

[5] Friedman, J. L. and Witt, D. N. : *Problems on diffeomorphisms Arising from Quantum Gravity*, Comtemp. Math. Vol. 71 (1988).

[6] Alvarez–Gaumé, L. and Witten, E. : *Gravitational Anomalies*, Nucl. Phys. B234 (1983).

[7] Witten, E. : *Topological Quantum Field Theory and the Jones Polynomial*, Commun. Math. Phys. 121 (1989), 351.

[8] Axerold, S.; Della–Pietra, S; Witten, E.: *Geometric Quantization of Chern–Simons Gauge Theories*, Journal of Differential Geom. 33 (1991).

[9] Baadhio, R. A. : *Global Gravitational Anomaly–Free Topological Field Theory*, Physics Letters B299 (1993), 37–40.

[10] Baadhio, R. A. and Kauffman, L. H. : *Link Manifolds and Global Gravitational Anomalies*, June 1992. To be published in Review in Math. Physics.
– Kauffman, L. H. : *Link Manifolds*. Michigan Maths. Journal 21 (1974) 33-44.

[11] Witten, E. : 2+1 *Dimensional Gravity as an Exactly Soluble System*, Nucl. Phys. B311 (1988/89), 46.

[12] Milnor, J. W. : *Differentiable Manifolds Wich are Homotopy Spheres*, Princeton Lecture Notes (1959) (unpublished).

[13] Kervaire, M. and Milnor, J. : *Groups of Homotopy Spheres I*, Annals of Math. 11 (1963), 504-537.

A DIAGRAMMATIC THEORY OF KNOTTED SURFACES

J. SCOTT CARTER
Department of Mathematics
University of South Alabama
Mobile, Alabama 36688
f4t3@usouthal.bitnet

and

MASAHICO SAITO
Department of Mathematics
University of Texas
Austin, Texas 78712
saito@math.utexas.edu

Abstract

A survey of our results in the diagrammatic approaches to knotted surfaces is given in part I of this paper. Results obtained by knotted surface diagrams and by movie descriptions will be reviewed. Diagrammatic ways of solving generalizations of the Yang-Baxter equation are explained. In part II, we give smoothings of higher dimensional knots that topologically generalize the classical knot smoothings of crossings and Kauffman's bracket trick for resolving the Yang-Baxter equation.

Key Words: Knotted surfaces, projections, movies, smoothings, simplex equations.

Part I. Diagrammatic Approaches to Knotted Surfaces

Although the Jones polynomial historically arose from algebra [17], knot theoretical and diagrammatic approaches simplified the theory and provided many applications. Some possibilities of quantum invariants in dimension 4 were reported in this conference. Such invariants will be defined by using the diagrammatic aspects of topology. Diagrammatic coincidences between topology and algebra have produced such invariants in the classical case. A close connection between topological and algebraic objects via diagrams is the remarkable aspect of this subject. Therefore diagrammatic approaches are also important to higher dimensional knot theory.

Knot diagrams also play a key role in other aspects of classical knot theory. After all, diagrams present knots precisely and effectively. Proofs are often carried out on diagrams, as are experiments in search of theorems. To compute an invariant, one starts with a diagram. Thus, diagrammatic approaches to higher dimensional knot theory are important — not only as an attempt to find quantum invariants, but also to understand knots by visualizing them. Indeed, we have obtained new topological results in 4-dimensional knot theory from this approach.

In part I of this paper, we review these results. The main interest will be knotted surfaces: *i.e.* smoothly embedded surfaces in 4-space. *The classical case* always means the case of knotted and linked curves in 3-space.

1 Projections

1.1 Knotted surface diagrams. Knotted and linked curves are represented by nice (smoothly and generically immersed) circles on the plane. In this representation, the circles have isolated crossing points and no more than two arcs of the circle intersect at a point. Any knot can be represented in this way, and such circles can be lifted to a knot by indicating crossing information (over/under paths).

A *knotted surface diagram* consists of a projection together with a method of distinguishing over and under crossing information. The terms *over and under* refer to the relative distance from the knotted surface to the hyperplane of projection when measured in the direction of projection. Crossing information is depicted by breaking one of the two intersecting sheets along the double curve. In this way, the standard model for a broken surface would be the sets $\{(x,y,0) : |x|,|y| \leq 1\}$ and $\{(x,0,z) : |x| \leq 1, 1/4 \leq |z| \leq 1\}$. The second set represents the broken surface as it appears in 3-space; this is the sheet that is closest to the hyperplane onto which the surface is projected. Similarly, representations of neighborhoods of points that project to triple points and branch points can be parametrized, and the parametrization is clearly indicated

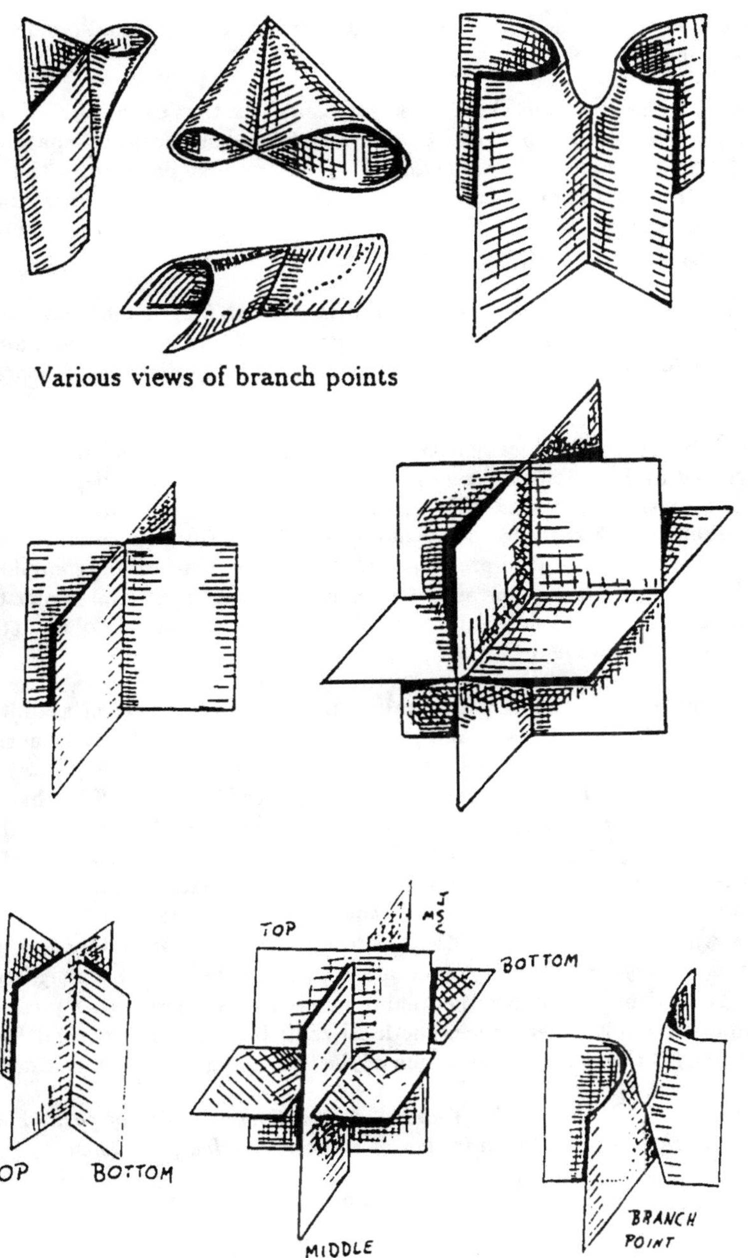

Various views of branch points

Figure 1

in Figure 1 where the local pictures of double points, branch points, and triple points are depicted.

Giller [13] observed that there are knotted surfaces that cannot be isotoped to have immersed projections, and there are immersed surfaces in 3-space that do not lift to embeddings in dimension 4. Thus the first problem in the diagrammatic theory of knotted surfaces is to determine the circumstances under which we can assume that the diagrams of knotted surfaces are sufficiently nice. Generalizing Giller's result, we proved:

1.1.1 Theorem [2]. *A knotted surface with normal Euler number n can be isotoped so that the projection has n branch points. This is the minimal number of branch points that can be achieved among all of the projections of isotopic embeddings.*

The proof utilizes the duality between characteristic classes and singularities first observed by Banchoff ([1], see the next section for more details).

On the contrary, it seems difficult to determine when we can lift a given immersed surface in 3-space to an embedding in 4-space. Combinatorial analyses on the double decker set (the preimages of the double point set) are employed in a rather *ad hoc* fashion. One asks if there is a lifting at each triple point to a top, middle, and bottom sheet such that the choices make sense globally (that is, over the entire double decker set.)

1.2 Formulas for normal Euler number. If the projection of a knotted surface is an immersion, it has a vanishing normal Euler number since the normal bundle has a trivial summand. Therefore the singularities (branch points) are related to the normal Euler number. It was first noticed by Banchoff [1] that the normal Euler number of a surface embedded in 4-space could be computed by means of indices associated to the branch points. When we fix a height function on 3-space, a branch point is described by a birth or a death of a small kink in an arc (type I Reidemeister move). A branch point is *positive* if it is the birth of a kink with negative writhe or the death of positive writhe. Otherwise, the branch point is *negative.* The key observations in understanding this assignment are: (1) The writhe determines the linking number of a curve with a push-off; (2) this linking number determines the local contribution to the normal Euler number; and (3) the induced orientation of the ball changes at birth and death.

1.2.1 Theorem [1] [2]. *Let $b(+)$ and $b(-)$ denote the number of positive and negative branch points of the projection of a knotted surface, F. Then*

$$b(+) - b(-) = n,$$

where n denotes the normal Euler number of the surface.

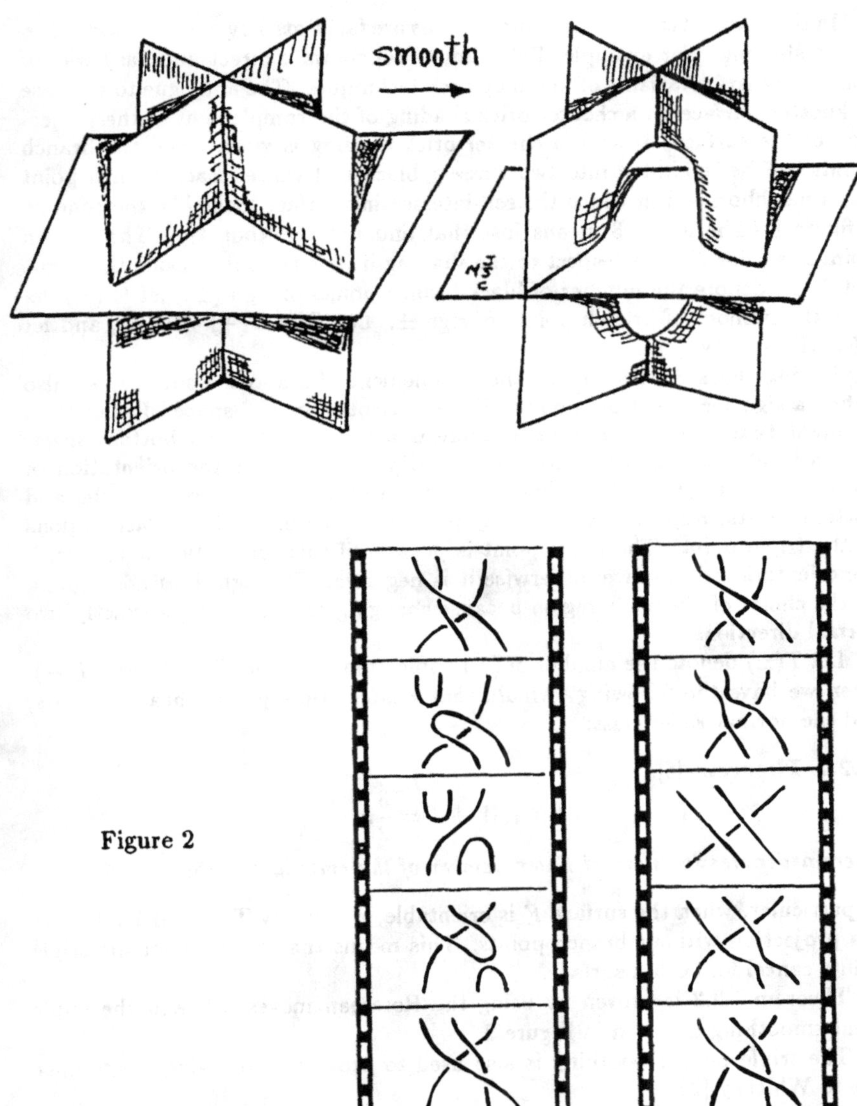

Figure 2

In the classical case, many computations are facilitated by means of a checker-board shading. For example, Pat Gilmer [14] shows a direct and easy way of computing Arf invariants of knots by such techniques. The analogue to the case of knotted surfaces is a checker-brick shading of the complement of the projection of the surface. If such a checker-brick shading is given, then the branch points can be separated into two classes: black and white. Each branch point has a neighborhood in which the self-intersecting surface looks like the cone on a figure 8 (The figure 8 means just that and not the knot 4_1). The branch point has color X with respect to the shading if X is the color inside this cone. Let $B(\pm)$ denote the number of black branch points of sign $(\pm)$, let $W(\pm)$ denote the number of branch points of sign $\pm$. Let $B = B(+) - B(-)$, and let $W = W(+) - W(-)$.

Having checker-brick shaded the complement of a knotted surface, we also define a sign for the triple points. Fix an orientation of 3-space. Recall from Figure 1 that at each triple point there is a top, middle, and bottom sheet. Pick a local (normal) orientation of the triple point to be the orientation of the triple (v_t, v_m, v_b), where these vectors are normal to the top, middle, and bottom sheets, respectively, and they point into one of the four black regions at the triple point. The triple point is positive if this orientation agrees with the orientation of 3-space; otherwise it is negative. The sign does not depend on the choice of the black region because changing regions changes exactly two normal directions.

Let $T(\pm)$ denote the number of $(\pm)$ triple points, and let $T = T(+) - T(-)$. Then we have the following formula that relates triple points, branch points, and the normal Euler class:

1.2.2 Theorem [6].

$$2B + W + T = \frac{3}{2}n,$$

where n denotes the normal Euler number of the embedded surface.

In particular, when the surface F is orientable, $n = 0$. By Theorem 1.1.1 there is a projection without branch points. This means that the signs of the triple points cancel for such a surface.

Theorem 1.2.2 is proven by using the Roseman moves [24] and the triple point smoothing depicted in Figure 2.

The triple point smoothing is also used to prove the following congruence due to Whitney [25]:

1.2.3 Theorem [2]. *The normal Euler number, n, of an embedded surface in 4-space satisfies*

$$\frac{1}{2}n \equiv \chi \pmod 2,$$

where χ denotes the Euler characteristic of the surface.

1.3 Seifert algorithms. In classical knot theory, the Alexander matrix is computed by using Seifert surfaces bounded by a given knot. Therefore it is important to have a visual construction of such surfaces for a given knot. The Seifert algorithm is such a construction. It constructs a surface from a given knot diagram (see for example [23]). First, crossings are smoothed, and the resulting disjoint nested circles bound nested embedded disks. The surface is obtained by attaching twisted bands that recover the original knot as the boundary. We generalized this method to knotted surfaces:

1.3.1 Theorem [6]. *Given an immersed knotted surface diagram of an orientable surface, there is an algorithm to construct an orientable 3-dimensional solid that is bounded by the represented surface.*

The construction is similar to classical case. Triple points are smoothed by means of the method depicted in Figure 2. Then the double point set is smoothed by the product of classical smoothings of crossings with an interval. A nested collection of embedded surfaces results in 3-space. Let them bound solids that are interiors in 3-space, and lift these cubes-with-knotted-holes-and-handles to disjoint levels in 4-space. Twisted bricks are attached along the double point curves, and then 2-handles are attached to recover the triple points at the boundary.

Furthermore, (after certain finger moves if necessary), we can get Heegaard diagrams for the solids thus obtained. As an application of this technique we gave a direct computation that Fox's Example 12 [11] bounds the connected sum of a lens space and copies of $S^2 \times S^1$. The proof is direct in the sense that we don't need to know at the outset that the knot is a twist spun trefoil (which was observed in [22]).

1.4 Diagrammatic algebras — diamonds. Classical knot projections give Wirtinger presentations of knot groups whose relations are conjugations. This formality was abstracted by Joyce to define algebraic objects called quandles [18]. Quandles were further generalized to crystals by Kauffman [21] and to racks by Fenn and Rourke [10]. Roughly speaking these objects consist of sets described by generators and relations defined in terms of certain binary operations. The relation set of a crystal that is associated to a knot comes directly from a representative diagram. The axioms among the operations are extracted from the Reidemeister moves. The generators are assigned to over-crossing arcs in a knot diagram, and each crossing gives a relation.

On the other hand, we assign different spins to each segment of a knot projection in a statistical mechanical model of the Jones polynomial.

This lead us to the idea of defining an abstract algebraic object by assigning generators to each arc in a knot projection. The axioms of this algebraic object (which we called a diamond [8]), are again derived from Reidemeister moves. These algebraic objects satisfy:

(a) To each knotted oriented surface there is an associated diamond that can be defined in terms of the knotted surface diagram.
(b) Finite diamonds can be used to define numerical invariants via partition functions. These give invariants of regular isotopy classes (which will be defined in the next section) of surface diagrams.
(c) By abelianization, there are modules over certain rings that have 0-divisors. These modules are invariants of knots that generalize the Alexander modules.

2 Movies

2.1 Visualizations by movies. In proving the results described in the previous sections, we have to ensure that the geometric techniques used are actually realizable as embeddings in 4-space. Because projections don't always lift and because one cannot always see all the information found in a knotted surface diagram (some regions are veiled by the surface), movies provide a more effective visualization tool. The cimematic technique has been known for years [11]. Here is a brief description.

Given a surface embedded in 4-space, consider its generic projection onto a 3-dimensional hyperplane. Pick a generic direction vector in this 3-space so that all but a finite number of the planes perpendicular to this direction intersect the image of the projected surface in a family of smooth generically immersed circles. The exceptional planes intersect at critical points of the (projection of the) surface, critical points of the double point set, triple points, or branch points on the projection. Therefore two planes — one on either side of an exceptional cross section — differ by the projection of one of the Reidemeister moves, a birth or death of a circle, or a surgery between arcs.

Each generic cross section is called *a still from the movie.* Since crossing information is encoded in a knotted surface diagram, this information is encoded in the movie by replacing the immersed circle in a still by a classical link diagram.

The *elementary string interactions (ESIs)* are the set of Reidemeister moves together with the birth or death of a simple closed curve and a surgery between two arcs. ESIs are depicted in Figure 3. Usually, in our movies pairs of successive stills differ by at most an elementary string interaction. In Figures 2 and 8 through 12, smoothings of crossings occur; these can be rewritten as a combination of ESIs. Moreover, from such a movie a knotted surface diagram is easy to reconstruct.

2.2 Movie moves. The stills in the movie description of a knotted surface differ by one of the ESIs. The problem of finding Reidemeister moves in this set-up is thus stated as follows: What is a finite set of moves among sequences of ESIs that suffice to produce all the equivalent sequences in ESIs? This question was posed to us by Kauffman. The solution is the topological part of the statistical mechanical approach to the Jones polynomial in dimension 4.

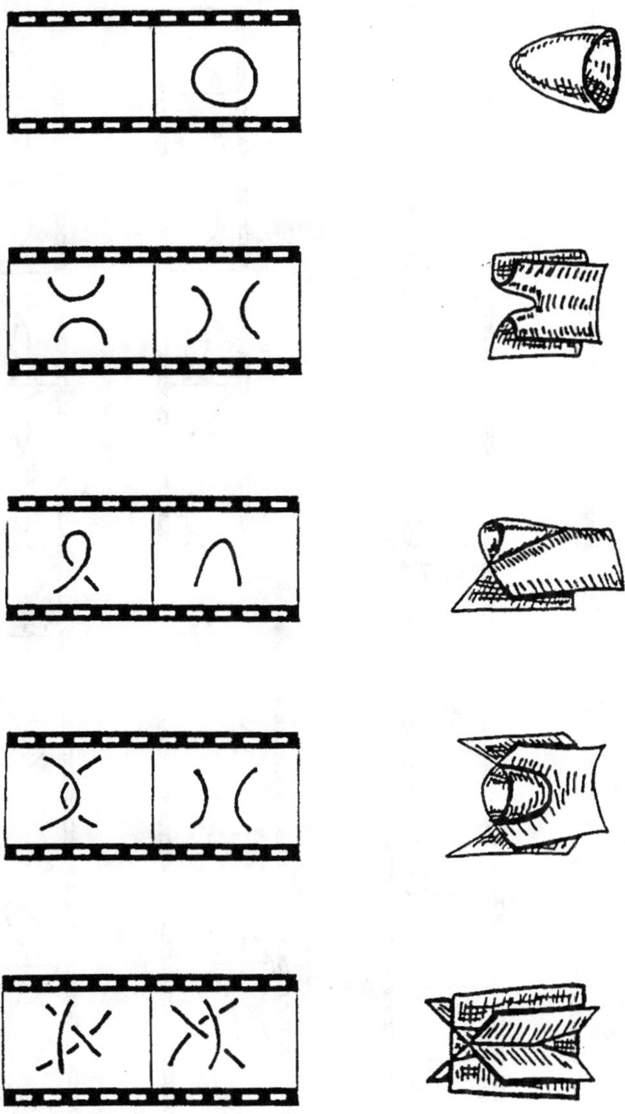

Figure 3

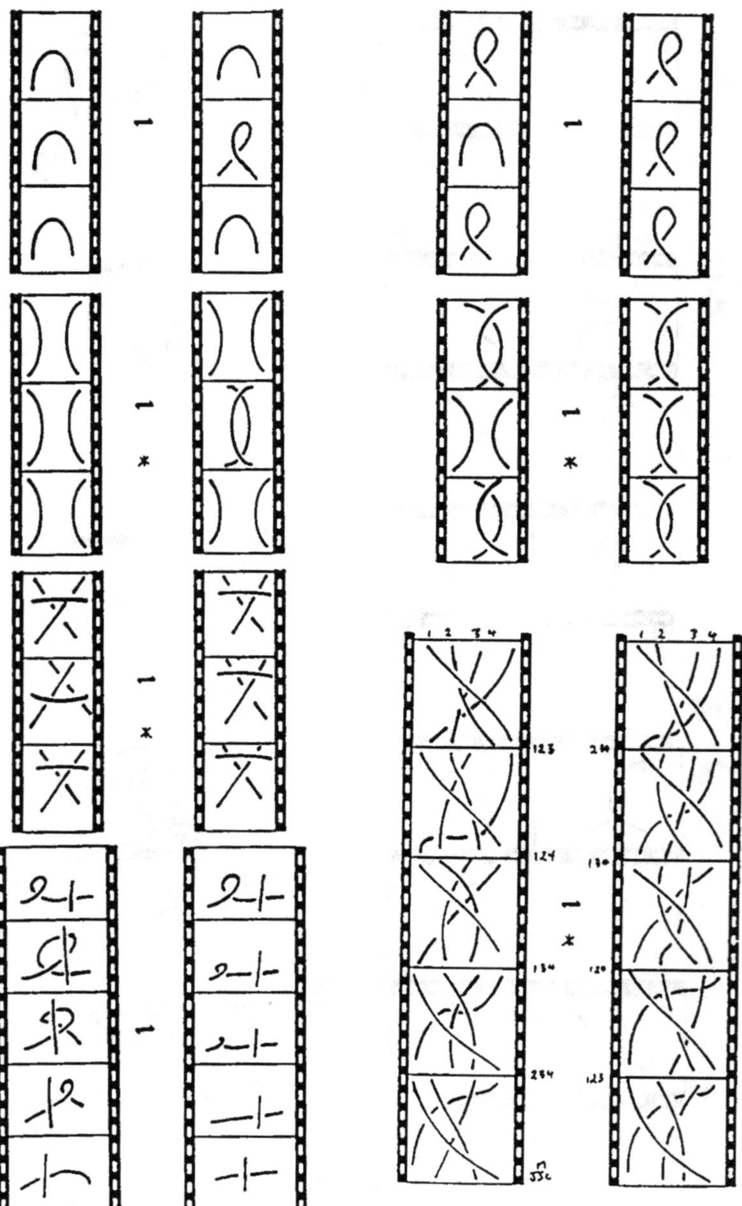

Figure 4

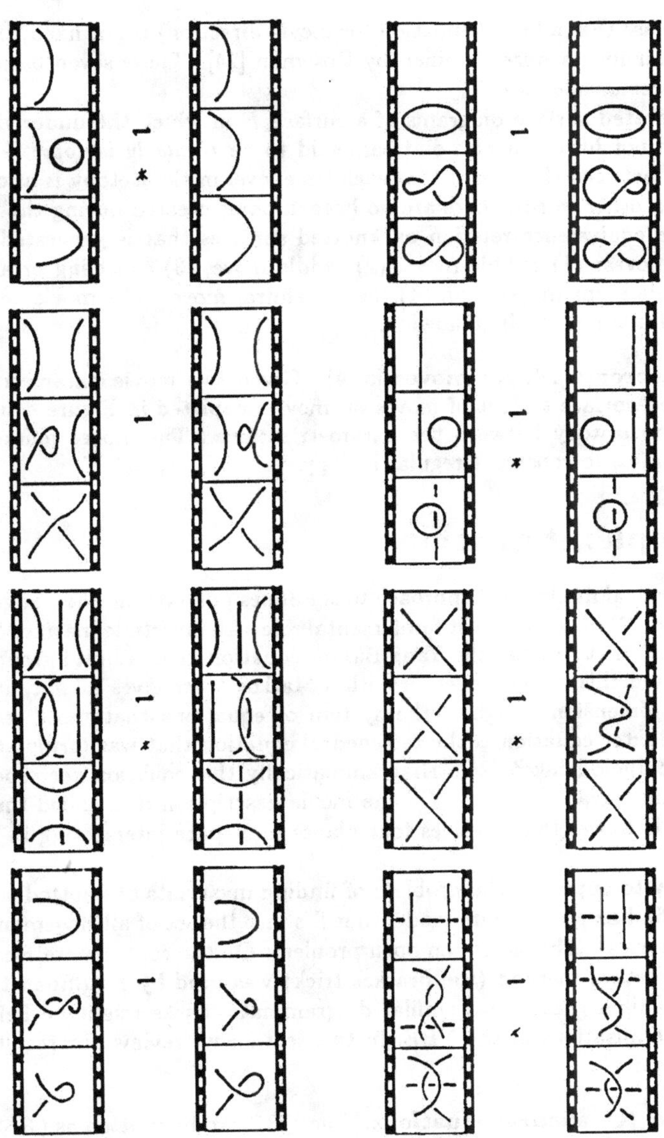

Figure 4 (continued)

In the case that a height function (or movie direction) is not fixed, the set of Reidemeister moves were obtained by Roseman [24]. These seven basic moves are ***the Roseman moves.***

Two knotted surface diagrams of a surface F in which the underlying projections do not have branch points are said to be ***regularly isotopic*** if there is an isotopy between them such that each time level in the isotopy is also an immersion. In other words, there are no branch point created during the isotopy. This is the equivalence relation on knotted surfaces that is generated by the Roseman moves: (1) bubble move, (2) saddle move, (3) canceling or adding a pair of triple point moves, and (4) the tetrahdral move. The movie version of these are illustrated in the following:

2.2.1 Theorem: Movie moves [3, 4]. Given two movie parametrizations of a knotted surface the set of moves on movies depicted in Figure 4 suffice to construct an isotopy between the parametrizations. The moves marked with asterisks suffice to construct regular isotopy.

3 Simplex Equations

In the statistical mechanical approach to the Jones polynomial, the Yang-Baxter equation (YBE) plays the most fundamental role. An invertible matrix R (called an R-matrix) that satisfies the Yang-Baxter equation gives a braid group representation, and this matrix together with a Markov trace gives a knot invariant.

In one dimensional higher, the system of equations that most resembles the Yang-Baxter equation is the tetrahedral equation that was formulated and solved by Zamolodchikov [26]. Diagrammatically this equation corresponds to the Roseman tetrahedral move (in the movie description it is called the tetrahedral movie move) that involves four planes in 3-space intersecting in general position.

One way to approach the problem of finding invariants of knotted surfaces, then, is to find an appropriate model that fits into the set of all Roseman moves (or movie moves). This is still an open problem. On the contrary we recall that a knot theoretical method (the bracket trick) was used by Kauffman to solve the YBE. This suggests that similar diagrammatic tricks might be helpful in solving generalizations of the YBE. In this section we review our results from this viewpoint.

3.1 Planar tetrahedral equations. The YBE can be written as $(R\otimes I)(I\otimes R)(R\otimes I) = (I\otimes R)(R\otimes I)(I\otimes R)$, where $R \in \mathrm{End}(V\otimes V)$ and I is the identity map on a vector space V over the complex numbers. This corresponds to the braid group relation, and thus soultions to the YBE give a braid group representation. Natural generalizations of this equation were formulated by repeating $(R\otimes I)$ and $(I\otimes R)$ alternating for a number of times where R :

$V^{\otimes a} \to V^{\otimes a}$ and where I denotes the identity operator on the bth tensor power of V. In particular, by the *planar tetrahedral equation* we mean the equation

$$(R \otimes I)(I \otimes R)(R \otimes I)(I \otimes R) = (I \otimes R)(R \otimes I)(I \otimes R)(R \otimes I).$$

In [5] we showed that there are solutions to the planar tetrahedral equation in case R acts on $V \otimes V \otimes V$ and I acts on V. The solutions are found in a representation of the Temperly-Lieb algebra. Specifically, we have the following:

3.1.1 Theorem. *The matrix*

$$R = A(H \otimes 1)(1 \otimes H) + B(1 \otimes H)(H \otimes 1) + C(H \otimes 1) + D(1 \otimes H) + E\,1$$

satisfies the planar tetrahedral equation provided that

$$\begin{aligned} A &= B && (1)\\ C &= D && (2)\\ A + E &= -\delta C && (3)\\ AE &= C^2 && (4) \end{aligned}$$

Here, H is the image of $\genfrac{}{}{0pt}{}{\cup}{\cap}$ *, under a representation of the Temperly-Lieb algebra.*

These solutions are found by a diagrammatic method utilizing the diagrams for the Temperly-Lieb algebra.

3.2 Double point equations. Suppose that R is a solution to the quantum Yang-Baxter equation. Thus $R_{12}R_{13}R_{23} = R_{23}R_{13}R_{12}$, each R is an endomorphism of $V \otimes V$ and R_{ij} acts on the ith and jth factors of $V \otimes V \otimes V$ while acting as the identity on the remaining factor. The Frenkel-Moore version [12] of the tetrahedral equation is

$$S_{123}S_{124}S_{134}S_{234} = S_{234}S_{134}S_{124}S_{123}.$$

We formulated a variant of the Frenkel-Moore equation as follows:

$$S_{124}S_{135}S_{236}S_{456} = S_{456}S_{236}S_{135}S_{124}.$$

This formulation is similar to one given in [20] and [19]. Each S is an endomorphism of $V \otimes V \otimes V$, and each S acts on the 6-fold tensor power of a vector space V, acting as the identity on the indices not mentioned.

This *double point equation* was formed by taking the set of pairs of indices in the Frenkel-Moore equation, lexicographically ordering these pairs, and reindexing the tensors with the lexical index. For example, in S_{134} each pair $\{1,3\}$,

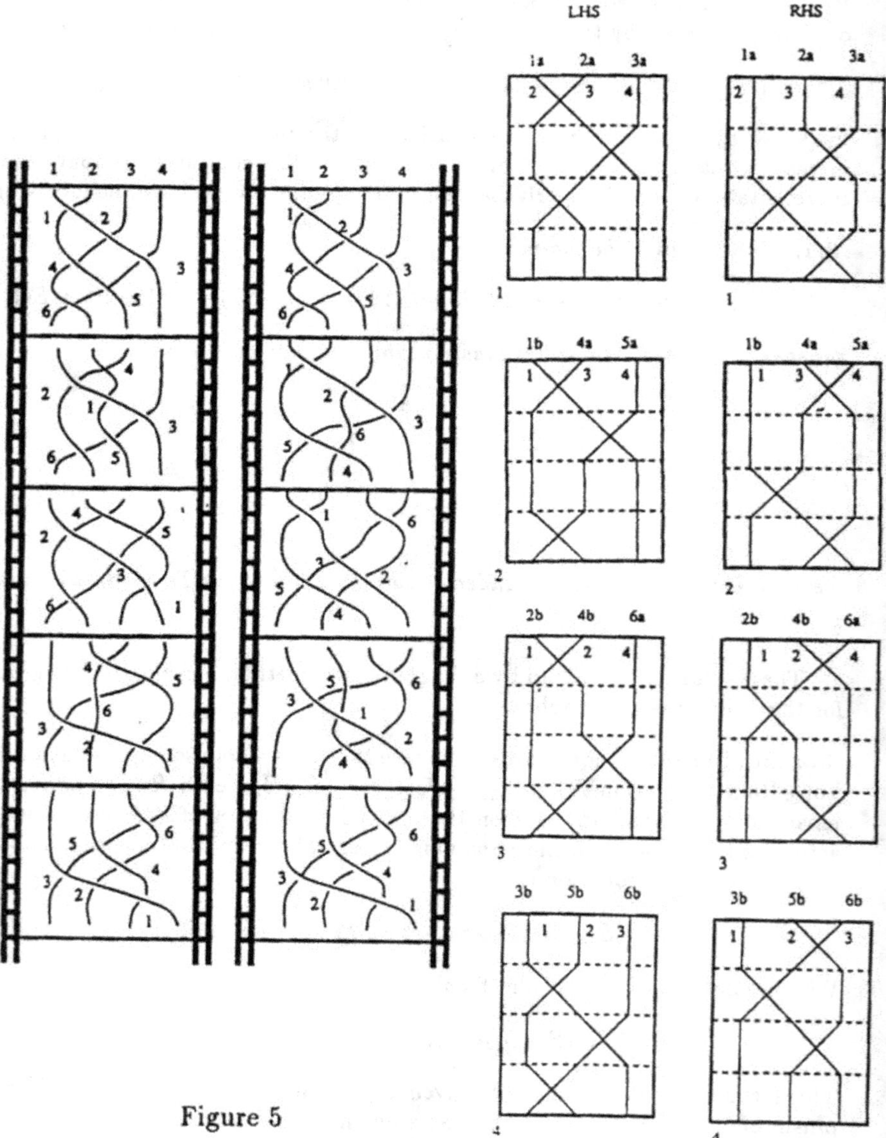

Figure 5

$\{1,4\}$, and $\{3,4\}$ appears and these are the lexically 2nd, 3rd, and 6th pairs in the set of all 2-fold subsets of $\{1,2,3,4\}$. These 6 pairs of integers correspond to the numbering of crossings in Figure 5.

Based on an idea due to Jacob Towber, we found the following solution of the double point equation.

3.2.1 Theorem [7]. *Let $V = W \otimes W$, where W is a vector space over which there is a QYB solution R: $R_{12}R_{13}R_{23} = R_{23}R_{13}R_{12}$. For given V give a and b as subscripts of W: $V_i = W_{ia} \otimes W_{ib}$ for $i = 1, \cdots 6$. For $S = S_{123}$ acting on $V_1 \otimes V_2 \otimes V_3 = W_{1a} \otimes W_{1b} \otimes W_{2a} \otimes W_{2b} \otimes W_{3a} \otimes W_{3b}$ set*

$$S_{123} = R_{1a2a}R_{1b3a}R_{2b3b}.$$

Then S is a solution to the double point equation.

It is not difficult to compute that this S is a solution, but the inspiration for the solution was given by the diagram in Figure 5. This figure depicts the preimage of the 4 planes that are intersecting generically in 4-space before and after the tetrahedral move. In these preimages a picture of Reidemester type III move that corresponds to QYBE appears. The labels a and b denote above and below to indicate which sheet (above or below with respect to crossing information) the preimages of the intersection arcs lie on.

Moreover, in [7] we formulated generalizations of these equations that correspond to the intersection sets of hyperplanes, and we showed that solutions to these simplex equation could be used to give solutions to an associated equation in one dimension higher. Thus whole classes of generalized n-simplex equations have solutions (starting from solutions to generalized Frenkel-Moore equations [9]).

3.2.2 Remarks. Hietarinta [15] generalized this method to yield two-variable solutions to the tetrahedral equation formulated above. Ruth Lawrence has told us that in a similar fashion, solutions to her system of equations can give solutions to all of the higher dimensional analogues.

Part II. Smoothings of Higher Dimensional Knots

Smoothing of crossings on classical knot diagrams played an important role in the study of Jones type invariants. The most remarkable result in this regard is Kauffman's bracket. This not only simplifies the definition of the Jones polynomial but also gives a solution to the Yang-Baxter equation in a simple, diagrammatic way.

Kauffman's trick can be seen as a smoothing of a triple point in the projection of isotopies. By two types of smoothing he can replace a type III Reidemeister move with other moves (type II and an extra move involving maximal/minimal points in the case where a height function on the plane is specified). The Reidemeister type III move projects to a triple point in the isotopy (see the figures of the ESIs). Therefore we can regard this trick as a method of resolving a triple point by means of smoothing of crossings, or a smoothing of a triple point by a smoothing of crossings.

We recall that this smoothing was used in the proofs Theorems 1.2.3 , 1.3.1 of part I.

Considering the significance of these smoothings in classical knot theory, we expect similar consequences in higher dimensions. The first question in this direction is: Can the triple point smoothing be used to resolve quadruple points? In other words, can triple point smoothings be used to resolve the Roseman move that generalizes the type III move?

Yes! Furthermore, the resolution of the quadruple point is used to resolve the crossings in one dimension higher as well. And, inductively, smoothing is generalized to any higher dimensions. Consequently, applications of smoothings in all dimensions that are similar to the applications sketched in Part I are expected and worth pursuing.

4 Smoothings of Quadruple Points

As we have observed, the continuous trace of Kauffman's bracket trick of resolving the type III move by smoothing crossing is the triple point smoothing we used to prove various theorems (Fig. 2). The goal of this section is to use this smoothing to resolve the Roseman tetrahedral move that involves moving one of four generically intersecting planes in 3-space past the intersection of the other three. There are four triple points of intersection that correspond to the four ways of choosing three sheets out of four planes. Our question is whether we can move the plane is a way that avoids a quadruple point via a smoothing of one of these four triple points.

4.1 Theorem. *We can smooth the quadruple point by means of the triple point smoothing. Furthermore, this can be realized as an isotopy.*

This is better explained and better proved by pictures. First we explain Figure 6. The top two pictures are before/after the terahedral move. Two pictures in the second row show the pictures obtained by smoothing one of four triple points. The first statement of the theorem says that we can homotop one to the other without quadruple point in the course of this homotopy. The rest of the Figure shows the proof. We leave it to the reader as an exercise to read from the pictures what types of simpler moves were used in this homotopy. Figure 7 depicts the case of the other possible triple point smoothing.

Recall that our triple point smoothing is realized by an embedded annulus and a plane. The four generic planes can be lifted to 4-space as embeddings.

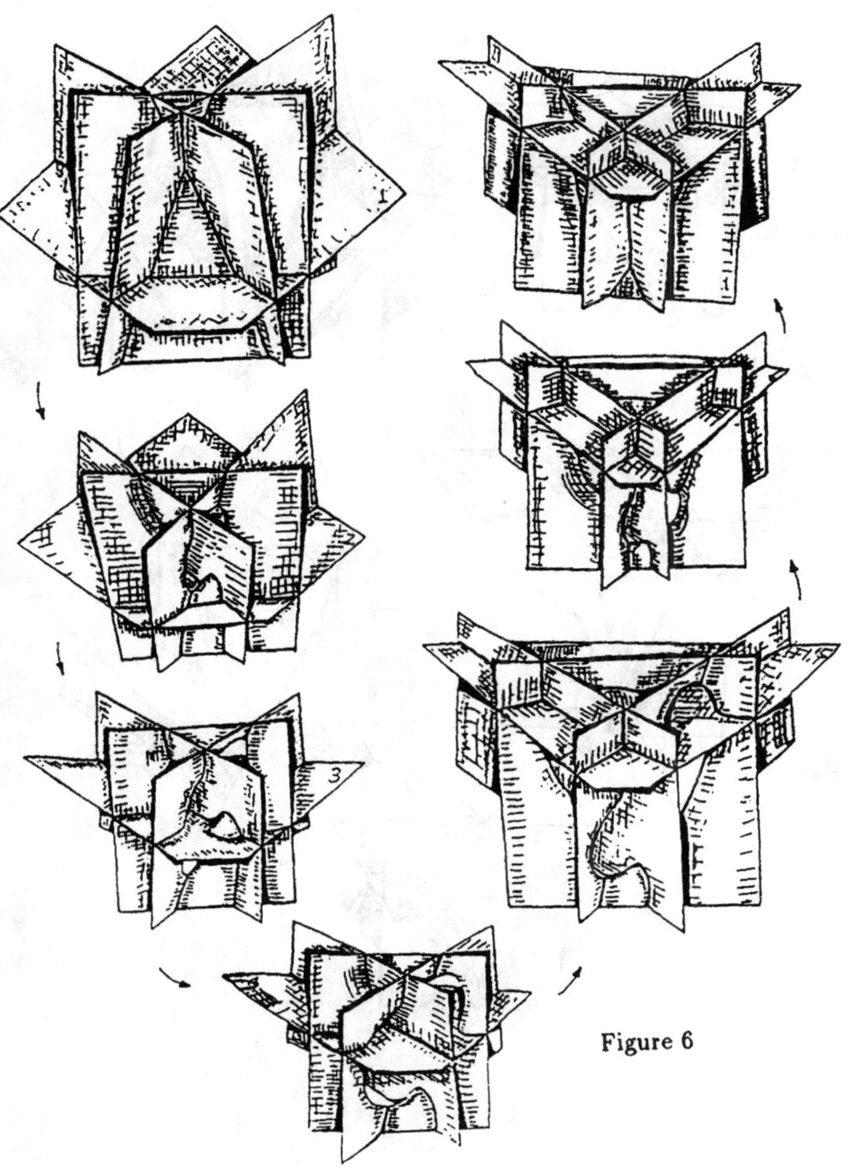

Figure 6

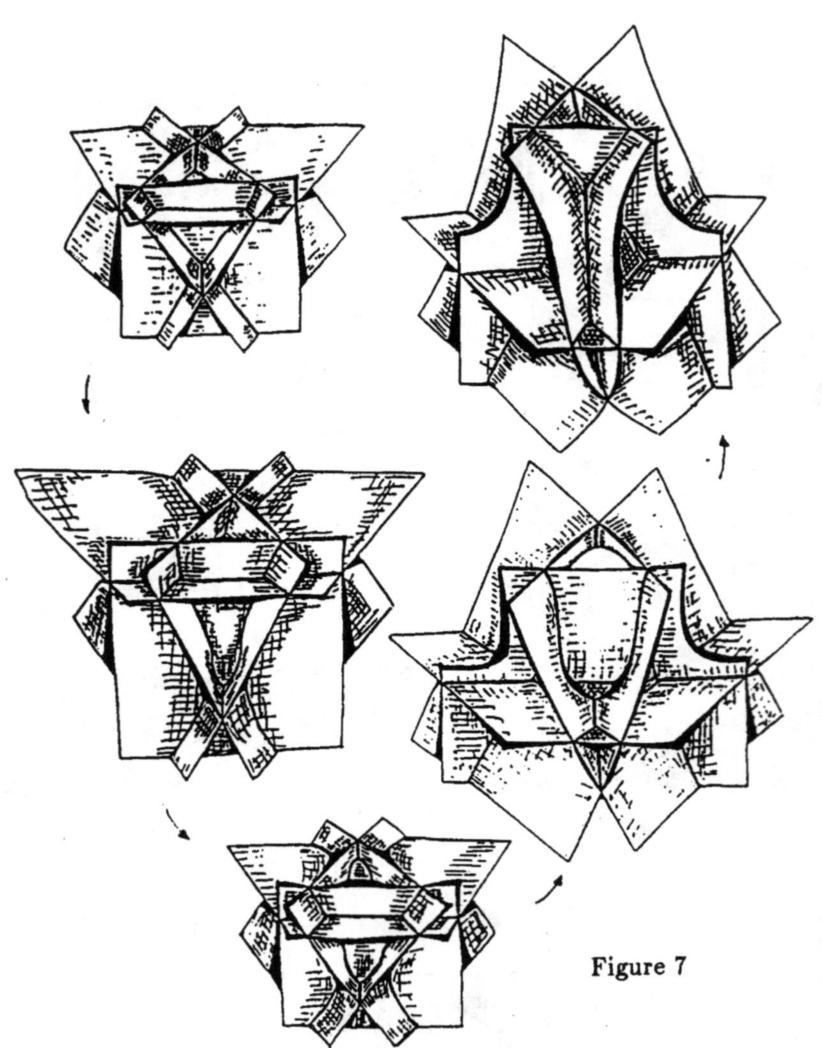

Figure 7

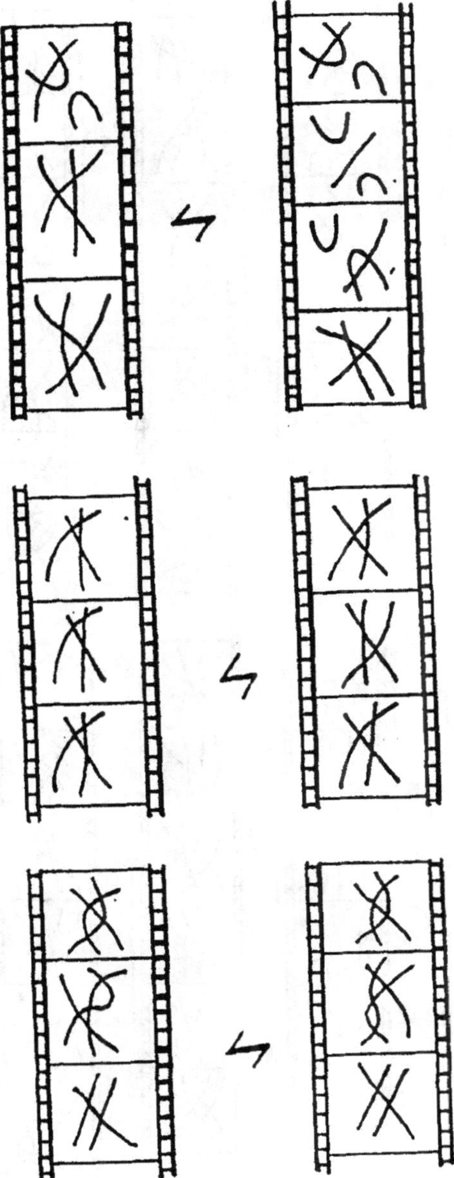

Figure 8

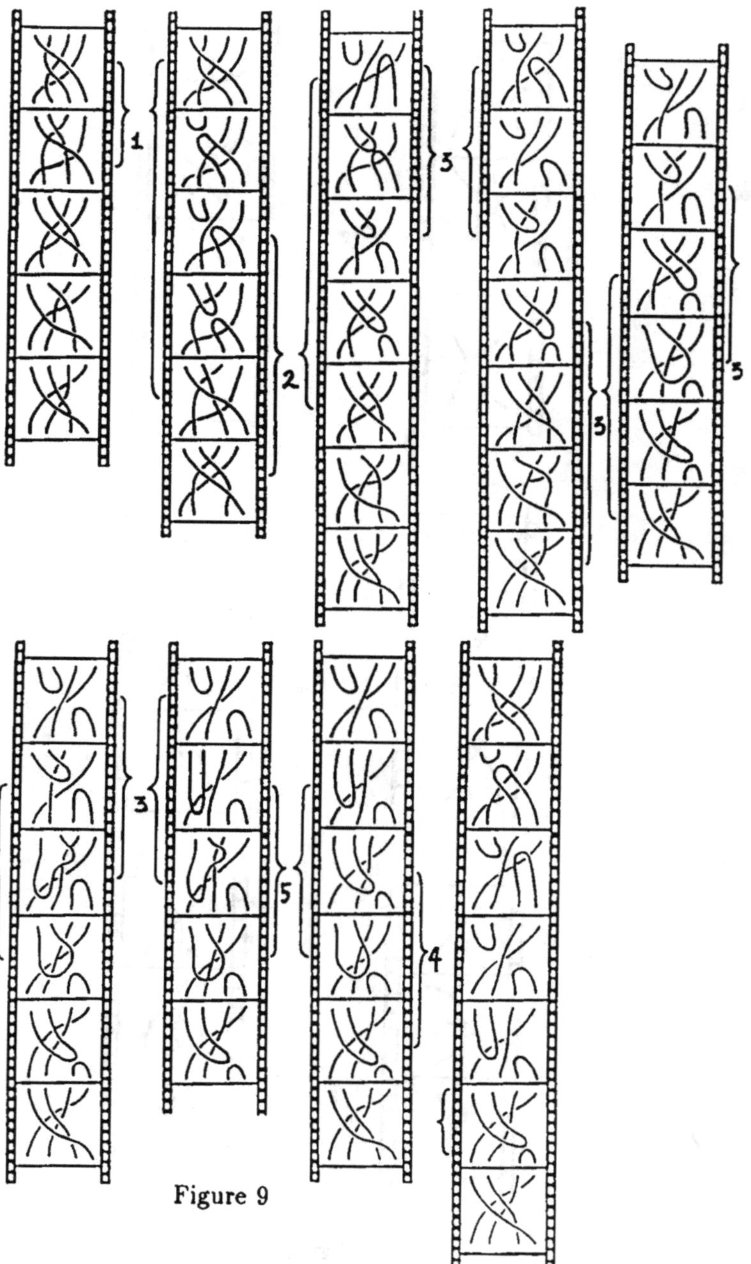

Figure 9

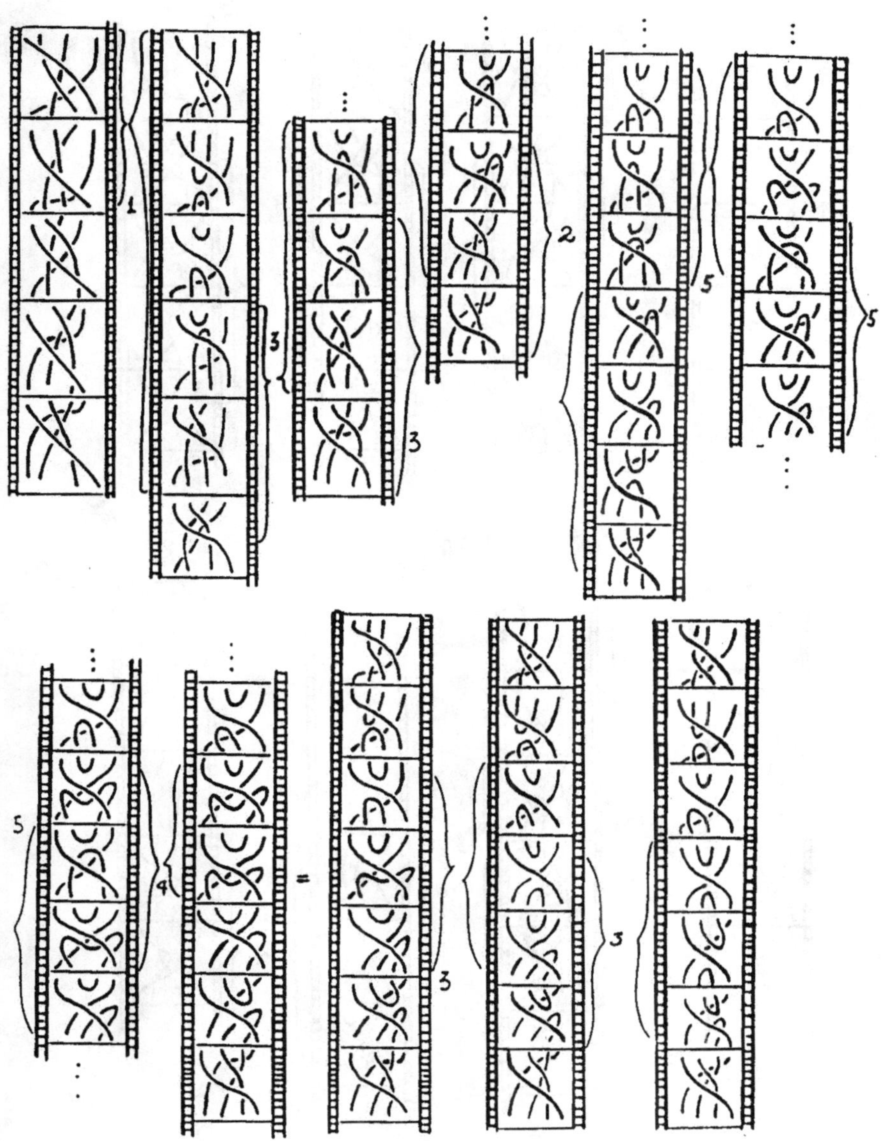

Figure 10

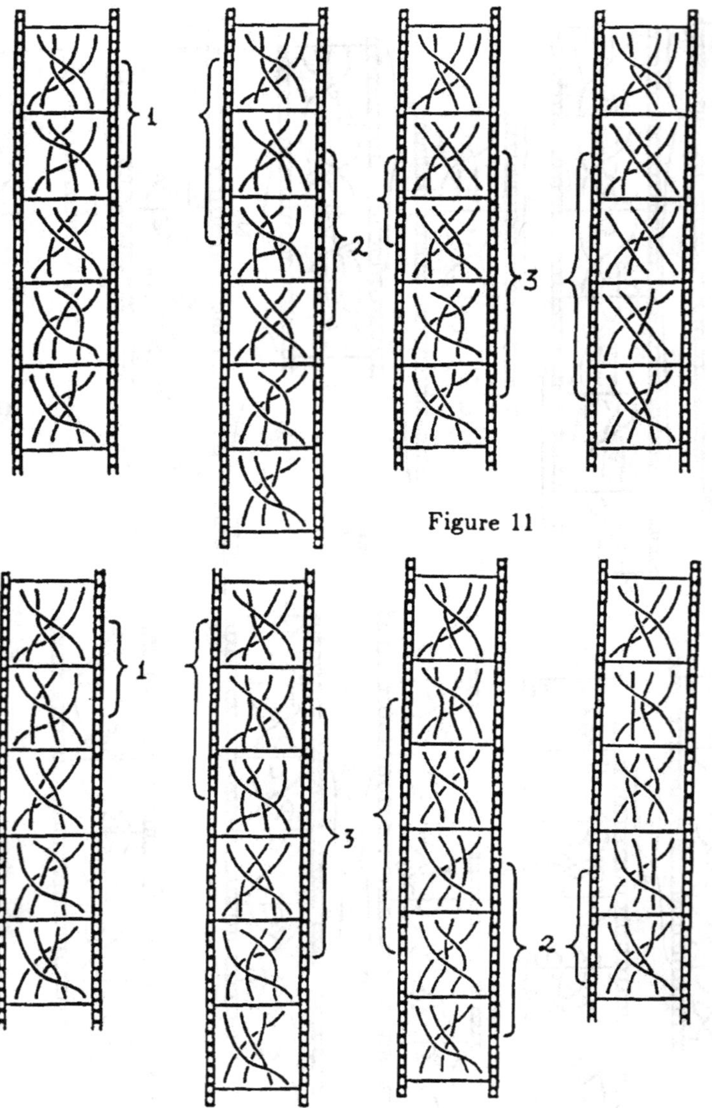

Figure 11

Figure 12

Therefore, up to the second row of Figure 6 these pictures can be realized as projections of embedded 2-manifolds in 4-space. The second half of the statement of the theorem says that the homotopy described by these pictures can be realized as an isotopy among these embedded surfaces in 4-space. To prove this part of the theorem we use movie methods. The changes from the first film strip to the second film strip in each of Figures 9, 11, 10, 12 depict four different movie descriptions of the triple point smoothing (these can be directly seen as Kauffman's bracket trick). Figures 9, 11, 10 and 12 prove the second half of the theorem. After each of these smoothings, the figures indicate isotopic moves to realize the homotopy depicted in Figures 6 and 7.

In these figures two successive movie films differ only by one of the following moves:

(1) The smoothings of the triple points,
(2) A branch point passing through a sheet,
(3) Distant moves commute,
(4) Cancellation/creation of a pair of triple points, and
(5) A movie move involving type II and III Reidemeister move.

See Figure 8 for the movies of the moves (2), (4) and (5). Notice here that the moves (3) and (5) are the moves we have to take into account because a time direction was fixed to give the movie descriptions. This is analogous to the fact in the classical case that we needed two extra Reidemeister moves when we fix a height function. (See the movie move section for more details.)

5 Higher Dimensional Generalizations

The smoothing of a quadruple point can be interpreted as a movie of a 3-manifold. When an embedding is depicted in each figure (via a broken surface diagram), this 3-dimensional movie depicts an embedding of a 3-manifold in 5-dimensional space. Furthermore, it can be shown that the smoothing of the quadruple point described in the previous section can be used to smooth a quintuple point in the projection of the isotopy of a 3-manifold in 5-space. Thus Kauffman's smoothing technique fits into an inductive scheme of smoothing the 0-dimensional crossing point of a knot projection in all dimensions.

Acknowledgements

We thank Lou Kauffman for organizing this exciting conference and for inviting us as speakers. Thanks also go to other participants for valuable conversations. These works were done while the second named author was a postdoctoral fellow at the University of Texas at Austin and the University of Toronto. He is grateful to professors C. McA. Gordon and K. Murasugi for their helpful conversations and support during this time. We also thank the following people for information and conversations: J. Fischer, D. Flath, R. Hitt, M. Kapronov, R. Lawrence, D. Silver, and J. Towber.

References

[1] Banchoff, T. F., *Double Tangency Theorems for Pairs of Submanifolds*, in Geometry Symposium Utrecht 1980 ed. Looijenga, Seirsma, and Takens, LNM v. 894, Springer-Verlag (1981), 26-48.

[2] Carter, J. Scott and Saito, Masahico, *Canceling Branch Points on Projections of Surfaces in 4-Space*, Proc. of the AMS. 116, No 1. (Sept 1992), 229-237.

[3] Carter, J. Scott and Saito, Masahico, *Syzygies among Elementary String Interactions in Dimension 2+1*, Letters in Mathematical Physics 23 (1991), 287-300.

[4] Carter, J. Scott, and Saito, Masahico, *Reidemeister Moves for Surface Isotopies and Their Interpretation As Moves to Movies*, Preprint.

[5] Carter, J. Scott, and Saito, Masahico, *Planar Generalizations of the Yang-Baxter equation and Their Skeins*, Journal of Knot Theory and its Ramifications Vol 1, No. 2 (1992), 207-217.

[6] Carter, J. Scott, and Saito Masahico, *A Seifert Algorithm for Knotted Surfaces*, Preprint.

[7] Carter, J. Scott, and Masahico Saito, *On Formulations and Solutions of Simplex Equations*, Preprint.

[8] Carter, J. Scott, and Masahico Saito, *Diagrammatic Invariants of Knotted Curves and Surfaces*, Preprint.

[9] Ewen, H. and Ogievetsky, O., *Jordanian Solutions of Simplex Equations*, preprint.

[10] Fenn, R., and Rourke, C., *Racks and Links in Codimension Two*, preprint.

[11] Fox, R.H., *A Quick Trip Through Knot Theory*, in Topology of Manifolds, Prentice Hall (1962).

[12] Frenkel, Igor, and Moore, Gregory, *Simplex Equations and Their Solutions*, Comm. Math. Phys. 138 (1991), 259-271.

[13] Giller, Cole, *Towards a Classical Knot Theory for Surfaces in* $\mathbf{R}^4$, Illinois Journal of Mathematics 26, No. 4, (Winter 1982), 591-631.

[14] Gilmer, P., *A Method of Computing Arf Invariants*, preprint.

[15] Hietarinta, J., *Some Constant Solutions to Zamolodchikov's Tetrahedron Equations,* preprint.

[16] Homma, T. and Nagase, T., *On Elementary Deformations of the Maps of Surfaces into 3-Manifolds I*, Yokohama Mathematical Journal 33 (1985), 103-119.

[17] Jones, V. F. R., *A Polynomial Invariant for Knots and Links via von Neumann Algebras,* Bull. AMS 12 (1985) 103-111. Reprinted in Kohno "New Developments in the Theory of Knots," World Scientific Publishing (1989).

[18] Joyce, David, *A Classifying Invariant of Knots, the Knot Quandle*, J. of Pure and Applied Alg., 23, (1982) 37-65.

[19] Lawrence, R., *Algebras and Triangle Relations*, preprint.

[20] Kapranov, and Voevodsky, *Braided Monoidal 2-Categories, 2-Vector Spaces and Zamolodchikov's Tetrahedra Equations,* (first draft) Preprint.

[21] Kauffman, Louis, "Knots and Physics," World Science Publishing (1991).

[22] Litherland, R., A letter to Cameron Gordon.

[23] Rolfsen, Dale, "Knots and Links," Publish or Perish Press, Berkley, 1976.

[24] Roseman, Dennis, *Reidemeister-Type Moves for Surfaces in Four Dimensional Space,* Preprint.

[25] Whitney, H., *On the Topology of Differentiable Manifolds*, in Lectures in Topology, (Wilder and Ayres, eds.), University of Michigan Press (1941), 101-141.

[26] Zamolodchikov, A. B., *Tetrahedron Equations and the Relativistic S-Matrix of Straight-Strings in* $2+1$*-Dimensions*, Commun. Math. Phys. 79, 489-505 (1981), Reprinted in Jimbo, "Yang- Baxter Equation in Integrable Systems," World Scientific Publishing Co., Singpore, (1989).

FOUR DIMENSIONAL TQFT; a Triptych

by Louis Crane[1]
Mathematics Department
Kansas State University
Manhattan KS, 66506

ABSTRACT: We discuss three possibly interrelated ideas. One idea is a suggestion for a categorical approach to quantizing gravity, one tells us how to lift a TQFT from D=3 to D=4, and one tells us how to refine some modular tensor categories into 2-categories. Relations between the ideas are conjectured.

Since it is impossible to state one idea adequately in 20 minutes, I shall outline three ideas sketchily instead. I am submitting to this conference, as joint work with David Yetter, a preliminary version of a paper which details one of the three ideas [1]. A second is written out in [2], and a third in [3].

The three ideas I describe begin in different places, but ought to connect with one another. They all revolve around the problem of constructing Topological Quantum Field Theory [TQFT] in dimension 4.

1 Quantum Gravity and augmented TQFT

There seems to be a connection between 3D TQFTs and the quantum theory of gravity, as formulated in the Ashtekar variables. There are several indications of this. One is that the Chern-Simons functional, which Witten [4] used to formally construct 3D TQFT, is (also formally), a solution of the constraints for general relativity in the Ashtekar variables [5]. Another indication is the deformed spin network interpretation of the CSW TQFT, [6] which shows that it is closely related to the spin network formulation of 3D quantum gravity due to Regge and Ponzano [7].

The deformed spin network picture gives us a picture of a 3D TQFT as a rule which assigns to a 3-manifold with boundary a number, which can be interpreted as an average over discretized 3D Riemannian metrics, when we specify a state in the space of conformal blocks of the boundary surface. This goes a long way towards giving an interpretation of the CSW TQFT as a state for the quantum theory of gravity.

In order to create an interpretation for the CSW state which would be related to physical experiments, it is necessary to reintroduce time into our picture, i.e. to make a 4-dimensional structure which is some sort of extension of a 3d TQFT. The word I coin for the new structure is "augmented."

An analysis of what would be physically useful for a 4D interpretation suggests a mathematically elegant form for what an augmentation should be. A

[1] supported by NSF grant DMS-9106476

TQFT can be described as a functor from a cobordism category to Vect (the category of finite-dimensional vector spaces) It is easy to restrict a TQFT to subsets of some 3-manifold M and get a functor on a relative cobordism category, of surfaces and cobordisms embedded in M.

An augmentation of a TQFT then amounts to a rule, which assigns, to any 4D cobordism between two 3-manifolds a natural transformation between the two functors associated to the relative categories of the two 3-manifolds. This translates physically into time evolution operators on relative states, together with appropriate consistency relations when observers observe one another evolving.

This leads us to the problem of trying to augment a 3D TQFT. Since CSW theory can be constructed from a modular tensor category (MTC) [8], we are led to consider two separate mathematical problems :

1) augmenting a 3D TQFT directly up to D=4, and

2)augmenting a modular tensor category into a 2-category.

I do not, at this point know how to use either of these ideas to construct an augmentation in the sense described above, but there is considerable progress to report on both, purely as mathematical problems.

2. Categorical construction of 4D TQFT.

This construction is discussed in the short paper submitted to the conference jointly with D. Yetter.Consequently, I confine myself here to a few simple remarks.

One important point is that we have a construction of a 4D TQFT from a type of tensor category we are already familiar with, rather than a hypothetical one which solves some list of axioms.

It is rather surprising that we get a 4D theory from a modular tensor category, since the general intuition in the field is that 4D invariants should come from 2-categories [9, 10]. The construction has another rather surprising feature, which is that we put labels of spins (or, more generally, of representations) on the 2-simplices of the triangulation rather than on the edges.

We believe that these two facts suggest that our invariant is a special case of one constructed from 2-categories. The MTC is playing the role of a 2-category with one object, so that the edges are labeled trivially, and the representations on the surfaces are really to be thought of as morphisms.

As I shall mention below, this suggests that the 2-category I describe in section 3 may have some relationship to 4D topology.

3. The Canonical Basis and 2-Categories

This part of my talk represents joint work with Igor Frenkel [3].

There are several ways to construct the MTCs which are related to CSW theory. One is to look at the representation theory of quantum groups when q is a root of unity. Quantum groups admit a rigid and elegant layer of structure related to the canonical basis of Lusztig [11], in which the structure constants and tensor operators between representations have purely integral coefficients, and with slight modification of the algebra, even purely positive integral coefficients.

At first glance, the structure related to the canonical basis seems unrelated to the topological applications of quantum groups. We do not know if we can use the canonical basis to produce 4D invariants, but we do have a result which strongly suggests that possibility.

Our result is that the quantum groups can be "categorified", ie. represented as the Grothendieck rings of tensor categories. The categories of representations of the quantum groups can then be recategorified into 2-categories.

We believe that this "categorification" is indicative of possible 4D topological significance because of an analogous phenomenon which relates 2D and 3D TQFTs via categorification.

(A detailed discussion of the notion of categorification can be found in [3]. Let me just explain, briefly, that if we have a category with direct sum and tensor product, we can make a ring of the formal linear combinations of objects in the obvious way. This we call the Grothendieck ring. The reverse operation, i.e. from ring to category, is called "categorification". It can be nonexistent or many-valued. Clearly, a ring needs a basis in which its operations have positive integral structure constants in order to be categorified. That is the point of departure for the connection between the canonical basis and categorification.)

2D TQFTs are constructible from commutative associative algebras with an inner product and suitable compatibility conditions. On the other hand, 3D TQFTs are constructed from modular tensor categories.

It turns out that the most physically interesting 2D TQFTs, namely the G/G models, are constructed from the algebras constructed from the chiral vertex operators of WZW models, called Verlinde algebras. The Verlinde algebras are precisely the Grothendieck algebras of the modular tensor categories which are related to CSW theories. Thus the 3D CSW theories are categorifications of the G/G models.

As yet, it is only a matter of conjecture that the 2-category we construct has any relationship with the 4D construction in part 2.

4. Connections

It is not yet clear if the three ideas outlined above connect. Recently, L. Smolin [12] has suggested that in order to make a 4- dimensional interpretation of states for general relativity in the Ashtekar variables, it is necessary to fill up space with a family of surfaces on which we could locate clocks.

Since the 4D construction described in 2 above is really built up out of 3D TQFT on a family of 2-surfaces in the 3-skeleton, it may be possible to interpret our 4D construction as implementing Smolin's suggestion under the assumption that the universe is in the CSW state. If so, and if a steepest descent approximation exists for our 4D theory analogous to Regge-Ponzano theory in D=3, then the relationship between ideas 1 and 2 in this paper may be close.

Acknowledgements:
Much of this work is joint work with D. Yetter and I Frenkel. the ideas about quantum gravity grow out of years of work with L. Smolin and C. Rovelli. I wish to thank J. Baez, J. Pullin, and G. Zuckerman for helpful conversations.

References:

1.L. Crane and D. Yetter, A Categorical Construction of 4D Topological Quantum Field Theories, this volume

2. L. Crane, Categorical Physics, preprint

3. L. Crane and I. Frenkel, Categorification and The Construction of Topological Quantum Field Theory, to appear in Conference Proceedings AMS conference on Geometry, Symmetry and Topology, Amherst MA, 1992

4. E. Witten, Quantum Field Theory and The Jones Polynomial, IAS preprint HEP 88/33

5. B. Brugman, R Gambini and J. Pullin, Gen. Rel. Grav. in press

6. L. Crane, Conformal Field Theory, Spin Geometry, and Quantum Gravity, Phys. Lett. B v259 3 (1991)

7. G. Ponzano and T. Regge, Semiclassical Limits of Racah Coefficients in Spectroscopic and Group theoretical methods in Physics. ed F. Bloch (North Holland, Amsterdam)

8. L. Crane, 2-d Physics and 3-d Topology, Commun. Math. Phys. 135 615-640 (1991)

9. D. Kazhdan, personal communication

10. M. Kapranov and V. Voevodsky, Braided Monoidal 2-Categories, 2-Vector Spaces and Zamolodchikov's Tetrahedra Equation, preprint

11. G. Lusztig, Canonical Bases in Tensor Products, MIT preprint 1992, and references therein

12. L. Smolin, Time, Measurement, and Information Loss in Quantum Cosmology, SUHEP preprint, 1992

A CATEGORICAL CONSTRUCTION OF 4D TOPOLOGICAL QUANTUM FIELD THEORIES

Louis Crane[1] and David Yetter

Department of Mathematics
Kansas State University
Manhattan, KS 66506

1. Introduction

In recent years, it has been discovered that invariants of three- dimensional topological objects, such as links and three-dimensional manifolds, can be constructed from various tools of mathematical physics, such as Von Neumann algebras [1], Quantum Groups [2], and Rational Conformal Field Theories [3].

Since these different structures lead to the same 3D invariants, it is natural to wonder how they are related. A fundamental connection is that they all give rise to the same special tensor categories, which act as expressions of "quantum symmetry" in the very different physical settings [4]. It is therefore very important that the 3D invariants can all be reconstructed from a tensor category with the appropriate properties, called a "modular tensor category (MTC) [3,5]." The invariants have the property that they factorize nicely if the manifold or link is cut along a surface, which we express by saying that they form a Topological Quantum Field Theory, or TQFT.

Thus the theorem proven in [3] states that a modular tensor category gives rise to a 3D TQFT. This represents a remarkable convergence with [6], in which it was realized that a suitable categorical structure would give rise to 3D topological information, although without the examples from physics.

The original suggestion to look for TQFT's appears in (not mathematically rigorous) work of Atiyah [7] and Witten [8]. Atiyah devotes more attention to the 4D than the 3D situation, and poses the question whether the two are related; or more concretely, whether the invariants of Donaldson and Jones are related. Witten actually works in the 4D situation, and formally suggests that Donaldson's invariant can be fitted into a 4D TQFT.

At this point, there is no mathematical construction of the 4D TQFT envisioned in [7] and [8]. Donaldson-Floer theory has so far eluded the efforts of the strongest of analysts [9].

It is therefore clear that a 4D construction of a TQFT parallelling the 3D categorical one would have very great implications for mathematics. As we shall mention in the conclusion section of this paper, there may very well be important physical implications also.

The construction we describe in this paper is a considerable step in this direction. Following a suggestion of Ooguri [10], we offer an expression which

[1]Supported by NSF grant DMS-9106476

gives an invariant of a 4-manifold. The expression is reminiscent of the invariant of Viro and Turaev [11], and depends on a triangulation of the manifold. The relative form for a manifold with boundary, and hence the TQFT, follow directly.

We believe that the invariants, which are constructed from a modular tensor category within the 3-skeleton of the 4-manifold, are not the most general, and that the construction should be extended to a more general one, in which the MTC is replaced by a suitable 2-category.

For the purposes of this paper, all manifolds are compact, oriented, and smooth (=PL for D=4).

2. The Construction

In [10], Ooguri, motivated by some ideas in combinatorial quantum gravity, proposed a formal expression for an invariant for 4-manifolds. The expression he proposed was divergent, but he suggested that replacing the representations of a Lie group by those of a quantum group at a root of unity might regularize it.

Since the representations of a quantum group at a root of unity form a MTC, it is natural to recast Ooguri's suggestion in the context of [3]. We discovered that a rather small modification of Ooguri's expression sufficed to produce an invariant of 4-manifolds.

The expression which we found to produce a topological invariant is a sum of products of contributions for each simplex of the triangulation.

In order to define the invariant, we need to split each tetrahedron of the triangulation, so that the faces are grouped into two pairs. Geometrically, we think of this as cutting the four-holed sphere which results from thickening the 1-skeleton of the dual triangulation of the tetrahedron into two trinions (three-holed spheres). See Figure 1.

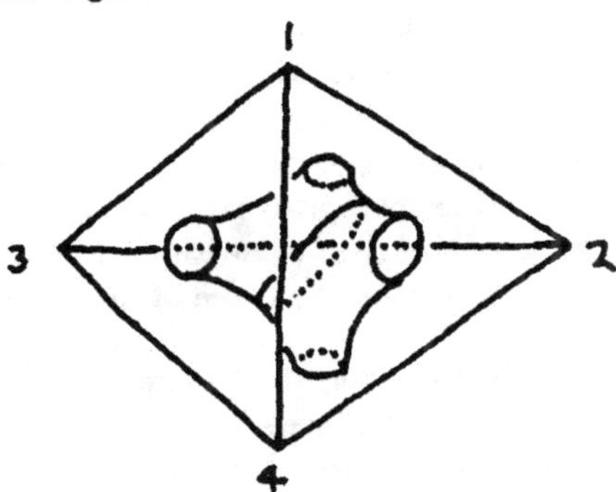

Figure 1

The surfaces which result from joining these 4-holed spheres together are very important in our definition. Since they reside in the 3-skeleton of the triangulation, we can often regard them as boundaries of 3-manifolds, so that the theorems about invariants of manifolds with boundary in [3] apply.

We can think of a labelling of the faces of the triangulation as contributing a labeling to the holes of the 4-holed spheres. In order to pick a basis for the vector space which the MTC attaches to the four-holed sphere, we must cut it along a circle into trinions, and sum over all labelings of the internal cut by irreducible objects of the MTC.(If we change cuts, a matrix called the fusion matrix relates the two bases we obtain.)

If we glue together all the 4-holed spheres from the tetrahedra on the boundary of one 4-simplex, we obtain a surface imbedded in S^3. See Figure 2.

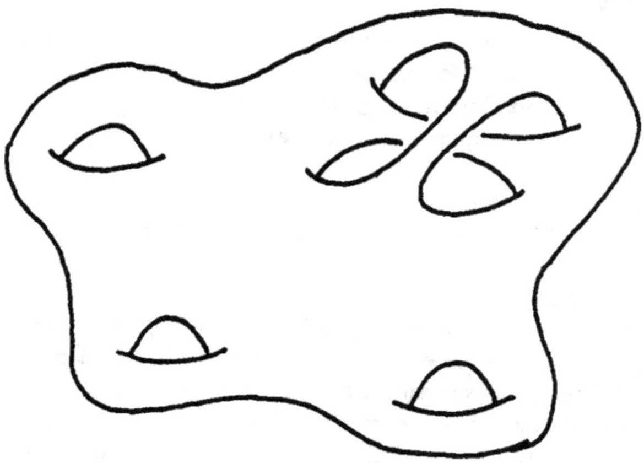

Figure 2.

If we pick a vector in the vector space which is assigned to this surface by the TQFT associated to the MTC, then, since the exterior of the surface in S^3 is a 3-manifold with boundary the surface, we can obtain an invariant number. In order to describe a vector in the space associated to the surface, we need to cut the surface into trinions and label each cut with an irreducible object. In a general MTC, we need to pick vectors in the space of intertwining operators as well.

Such a decomposition requires 15 cuts, 10 at faces and 5 inside tetrahedra. We call the invariant we obtain a generalized 15J symbol. Such an invariant can also be written by combining generalized Clebsch-Gordan coefficients, analogously to the classical 15J symbols, except that care must be taken to describe the embedding of the surface (thought of as a thickened graph) correctly, by using the braiding of the category.

Thus, for each 4-simplex, we have a set of numbers, called generalized 15J symbols, corresponding to labellings of the faces and tetrahedra of the triangulation with irreducibles in the category, and a choice of two intertwining operators for each tetrahedron.

Our recipe for an invariant is as follows: for each labelling of the faces of the triangulation by an irreducible object of the category, each labelling of the cut in each tetrahedron by an irreducible, and each labelling of all the resulting trinions by intertwining operators, we take the product of all the 15J symbols of the 4-simplices of the triangulation, and correction factors corresponding to lower dimensional simplices. We then sum over all labellings.

The result is a topological invariant of the 4-manifold.

This expression is very similar to Ooguri's, except that the generalization of the 15J symbols to a MTC requires topological data, because of the braiding in the category. (As we explain below, our more technical definition of the invariant, which uses an ordering of the vertices of the triangulation, makes Ooguri's 6J symbols trivial.)

3. Proof of Topological Invariance

In the special case of quantum SU(2), there are (projectively) unique intertwiners, and the various symbols admit explicit expressions in closed form [12]. In the interest of clarity, we complete the proof in that case only. We believe the generalization should work, because the proof for the case of $U_q(sl_2)$ essentially amounts to factorization on a corner for the 3D TQFT, which works for the other quantum groups as well.

In order to prove topological invariance, we need a set of moves which relate any two triangulations of the same PL manifold. We are fortunate to possess a very convenient set of moves, as a result of recent work by Pachner, in which he showed that triangulated manifolds are equivalent if and only if they are bispherically equivalent [13]. (We wish to thank W.B.R. Lickorish for bringing this to our attention.)

The translation of our formula from 3D TQFT into the recombination diagrams of quantum SU(2), as defined in [12], requires a great deal of delicacy. We have found that the best way to define an invariant of 4-manifolds in terms of the expressions in [12] is to order the vertices of the triangulation and to make a consistent choice of splittings, where each tetrahedron has its odd and even faces joined separately, then connected in the middle. (By odd (even) faces, we mean the faces which result from removal of the first and third (second and fourth) vertices of the tetrahedron under the ordering). It is then necessary to normalize the diagram with an appropriate product of quantum dimensions.

In translating the trivalent graph associated to the 4-simplex into a recombination diagram, we make the choice that the odd half of each tetrahedron always comes from above. Thus, we associate to each ordered 4-simplex the following graph, or its reflection depending on whether the orientation induced by the ordering agrees with the orientation on the 4-manifold.

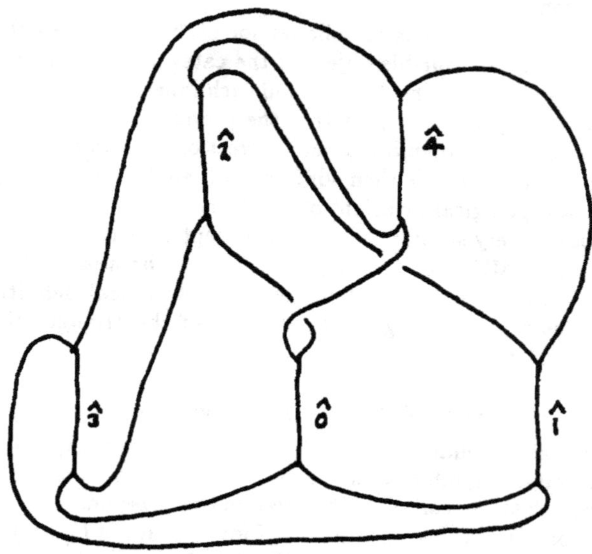

Figure 3

With this choice,Ooguri's 6J symbols become trivial, and the formula, with the proper normalization, takes the following form.

$$\sum N^{\#vertices-\#edges} \prod_{faces} dim_q(j) \prod_{tetrahedra} dim_q^{-1}(p) \prod_{4-simplices} 15J_q \qquad (*)$$

where the sum ranges over all assignments of spins to the faces and tetrahedra of the triangulation and j represents the spin labelling a face, p represents the spin labelling the cut interior to a tetrahedron, dim_q is the quantum dimension [12], and N is the sum of the squares of the quantum dimensions.

It is now a direct matter of computation to check that our expression is invariant under Pachner's moves. Basically, the moves work because each move consists in replacing one half of the boundary of a 5-simplex by the other half.

When computing a sum of products of diagrams, in which an object in one diagram is identified with one in another diagram, we can use some simple recombination rules, which were originally discovered in the representation theory of SU(2), but hold equally well for quantum SU(2) (and have analogues for any

MTC). Graphically, these rules are as follows:

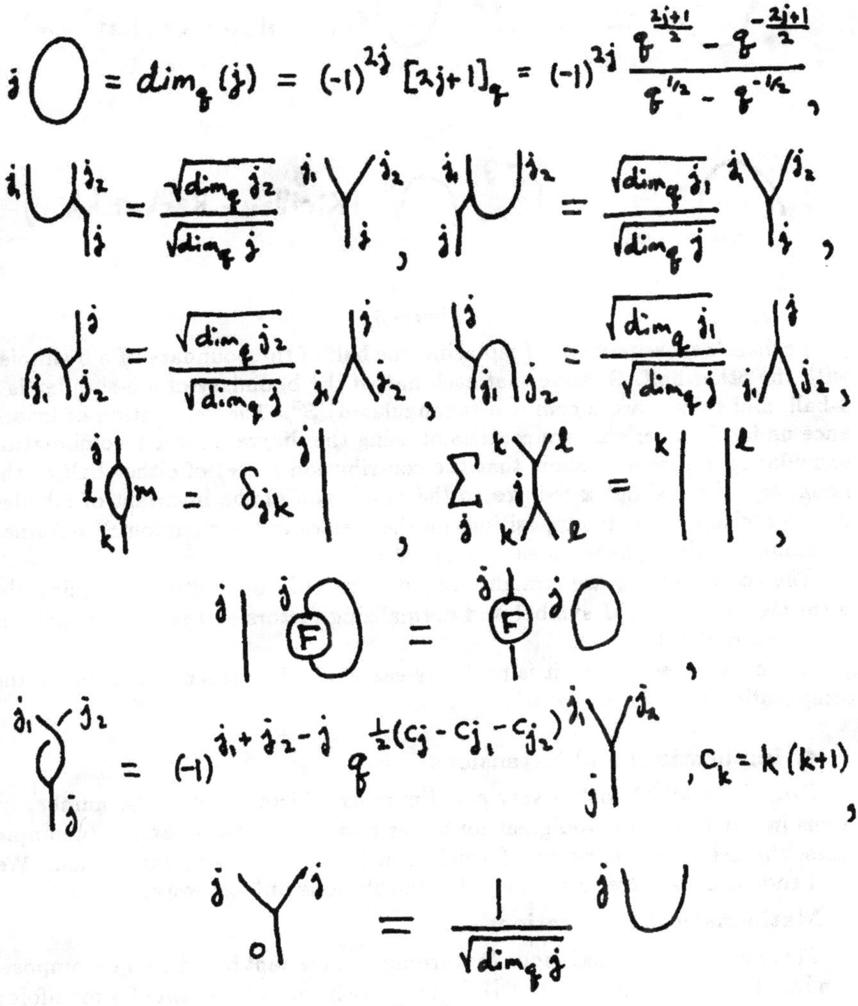

$$j\bigcirc = \dim_q(j) = (-1)^{2j}\,[2j+1]_q = (-1)^{2j}\,\frac{q^{\frac{2j+1}{2}} - q^{-\frac{2j+1}{2}}}{q^{1/2} - q^{-1/2}},$$

$$= \frac{\sqrt{\dim_q j_2}}{\sqrt{\dim_q j}} \quad , \qquad = \frac{\sqrt{\dim_q j_1}}{\sqrt{\dim_q j}} \quad ,$$

$$= \frac{\sqrt{\dim_q j_2}}{\sqrt{\dim_q j}} \quad , \qquad = \frac{\sqrt{\dim_q j_1}}{\sqrt{\dim_q j}} \quad ,$$

$$= \delta_{jk} \quad , \qquad \sum_j \quad = \quad ,$$

$$= \quad ,$$

$$= (-1)^{j_1+j_2-j}\, q^{\frac{1}{2}(c_j - c_{j_1} - c_{j_2})} \quad , \; c_k = k(k+1) \quad ,$$

$$= \frac{1}{\sqrt{\dim_q j}}$$

Figure 4

Notice that in the preceeding figure, we have modified the convention of Kirillov and Reshetikhin, and use maxima and minima to denote mutiples of the usual contraction of indices maps. Our modified normalizations are related

to those in [12] as follows

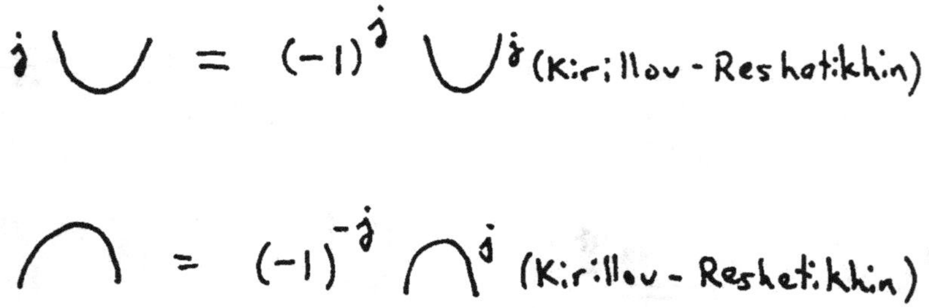

$$j \cup = (-1)^{j} \cup^{j} \text{(Kirillov-Reshetikhin)}$$

$$j \cap = (-1)^{-j} \cap^{j} \text{(Kirillov-Reshetikhin)}$$

Figure 5

Pachner's moves consist of replacing one half of the boundary of a 5-simplex with the other half. Observe that each half of the boundary of a 5-simplex is a 4-ball, and they share a common (triangulated) S^3. The verification of invariance under Pachner's moves consists of using the diagrammatic recombination formulas of Figure 4 to show that the contribution to $(*)$ of either half of the boundary of a 5-simplex reduces to the evaluation of the invariant of labelled surfaces with trinion decompositions on the surface in the common S^3 obtained by gluing 4-holed spheres in each tetrahedron.

The computations are straightforward. The only difficulty was finding the correct generalized 15J symbol and normalizing factors so that the reductions can be carried out.

Given this description, it is probably easier for the reader to reproduce the computation than for us to write it down.

4. Implications and Extensions

Even for a MTC with a very small number of irreducibles, the number of terms in our formula is too great for paper and pencil computation. In simple cases, the use of diagrammatic formalism will allow for hand calculation. We shall endeavor to make some computer calculations in later work.

Mathematical Implications

There is a standard technique for turning an invariant based on a decomposition into tetrahedra into a TQFT [14]. Thus, we have an invariant of 4-manifolds of the same algebraic structure as the one conjectured for Donaldson-Floer theory. We are led naturally to two questions: what do our new invariants tell us about smooth 4-manifolds? and how closely are they related to Donaldson-Floer theory?

Witten [8], gives a gauge fixed form of a formal Lagrangian for Donaldson-Floer theory, which is supersymmetric. We could attempt to imitate this by using the representations of quantum supergroups to construct a supersymmetric MTC, which might bring us closer to Donaldson-Floer theory.

It would be interesting to compute our invariants, for the MTCs at small roots of unity, on the examples of 4-manifolds which are known to be homeomorphic but not diffeomorphic [15]. This calculation would require nothing difficult, except extensive computation.

Mathematical Extensions

It is rather surprising that a 4D TQFT can be constructed from a MTC. The form of the invariant, with corrections associated to simplices of different dimensions, is suggestive of a picture in which one associates categorical structures of different levels to simplices of different dimensions. One is tempted to think that the MTC is functioning here as a 2-category with one object, or even as a 3-category with 1 object and one morphism. This would explain the most surprising aspect of Ooguri's suggestion, namely that spins appear on faces, rather than on edges, as in a gauge theory.

We are therefore led to conjecture that the invariant we calculate here is a special case of one calculated from a 2-category.

It is interesting to wonder whether the new 2-categories related to the canonical basis for quantum groups [16, 17] yield any interesting data in 4D topology.

The technique of attaching surfaces to triangulations we use is similar to the standard technique for producing a Heegaard splitting of a 3-manifold from a triangulation [22]. A Heegaard splitting can be thought of as a handlebody decomposition, in which the attaching of the 2-handles to the 1-handles is the surface map. One has the feeling that the formulation of our 4D invariant in terms of triangulations could be more elegantly restated in terms of handlebodies.

Physical Implications

The invariants of knotted graphs which are constructed from MTCs are closely related to spin networks [18, 19]. Spin networks are an old idea due to Penrose, which admit an interpretation as a discretized form of 3D quantum gravity. Furthermore, the Chern-Simons lagrangian, which Witten used to formally produce 3D TQFT [20], is also a state for quantum general relativity in the Ashtekar formalism [21]. It is a natural goal to try to relate the spin network picture to a geometric interpretation of Chern-Simons-Witten theory as a state for quantum general relativity.

The spin network picture allows us to rewrite the CSW invariant as a sum over labelings of a triangulation of a region in a 3-manifold, which can be interpreted as a sum over discretized 3-geometries[18, 21].

One is faced with the thorny problem of reinterpreting the CSW state as a 4D picture, ie, of reintroducing time into the picture. It is very suggestive to note that the construction outlined in this paper produces a 4D TQFT which is so closely related to a 3D one, and which comes from labelling a triangulation of the 4-manifold, thus suggesting a sum over discretized geometries of the 4-manifold.

In the best of possible worlds, Einstein's equation would appear in a classical limit of a 4D summation, just as flat geometries appear in [21].[2]

---

[2]note added in proof: A. Ocneanu has published a note in which he argues that our invariant is a constant. We have calculated our invariant for the two levels r=3,4 for $S^3 x S^1$ and for S^4, and get different results, namely, 1 in both cases for $S^3 x S^1$ versus 2 and 4 for S^4. The resolution of this contradiction is that Professor Ocneanu is calculating a different formula from ours, despite a very suggestive analogy.

REFERENCES

[1] V. Jones, *A Polynomial Invariant Of knots via Von Neumann Algebras* Bull. Am. Math. Soc. **12** (1985) 103-111.

[2] N. Reshetikhin and V. Turaev, *Ribbon Graphs and Their Invariants Derived From Quantum Groups*, Comm. Math. Phys. **127** (1990) 1-26.

[3] L. Crane, *2-d Physics and 3-d Topology*, Commun. Math. Phys. **135** (1991) 615-640.

[4] L. Crane, *Quantum Symmetry, Link Invariants and Quantum Geometry*, in Proceedings XX International Conference on Differential Geometric Methods in Theoretical Physics, Baruch College (1991)

[5] G. Moore and N. Seiberg, *Classical and Quantum Conformal Field Theory*, Commun Math. Phys. **123** (1989) 177-254.

[6] P. Freyd and D. Yetter, *Braided Compact Closed Categories with Applications to Low-Dimensional Topology*, Adv. in Math. **77** (1989) 156-182.

[7] M. Atiyah, *New Invariants For Three And Four Manifolds* in The Mathematical Heritage Of Hermann Weyl, AMS (1988).

[8] E. Witten, *Topological Quantum Field Theory*, Comm. Math. Phys. **117** (1988) 353-386.

[9] C. Taubes, *L^2 Moduli Spaces On 4-Manifolds With Cylindrical Ends, I*, Harvard preprint.

[10] H. Ooguri, *Topological Lattice Models in Four Dimensions*, Kyoto University preprint.

[11] V. Turaev and O. Viro, *State Sum Invariants of 3- Manifolds and Quantum 6-J Symbols*, Topology **31** (1992) 865-902.

[12] A. Kirillov and N. Reshetikhin, *Representations Of The Algebra $U_q(sl(2))$, q-Orthogonal Polynomials And Invariants of Links* in Infinite Dimensional Lie Algebras and Groups, World Scientific (1989).

[13] U. Pachner, *P.L. Homeomorphic Manifolds Are Equivalent by Elementary Shelling*, Europ. J. Combinatorics **12** (1991) 129-145.

[14] D. Yetter, *Topological Quantum Field Theories Associated To Finite Groups and Crossed G-Sets*, Journal of Knot Theory and its Ramifications **1** 1 (1992) 1-20.

D. Yetter, *Triangulations and TQFT's*, to appear.

[15] S.Donaldson, *Irrationality and The h-Cobordism Conjecture*, J. Diff. Geo. **26** (1987) 141-168.

[16] L. Crane and I. Frenkel, *Hopf Categories and Their Representations*, to appear.

L. Crane and I. Frenkel *Categorification and the Construction of Topological Quantum Field Theory*, to appear in Conference Proceedings AMS Conference on Geometry Symmetry and Physics, Amherst, MA, Summer 1992.

[17] D. Kazhdan and Y. Soibelman, *Representations of The Quantized Function Algebras, 2-Categories and Zamolodchikov Tetrahedra Equation*, Harvard preprint.

[18] L. Crane, *Conformal Field Theory, Spin Geometry, and Quantum Gravity*, Phys. Lett. B **259** 3 (1991).

[19] R. Penrose, *Angular Momentum; an Approach to Combinatorial Space Time*, in Quantum Theory and Beyond, ed T.Bastin, Cambridge.

[20] E. Witten, *Quantum Field Theory and The Jones Polynomial*, Comm. Math. Phys. **121** (1989) 351-399.

[21] G. Ponzano and T. Regge, *Semiclassical Limits of Racah Coefficients*, in Spectroscopic and Group Theoretical Methods in Physics, ed. F. Bloch, North-Holland.

[22] D. Rolfson, *Knots and Links*, Publish or Perish, (1976).

Evaluating the Crane-Yetter Invariant

Louis Crane
Kansas State University
Manhattan, Kansas 66506

Louis H. Kauffman
University of Illinois
Chicago, Illinois 60680

David Yetter
Kansas State University
Manhattan, Kansas 66506

I. Introduction

The purpose of this paper is to give an explicit formula for the invariant of 4-manifolds introduced by Crane and Yetter in [CR93]. For a closed 4-manifold W, this invariant will be denoted herein by CY(W). Our main result is the following theorem, proved in section 4.

Theorem. Let W be a closed 4-manifold. Let $\sigma(W)$ denote the signature of W, $\chi(W)$ denote the Euler characteristic of W and CY(W) denote the Crane-Yetter invariant of W. Let values N and κ be defined as in the beginning of section 4. Then $CY(W) = \kappa^{\sigma(W)} N^{\chi(W)/2}$.

This result is of general interest because *it expresses the signature of a 4-manifold in terms of local combinatorial data* (these data produce the state summation CY(W) in terms of a triangulation of W). Our result should be compared with the work of Gelfand and Macpherson [GM92] where the Pontrjagin classes (and hence the signature) are produced by a combination of subtle combinatorics and geometric topology. Here we give a formula for the signature in terms of a topological quantum field theory that is based on $SU(2)_q$ and on q-deformed spin networks.

It should also be mentioned that the invariant CY(W) is a rigorous version of ideas of Ooguri [OG92]. It is of interest to examine the implications of

our result for the physics that is inherent in Ooguri's work.
In section 2 we recall the definition of the Crane Yetter invariant. In section 3, CY(W) is reformulated in terms of Temperley-Lieb recoupling theory. The Theorem is proved in section 4.

Louis Crane wishes to acknowledge support by NSF Grant DMS-9106476. Louis Kauffman wishes to acknowledge support by NSF Grant DMS 9205277 and the Program for Mathematics and Molecular Biology of the University of California at Berkeley, Berkeley, CA.

II. A Concise Description of the Invariant CY(W)

Let $W = W^4$, a closed 4-manifold. Let CY(W) denote the Crane-Yetter invariant of W, as defined in [CY93]. The formula for this invariant, in terms of a triangulation of W, is a state summation over colorings from the index set $\{0,1,2, ..., r-2\}$ of the two dimensional faces and three dimensional simplices of the triangulation of W. Here we use integer labellings for convenience. In [CY93] the labellings are by half integers, but the formulas are equivalent. Thus the invariant is a function of the integers $r = 3,4, ...$.

To each colored face of the triangulation is assigned the quantum integer (quantum dimension)

$$\Delta[\text{face}] = \Delta(i) = (-1)^i[q^{i+1} - q^{-i-1}]/[q-q^{-1}] \quad \text{where} \quad q = \exp(i\pi/r).$$

where i denotes the color assigned to that face.

To each colored tetrahedron is assigned $\Delta[\text{tet}] = \Delta(i)$, where i is the color assigned to that tetrahedron.

To each 4-simplex is assigned the "15-j symbol" Φ(4plex) that is associated to the coloration of its boundary. This 15-j symbol is an evaluation of a network associated with the boundary of the 4-simplex, obtained by having two interconnected 3-vertices in each tetrahedron, forming a network with four free ends corresponding to the boundary of the tetrahedron. These nets are then interconnected in the pattern of the joinings of the faces of the tetrahedra in the boundary of the 4-simplex. In

[CY93] a specific convention for forming this net is given, and we refer to that paper for the details. The 3-vertices and network evaluations are done by Crane and Yetter in the Kirillov-Reshetikhin [KR88] diagrammatics for the recoupling theory of SU(2)q.

The formula for the invariant CY(W) is then given as shown below.

$$CY(W) = N^{n0-n1} \sum \prod \Delta(\text{face}) \prod \Delta(\text{tet})^{-1} \prod \Phi(\text{4plex})$$

The summation is over all colorings of the faces and tetrahedra from the index set {0,1,2,...,r-2}. The products are over all faces, tetrahedra and 4-simplices respectively. The values n0 and n1 are the number of 0-simplices and the number of 1-simplices in the triangulation. The value N is equal to the sum of the squares of the quantum dimensions , and it has the specific value

$$N = \sum_i \Delta(i)^2 = -2r/(q-q^{-1})^2.$$

This completes our description of the invariant CY(W).

III. Translating CY(W) into the Temperley-Lieb Format

In order to prove our result about the evaluation of CY(W) it is useful to translate the state sum into the language of the Temperley-Lieb version of the recoupling theory. This recoupling theory is explained in detail in [KL93], and expositions of it are given in [K91] and [K92]. We shall refer to Temperley-Lieb recoupling as the *TL theory,* and to Kirillov-Reshetikhin recoupling as the *KR theory*.

The TL theory is rooted in combinatorics of link diagrams, and it is a direct generalization (q-deformation) of the Penrose spin network theory. Its advantage for us here is that there is no dependence in the diagrammatics of the TL theory on maxima and minima or on the orientation of the diagrams with respect to a direction in the plane. Thus TL networks can be freely embedded in handlebodies and 3-manifolds.

The basic information needed to transform KR nets into TL nets is the relationship of their 3-vertices. This is given by the formula below where the subscripts KR and TL discriminate the vertices in question.

$$[\text{3-vertex}/a,b;c]_{KR} = (\sqrt{\Delta(c)}/\sqrt{\theta(a,b,c)})\ [\text{3-vertex}/a,b,c]_{TL}$$

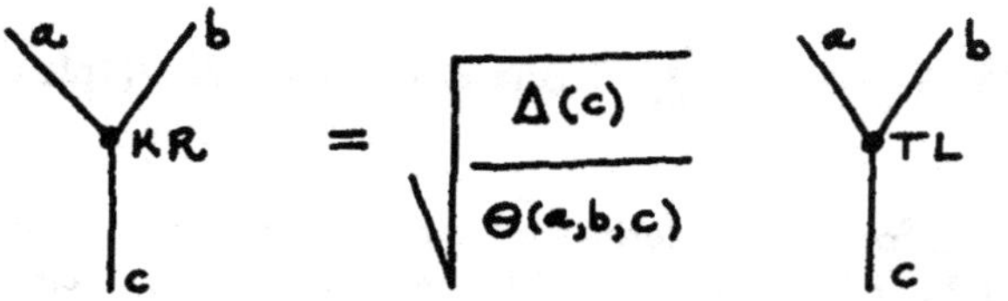

The $\theta(a,b,c)$ is the TL evaluation of a theta net.

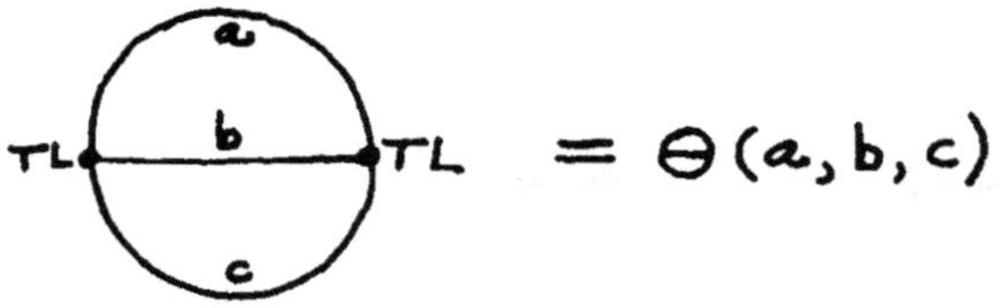

Note that the KR vertex is oriented with two legs up and one leg down. The TL vertex does not have a dependence on the leg placement. The value of a closed loop labelled $i = 0,1,\ldots,r-2$ is $\Delta(i)$ in both theories. We omit further details of the relationship between TL and KR

It is now a straightfoward matter to translate the CY(W) into the TL framework. The result is as shown below where ϕ denotes the TL 15-j symbol. The TL 15-j is described by exactly the same diagram as the KR 15-j, but all its 3-vertices are of TL type.

$$CY(W) = N^{n0 - n1} \sum \prod \Delta(\text{face}) \prod \Delta(\text{tet}) \theta_1(\text{tet})^{-1} \theta_2(\text{tet})^{-1} \prod \phi(\text{4plex})$$

In this formula, $n0$ = number of vertices, $n1$ = number of edges, $n2$ = number of faces, $n3$ = number of tetrahedra, $n4$ = number of 4 simplexes in the triangulation of W. The sum is over all colorings of the faces and tetrahedra of the triangulation.

$\theta_1(\text{tet})$, $\theta_2(\text{tet})$ are the two theta evaluations assigned to each tetrahedron. Each is of the form $\theta(a,b,c)$ where a,b and c are the colors assigned to one of the 3-vertices in the net associated with this tetrahedron. This means that we can take a and b to be the colors of two (paired) faces of the tetrahedron, and c to be the color associated with the tetrahedron itself.

$\phi(\text{4plex})$ is the TL evaluation of the 15-j net associated with each tetrahedron.

This completes our translation of the invariant CY(W) into the Temperley-Lieb format.

IV. A Formula for CY(W)

In this section we use results of Justin Roberts [R93] to give an explicit formula for CY(W) in terms of the Euler characteristic and the signature of the closed 4-manifold W.

(The reader should note that Roberts uses a notation CY(W), but that his CY(W) is not precisely the Crane Yetter invariant. It differs from it by a factor involving the Euler characteristic of the 4-manifold.)

We need the following values

$$\eta = (q - q^{-1})/i\sqrt{(2r)}$$

$$\kappa = \exp(i\pi(-3-r^2)/2r)\exp(-i\pi/4).$$

Note that $N = \eta^{-2}$, where N is as defined in section 2.

Roberts [R93] proves that

$$\kappa^{\sigma(W)} = \eta^{-n0+n1+n2-n3+n4} S(W)$$

where $S(W) = \sum \prod \Delta(\text{face}) \prod \Delta(\text{tet})\theta_1(\text{tet})^{-1}\theta_2(\text{tet})^{-1} \prod \phi(\text{4plex})$

and $\sigma(W)$ denotes the signature of the manifold W.

Roberts' state summation S(W) has exactly the same form, except for the power of N, as our TL version of the Crane-Yetter invariant. The 15-j evaluations of Roberts involve orientation conventions that are consistent with his use of TL networks in 3-dimensional handlebodies. This means that it follows from Roberts' work that different but consistent conventions for the 15-j symbols will lead to the same results. The Crane-Yetter convention in the TL format is one such choice.
Therefore $CY(W) = N^{n0-n1} S(W)$ by the results of section 3.

We can now prove the main theorem.

Theorem. Let W be a closed 4-manifold. Let $\sigma(W)$ denote the signature of W, $\chi(W)$ denote the Euler characteristic of W and CY(W) denote the Crane-Yetter invariant of W. Let values N and κ be defined as in the beginning of section 4. Then $CY(W) = \kappa^{\sigma(W)} N^{\chi(W)/2}$.

Proof.

$$\kappa^{\sigma(W)} = \eta^{-n0+n1+n2-n3+n4} N^{n1-n0} CY(W)$$

$$= \eta^{-n0+n1+n2-n3+n4} \eta^{-2n1+2n0} CY(W)$$

$$= \eta^{n0-n1+n2-n3+n4} CY(W)$$

$$= N^{-\chi(W)/2} CY(W).$$ Q.E.D.

Remark. Note that if we choose r greater than the number of 2-simplices in W, then $\sigma(W) < r$ and is therefore determined by $CY(W)$ and $\chi(W)$ via the formula in the theorem. Thus it is quite correct to say that the Theorem produces a combinatorial formula for the signature of a compact 4-manifold in terms of local data from the triangulation.

References

[CR91] L. Crane. Conformal Field Theory, Spin Geometry and Quantum Gravity. Physics Letters B. Vol. 259. No. 3. pp. 243-248. (1991).

[CR93] L. Crane and D. Yetter. A Categorical Construction of 4D Topological Quantum Field Theories. (In *Quantum Topology*, ed. by L.Kauffman and R.Baadhio).

[G92] I.M.Gelfand and R.D.Macpherson. A combinatorial formula for the Pontrjagin classes. Bulletin of the AMS. Vol. 26, No.2, April 1992. pp. 304-308.

[K90] L.H.Kauffman. Spin networks and knot polynomials. Intl. J. Mod. Phys. A. Vol. 5. No. 1. (1990). pp. 93-115.

[K91] L.H. Kauffman. **Knots and Physics** , World Scientific Pub. (1991).

[K92] L.H. Kauffman. Map Coloring, q-Deformed Spin Networks, and Turaev-Viro Invariants for 3-Manifolds. In the Proceedings of the Conference on Quantum Groups - Como, Italy June 1991 - edited by M. Rasetti - World Sci. Pub., Intl. J. Mod. Phys. B, Vol. 6, Nos. 11 & 12 (1992), pp. 1765 - 1794.

[KL93] L.H.Kauffman and S. Lins. Temperley Lieb Recoupling Theory and Invariants of 3-Manifolds. (to appear)

[KR88] A.N. Kirillov. and N.Y. Reshetikhin. Representations of the algebra $U_q(sl_2)$, q-orthogonal polynomials and invariants of links. In **Infinite Dimensional Lie Algebras and Groups.** ed. by V.G. Kac. Adv. Ser. in Math. Phys. Vol. 7. (1988). pp. 285-338.

[OG92] H. Ooguri. Topological Lattice Models in Four Dimensions. Mod.Phys.Lett.A, Vol. 7, No. 30 (1992) 2799-2810.

[PEN69] R. Penrose. Angular momentum: an approach to Combinatorial Space-Time. In **Quantum Theory and Beyond.** ed. T.A. Bastin. Cambridge Univ. Press (1969).

[R93] J. Roberts. Skein theory and Turaev-Viro invariants. (preprint 1993).

[TV92] V.G. Turaev and O. Viro. State sum invariants of 3-manifolds and quantum 6j symbols. Topology, Vol. 31, No. 4, pp. 865-902 (1992).

CANONICALLY QUANTIZED LATTICE CHERN-SIMONS THEORY

D. Eliezer
Physics Dept., L-412
Lawrence Livermore National Laboratories
PO Box 808
Livermore, CA 94551-9900
and
G. W. Semenoff

Department of Physics, University of British Columbia
Vancouver, British Columbia, Canada V6T 1Z1

ABSTRACT

We present canonical abelian lattice Chern-Simons theory as a model for anyons, and as a topological field theory and discuss the relationship between gauge invariance of Chern-Simons-like lattice actions and the topological interpretation of the canonical structure. We show that these theories are exactly solvable and have the same degrees of freedom as the analogous continuum theories. We also examine the description of lattice anyons as conventional particles coupled to Chern-Simons theory. Finally, we show how the problem is in general identical to that of finding a lattice-geometrical object which we call a local intersection form, which allows us to define the exterior algebra of lattice forms without benefit of a dual lattice, and how the partition function may be understood in the same way.

1. Introduction

Over the past few years physics and topology have cross-fertilized to produce exciting new results and deeper understanding over a range of topics, of which knot theory and electrodynamics in $2+1$ dimensions are two. Several years ago Ed Witten undertook a study of a class of (nondynamical) physical models and discovered that they contained topological information about knots, when the latter are viewed as externally imposed trajectories of charged particles.

The physical models are known as Chern-Simons theories, which exist as models within the kinematical framework of quantum field theory. Quantum field theories in general exist in something of a mathematical twilight, in that many of its operations are ill-defined. The work I will outline today resolves some of these ambiguities in the context of the simplest

This work is supported in part by the Natural Sciences and Engineering Research Council of Canada.

variety of Chern-Simons theory, namely abelian Chern-Simons theory. In the process, we shall observe a nice relation between the geometry of 2-manifolds and a geometry we may define on CW-complexes embedded therein.

2. Review of Canonical Quantization

2.1. Lagrangian Classical Particle Mechanics

Because most here are not physicists, I will conduct a brief review of canonical quantum mechanics and field theory, starting with Lagrangian classical mechanics, then outlining the mapping to the canonical reformulation, and finally giving the rules by which such a system is "quantized". Then I will try to show how a field, a physical variable obeying a Lagrangian partial differential equation, may be treated within the same canonical framework, and then quantized.

In physics, we typically define a classical Lagrangian theory describing the motion of a particle by defining a configuration space *Conf*, which is a manifold in which physical variables q may take values; a space of trajectories *Traj(Conf)*, i.e. maps $q(t) : \mathcal{I} \longrightarrow Conf$, $\mathcal{I}$ a finite or infinite interval on the real line; and a functional S on the space of trajectories S: $Traj(Conf) \longrightarrow \Re$. S is usually restricted to the form $S = \int L dt$ with L (the "Lagrangian") a function of q and $\dot{q}$, i.e. a function on the tangent bundle $T Conf$.

For example, let $Conf = \Re$, q a particle allowed to move on a line, for all time $-\infty < t < \infty$. *Traj(Conf)* is the set of maps $q(t)$: $\Re \longrightarrow \Re$. Let $L = (m/2)\dot{q}^2 - (k/2)q^2$, $S = \int Ldt$. This defines the harmonic oscillator, which describes the motion of a particle with mass m experiencing the force of a spring, with spring constant k. The final step in defining the classical theory is the extremizing of the action, which leads to n differential equations determining the dynamics, where $n = \dim(Conf)$. These are the famous Euler-Lagrange equations $\frac{d}{dt}\frac{\partial L}{\partial \dot{q}_j} - \frac{\partial L}{\partial q_j} = 0$, which in the particular example mentioned above work out to be $m\ddot{q} + kq = 0$. In general, because of the form of S, the equations must be second order ordinary differential equations, whose 2 initial data each correspond to the initial positions and initial velocities of the particles. In general, this defines the notion of a (classical) physical degree of freedom - a quantity evolved forward in time by a second order equation of motion, and having 2 freely chosen initial data (so that constraints among the initial data subtract out degrees of freedom from the system).

2.2. Canonical Classical Particle Mechanics, or Hamiltonian Particle Mechanics

The canonical formalism reformulates these equations into $2n$ first order differential equations, via a Legendre transformation which defines a new variable $p_i = \partial L/\partial \dot{q}_i$ to replace the velocities $\dot{q}_i$, and a function

$$H = \sum_j p_j \dot{q}_j(q_k, p_k) - L(q_i, \dot{q}_i(q_k, p_k)) \tag{2.1}$$

the Hamiltonian, to replace L (where the $\dot{q}$'s are regarded as functions of the p's and q's). We regard these p s and q s as coordinates on a $2n$-dimensional manifold, which is in fact the cotangent bundle on $Conf$: $T^* Conf$, known as "phase space", which is equipped with a natural symplectic (nondegenerate, skew-symmetric) form $\omega = \sum_i dp_i \wedge dq_i$, mapping vectors to 1-forms. This symplectic form defines an antisymmetric bracket $\{\cdot,\cdot\}$: $C^\infty(T^* Conf) \otimes C^\infty(T^* Conf) \longrightarrow C^\infty(T^* Conf)$ (the "Poisson Bracket") on functions on phase space: $\{f(q_j,p_j), g(q_j,p_j)\} = df(Idg)$ (with I mapping 1-forms to vectors, $\omega(I\alpha) = \alpha$) and makes these functions into a Lie Algebra, the Poisson Algebra. Explicitly

$$\{f(q_j,p_j), g(q_j,p_j)\} = \sum_j \frac{\partial f}{\partial q_j}\frac{\partial g}{\partial p_j} - \frac{\partial g}{\partial q_j}\frac{\partial f}{\partial p_j} \tag{2.2}$$

so that the coordinate functions satisfy $\{q_i,p_j\} = \delta_{ij}$, $\{q_i,q_j\} = \{p_i,p_j\} = 0$. Time evolution equations which reformulate the equations extremizing S are determined by ω and the Hamiltonian, as $\dot{x} = dx(IdH) = \{x, H\}$, where $x \in T^*Conf$, or $\dot{q}_i = \partial H/\partial p_i$, $\dot{p}_i = -\partial H/\partial q_i$. Pictorially, the Hamiltonian function on the phase space manifold, together with the symplectic form define a vector field IdH, whose integral curves define a flow on the manifold, and it is this flow which we identify with time evolution. In this picture, there are again $2n$ initial data, corresponding to the initial positions and momenta of the particles. In the example above, the Hamiltonian for the harmonic oscillator is $H = (1/2m)p^2 + (k/2)q^2$, and with the explicit form of the Poisson bracket given above, we may work out the equations of motion to be $\dot{q} = p/m$, and $\dot{p} = -kq$, which are equivalent to the original equation upon elimination of p.

2.3. Canonical Quantization of Particle Mechanics

2.3.1 Phase space is mapped into an algebra of operators on $\mathcal{L}^2(QConf)$ via the rule $\{\cdot,\cdot\} \longrightarrow -i[\cdot,\cdot]$

Canonical quantization is essentially a set of mathematical procedures mapping one mathematical system onto another, designed originally to mimic our intuition about the physics of the electron, and later abstracted to more general systems. Through a series of experiments physicists became aware that, among a collection of electrons, individual electrons are not distinguishable even in principle. Thus, for example (electron 1 at point x and electron 2 at point y) represents the same physical state as (electron 2 at point x and electron 1 at point y). Thus *Conf* should be replaced by *QConf*, defined as *Conf* modulo permutations of particle positions. In addition it was found that electrons could not be localized to a single point in configuration space, but rather were distributed in at least a finite (and possibly infinite) sized neighborhood, with distribution defined by an $\mathcal{L}^2$ function on *QConf*, $\psi(q)$, called the wave function. The wave function defines the physical state of the system, and is time-independent.

The phase space coordinates q and p are made into operators on $\mathcal{L}^2(QConf)$, and the set of functions on phase space is now an associative algebra. Specifically, it is the universal enveloping algebra of an extended affine Lie Algebra: the Heisenberg algebra. The Heisenberg algebra is defined by allowing the commutator bracket of defining basis elements (q_i, p_i) to be simply equal to the Poisson brackets for those same objects as elements of the Poisson algebra (times $-i$): $[\cdot,\cdot]$, $[q_i,p_j] = -i\delta_{ij}$, $[q_i,q_j] = [p_i,p_j] = 0$. These operators act on the Hilbert space $\mathcal{L}^2(QConf)$, usually so that the q_i act as multiplication operators, and the p_i as differentiation operators ($p_i = i\partial/\partial q_i$). The startling physical statement contained in all of this is that the momenta are no longer independent data, but rather depend on how the distribution function (the "wave function") is localized in configuration space. This is a crude statement of the uncertainty principle.

Under this mapping, all other observables, which are functions of the p s and q s, also become operators on the Hilbert space. In order to extract physical information about our system, i.e. the values of these observables, we must allow them to act on the physical state, the wave function ψ. This usually takes the form of computing expectation values of these operators. There is often more than one quantum operator corresponding to a classical observable, and these must be chosen to be Hermitian operators, so that expectation values are real. We shall see that, even in field theory, the fields are almost always taken to be Hermitian operators. The one exception to this rule is the fermion field operator, which appears as a complex representation of the spin algebra.

2.3.2. Classical time evolution is mapped to quantum time evolution via the rule $\{\cdot,\cdot\} \longrightarrow -i[\cdot,\cdot]$

Thus the Hamiltonian function, which determined the flow of time evolution, is also an operator on the Hilbert space. a function of the operators p_j, q_j. Time evolution of physical observables $X(q_j, p_j)$ is also translated directly over from the Poisson algebra using the same rule as above: $\dot{X}(p_j, q_j) = i[X(p_j, q_j), H(p_j, q_j)]$. If H has no time dependence, this equation may be solved formally, as $X(p_j, q_j, t) = \exp(iH(p_j, q_j)t)X(p_j, q_j, 0)\exp(-iH(p_j, q_j)t)$ i.e. time evolution acts by a unitary transformation on the operators. For our purposes, it is more helpful to invert this unitary transformation simultaneously on the operators and states, such that the operators are independent of time, and the states evolve in time instead. Now the evolution of the state may be given by a first order ordinary differential equation as

$$i\frac{\partial}{\partial t}\psi(q_i) = H(p_i, q_i)\psi(q_i), \tag{2.3}$$

which is known as the Schrödinger equation.

2.3.3. Statistics: Symmetries of the wave function under particle exchange

There is a further physical statement, that we must take as given by nature – the "statistics" of particles. Particles in quantum mechanics must be indistinguishable, as noted above. In addition, however, the wave functions must occur as specific representations of $\pi_1(QConf)$. Essentially, there are two known types of fundamental particles known to exist in the world, which are given the names "boson" and "fermion". The wave functions of the former are required to be totally symmetric representations of $\pi_1(QConf)$, while those of the latter must be totally antisymmetric. Among fundamental particles living in 3 spatial dimensions, these are the only statistics allowed. However, in 2 spatial dimensions, the $\pi_1(QConf)$ is richer (in the case of fermions only), and allows other possibilities, other 1-dimensional representations, realized in nature not as a fundamental particle, but as a composite of other, more fundamental particles. They are called "anyons", because their statistics interpolate between bosons and fermions, taking "any" value.

2.4.1 Classical Field Theory

We now attempt to make the transition between the mechanics of individual particles, described above, and the mechanics of fields. Suppose we are given a manifold $\mathcal{M}$, representing space. The fields which appear in physical theories may be thought of either as a section of a bundle over $\mathcal{M}$, or as a connection on that bundle. In the common examples this may be a tangent bundle, vector bundle , principal fibre bundle or line bundle. We identify the space of sections and/or connections with *Conf*, and choose convenient coordinates ϕ on *Conf*. We write down an action functional $S = \int L dt = \int dt \int_{\mathcal{M}} \mathcal{L}$, where $\mathcal{L}$, the Lagrangian density, is a function of the field ϕ, its time derivative $\dot{\phi}$, and its derivatives in $\mathcal{M}$, usually up to first order. By extremizing the action we may define Lagrangian dynamics exactly as before. Now however, the Euler-Lagrange equations are a partial differential equation with initial data $\phi(x, t = 0), \dot{\phi}(x, t = 0)$ specified at each point x of $\mathcal{M}$.

Hamiltonian dynamics also follows in analogy with classical particle mechanics as described above. The Lagrangian density $\mathcal{L}$ is usually a local function of $\phi(x)$, $\dot{\phi}(x)$, so that $\pi(x) = \partial\mathcal{L}/\partial\dot{\phi}(x)$ is a function of $\phi(x)$, $\dot{\phi}(x)$ and their spatial derivatives, but not of $\phi(y)$, $y \neq x$, or its derivatives. This implies that the Poisson Bracket is of the form $\int_{\mathcal{M}} dx d\phi(x) \wedge d\pi(x)$, so that $\{\phi(x), \phi(y)\} = \{\pi(x), \pi(y)\} = 0$ and $\{\phi(x), \pi(y)\} = \delta(x - y)$. This gives us the picture that $\phi(x)$ and $\phi(y)$ are to be considered as independent degrees of freedom in much the same way as q_1 and q_2, back in the section on particle mechanics.

2.4.2 Canonical Quantum Field Theory

The program of canonical quantum field theory is to write down a theory describing the dynamics of entire field configurations using the same prescription as for the quantum mechanics of particles. Having defined a Poisson algebra, the rule is the same as before. We map the Poisson algebra into the Heisenberg algebra (now with infinite basis) on $\mathcal{L}^2(QConf)$ via the rule $\{\cdot,\cdot\} \longrightarrow -i[\cdot,\cdot]$ i.e. $[\pi(x), \phi(y)] = 0, \quad x \neq y$, $[\phi(x), \phi(y)] = [\pi(x), \pi(y)] = 0$. Also, the wave functions $\psi(q)$ become simply wave-functionals $\psi[\phi]$, and as before, the operator ϕ acts on this state by multiplication, and its canonical momentum acts by functional differentiation $\pi(x)\psi[\phi] = i\delta/\delta\phi(x)\psi[\phi]$.

However, there is a great conceptual difference. Whereas in quantum mechanics the configurations are the possible positions of the particles living in our system, in quantum field theory they may be understood as the *existence* of particles (or the number present) at given positions.[‡] In other words, the actual number of particles in the system at each point of $\mathcal{M}$ is a quantum variable which we may compute within quantum field theory. And the operators on the Hilbert space actually create and destroy particles at various positions or momenta from the physical state, whose role is now partly to act as a bookkeeping device for these quantities. The imposition of statistics within quantum field theory then devolves to the commutation relations between operators – for fermions the above rule actually changes the bracket from a commutator $[a, b] = ab - ba$ to an anticommutator $[a, b]_+ = ab + ba$ †. And for anyons we shall seek the commutator $[a, b]_\alpha = ab - e^{\pm i\alpha} ba$. •

---

‡ see footnote, pg. 13.

† Thus fermions do not actually derive from the picture of classical physical variables living on a symplectic manifold described above. In fact, they may be reincorporated into this picture through the use of Grassmannian variables.

• α is not quite independent of x and y. It's form (see ahead) is dictated by the $\pi_1(QConf)$ and similar requirements on its Hermitian conjugate algebra.

2.5 Abelian Gauge Theories

2.5.1 A system defined on a space of connections and sections of a line bundle

We shall use the foregoing to study a theory of (abelian) gauge fields coupled to fermion fields. An abelian gauge field is a connection on a line bundle. We will introduce coordinates on the space of connections as a choice of 1-form A. Two 1-forms define the same connection if their difference is exact. Let the Hodge decomposition of A be $A = d\chi + *d\xi$. In the terminology of physicists, we say that this 1-form is an "abelian gauge field", and that the action is "gauge invariant", when it is shown that the action is independent of the exact part of A, i.e. of χ. When this field is coupled† to fermions (the fermions are Grassman-valued sections of the line bundle), we say that the resulting theory is a gauge theory if the action is gauge invariant when the phase of ψ is identified with χ, so that the gauge transformation under which the action must be invariant is now $A \longrightarrow A + d\chi$, $\psi \longrightarrow e^{i\chi}\psi$. These gauge transformations form a group, called the gauge group, and the exact part of A is called the gauge function, or the gauge part of A. This group is a group of symmetries among the configuration variables (or, in the case of Chern-Simons theory, the phase space). Of particular interest in constructing actions for a gauge theory is a knowledge of the possible gauge invariant expressions out of which one can build an action functional. These are the 2-form $F = dA$, and the covariant derivative $\Psi^\dagger(i\partial - A)\Psi$. Related to this last expression is the charge transporter $\Psi(x)^\dagger \exp\left(i\int_{C_{x,y}} A\right)\Psi(y)$.

2.5.2. Abelian gauge theories have degenerate canonical structure, and constraints on the initial data

Within the canonical formalism, however, gauge invariance takes on a peculiar aspect. This is because, in the normal run of gauge theories, there is no momentum for one of the components (the one pointing in the time direction) because the Lagrangian actually does not depend on its velocity. In this special case there is nothing to do but eliminate this component from the dynamics of the theory, and rebuild the canonical formalism again without it, or equivalently, enforce a constraint that sets this momentum to zero for all time. Because our set of canonical variables is now smaller, our symmetry group is smaller, a subgroup of the group of gauge transformations, namely those gauge transformations that are independent of time. We thus have a dynamical model whose time evolution includes an unconstrained function on $\mathcal{M}$ (the exact part of the gauge variable) which cannot evolve in time. And this means that its momentum (which determines the time development) must be set to zero. This constraint is known as Gauss' law, and is responsible for the famous $1/r^2$ repulsion between stationary charged particles, in ordinary electrodynamics.

Chern-Simons Theory in the Continuum

We are now ready to discuss Chern-Simons theory. Chern-Simons theory is a theory of gauge fields which transmutes the statistics of the charged particles to which it is coupled[3-22]. It allots essentially no physical degrees of freedom to the gauge field, but instead only

† This means that the action is a function of both fields, and $S(\psi, A) \neq S_\psi(\psi) + S_A(A)$.

a long range force which shadows matter particles, such as the fermions introduced in the previous section. Chern-Simons theory, like fractional statistics, only makes sense in 2 spatial dimensions[23-30]. When considered in isolation, i.e. uncoupled to any matter, Chern-Simons theory is not quite trivial, but instead has a finite number of degrees of freedom – the zero modes of the connection 1-form A (there are two for each hole in the manifold)[31,32]. Consider as a configuration space *Conf* the space of 1-forms on a 2 dimensional Riemann surface $\mathcal{M}$. By a suitable embedding, we may take *Traj* as the space of 1-forms on the 3-manifold $\mathcal{M} \times \Re$.

3.1. Action for Chern-Simons Theory

In $2+1$ dimensions, aside from the normal kinetic gauge action $\int dt \int_{\mathcal{M}} (1/4) F^2$, there is another gauge invariant possibility (let $L = \int_{\mathcal{M}} \mathcal{L}$, also $H = \int_{\mathcal{M}} \mathcal{H}$)[†]

$$\mathcal{L}_{CS} = \frac{k}{4\pi} A \wedge dA = \frac{k}{4\pi} \epsilon_{\mu\nu\lambda} A_\mu(x,t) \partial_\nu A_\lambda(x,t) \tag{3.1}$$

This term is gauge invariant after an integration by parts, under transformations which vanish on the boundary. It also has the interesting property of being odd under time reversal $t \longrightarrow -t$, and parity $x_1 \longrightarrow -x_1, x_2 \longrightarrow x_2, \quad x = (x_1, x_2) \in \Re^2$. These two features allow interesting behavior, in that the states are now not required to be even under these two symmetries, in contrast to more ordinary gauge theories. For example, this property is believed to allow for a superconducting anyonic ground state in some cases.

3.2. Canonical Quantization of Chern-Simons Theory

3.2.1 The Lagrangian is degenerate, the phase space is identified with Conf

We perform a canonical split

$$\mathcal{L}_{CS} = \epsilon_{ij} A_i(x,t) \dot{A}_j(x,t) + A_0(x) B(x), \quad B(x) = dA(x) \tag{3.2}$$

and find that L_{CS} is a "degenerate" Lagrangian – meaning that A_1, A_2, $\pi_1 = \partial \mathcal{L}_{CS} / \partial \dot{A}_1$ and $\pi_2 = \partial \mathcal{L}_{CS} / \partial \dot{A}_2$ are not all independent. Because $\pi_1 = -A_2$ and $\pi_2 = A_1$, we find that the phase space is simply equal to the configuration space, rather than being twice the dimension. Thus we find that we are to impose the commutation relations on the configuration (now also phase space) variables

$$[A_1(x), A_2(y)] = \frac{2\pi i}{k} \delta(x-y) \tag{3.3}$$

Thus, the gauge symmetry is not a configuration space symmetry in this case, but a phase space symmetry. [†]

[†] Notation: Sums over Greek indices are to be read as $a_\mu b_\mu = -a_1 b_1 - a_2 b_2 + a_0 b_0$. Also, for latin indices: $a \cdot b = a_i b_i = a_1 b_1 + a_2 b_2$.

[†] The reader may be concerned by the appearance of Dirac δ-functions, which are not functions at all and a vice common to physicists. In fact, these are part of the motivation for studying the lattice versions of physical theories, treating the continuum models as a limit thereof. On the lattice, these δ-functions revert back to the Kronecker- δ's of particle mechanics.

3.2.2. Gauge constraint for Chern-Simons theory

Notice also that there is no dependence on $\dot{A}_0$. As we discussed earlier, this implies that $\pi_{A_0} = 0$, and in fact equivalence with the Lagrangian theory demands that π_{A_0} should vanish for all time, when acting on physical states. This is a nontrivial additional constraint we need to impose, and it cannot be enforced over the entire Hilbert space. Hence we must restrict ourselves to states in the Hilbert space which this operator annihilates, which we regard as the "physical subspace". This constraint (which we write as $\mathcal{G}$) appears as

$$\begin{aligned} 0 &= \dot{\pi}_{A_0}|\text{phys}\rangle \\ &= \delta H_{\text{CS}}/\delta A_0(x)|\text{phys}\rangle \\ &= B(x)|\text{phys}\rangle \end{aligned} \tag{3.4}$$

The gauge symmetry can be used to show that π_{A_0} is equal to the momentum of the exact part of A, and thus Eq. (3.4) is the Gauss Law constraint discussed above.

3.2.3. The Hamiltonian for Chern-Simons theory vanishes

Having defined the momenta, and found the expected constraint, we now seek the Hamiltonian. However, inspection of the Lagrangian, together with the above definitions for momenta, shows us that the time derivative parts of the Lagrangian Eq. (3.2) don't contribute to H_{CS}, and the only thing left is the constraint, which vanishes. Thus we find $\mathcal{H}_{CS} = 0$.

3.3. Properties of Chern-Simons Theory

3.3.1. The commutator of Wilson lines

The first interesting feature that Chern-Simons theories have is reflected in the properties of their Wilson line operators $\int_C A$. Using the commutator Eq. (3.3) it is easy to see that

$$[\int_C A, \int_{C'} A] \tag{3.5}$$

is a very simple quantity which merely counts the intersections of the curves (suitably well-behaved) C and C'. In fact, it counts the signed intersection number of the two curves, which is a fixed-endpoint homotopy invariant.

3.2.2. Coupling to external currents - fractional statistics

Some other elementary properties of Chern-Simons theory may be observed by coupling the gauge field to an external current, $\mathcal{L}_{CS} \longrightarrow \mathcal{L}_{CS} + A_\mu j_\mu$, where j is a c-number current, satisfying the current conservation equation $\partial_\mu j_\mu = 0$ (i.e. particles in the current are neither created or destroyed)[7,41]. Here let $\mathcal{M} = \Re^2$ for simplicity. With this addition the Gauss Law constraint $\mathcal{G}$ becomes $\mathcal{G} = B + (2\pi/k)j_0 = 0$. Now let us rewrite A, using its Hodge decomposition, as

$$\begin{aligned} A &= d\xi + *d\chi \\ &= d\xi + *d\frac{1}{\Delta}B, \end{aligned} \tag{3.6}$$

Note that $\mathcal{G} = \pi_\xi = B + (2\pi/k)j_0$ so that $[B(x), \xi(y)] = [\pi_\xi(x), \xi(y)] = \frac{2\pi i}{k}\delta(x-y)$, which we may represent as $B(x) = \frac{2\pi i}{k}\delta/\delta\xi(x)$.

Thus we may choose to use B and ξ rather than A_1 and A_2 as our canonical variables. The Hamiltonian consists only of the spatial part of the interaction with the external current

$$\begin{aligned} H &= \int A_i j_i \\ &= \int j \wedge d * \frac{1}{\Delta} B + j \cdot d\xi \\ &= \int j \times \partial \frac{1}{\Delta} B + j \cdot d\xi \end{aligned} \tag{3.7}$$

(where $a \times b$ means $a_i b_j \epsilon_{ij}$) and the Schrödinger equation is

$$\left[\int \left(\frac{2\pi i}{k}\right) j \times \partial \frac{1}{\partial^2}\frac{\delta}{\delta\xi} + j \cdot d\xi\right] \Psi[\xi, t] = i\frac{\partial \Psi[\xi, t]}{\partial t} \tag{3.8}$$

In addition, Gauss Law is

$$\left(i\frac{\delta}{\delta\xi} + j_0\right) \Psi[\xi, t] = 0 \tag{3.9}$$

whose solution is simply

$$\Psi[\xi, t] = \exp i\left(\int_{\mathcal{M}} j_0(x, t)\xi(x)\right) \tilde{\Psi}(t) \tag{3.10}$$

Plugging this result into the Schrödinger equation gives the result

$$\Psi[\xi, t] = \exp i\left(\int_{\mathcal{M}} j_0(x, t)\xi(x)\right) \exp\left(-\frac{2\pi i}{k}\int_{-\infty}^{t}\int_{\mathcal{M}} j \times \partial \frac{1}{\partial^2} j_0\right) \tag{3.11}$$

The latter term has an interesting interpretation, when j_0 and j_i are taken to be currents from two distinct point particle trajectories $j_i = \dot{x}_i\delta(r - x(t))$, $j_0 = \delta(r - y(t))$, (i.e. j_μ is such that $\int_{\mathcal{M}} A_\mu j_\mu = \int_C A$, where C is the particle trajectory). In this case the argument of the second exponential becomes $\theta(x(t) - y(t))$, where θ is a "multi-valued function" on the plane, $\theta(x) = \operatorname{Im}\ln(x_1 + ix_2)$, $x = (x_1, x_2)$ which we may make well-defined by the introduction of a cut. Since we will be making use of the properties of the angle function, we shall define it carefully.

3.4 Definition of Angle Function

The angle function may be defined by the following formula[42]

$$\theta_C(x, y) = \int_C f^{(y)} = \int_C d\ell_i f_i(\ell - y)$$

where $f^{(y)}(x) = f(x - y)$, and f is a 1-form defined as $f = *dg = \epsilon_{ij}\partial_j g$, where g is the Green's function of the Laplacian, $\partial^2 g(x) = *d * dg(x) = \delta(x)$. C is an integration contour that runs from a point at ∞ to x (we choose it to be along the positive $\hat{1}$ axis). Note f satisfies $df = \partial \times f = \delta$, i.e. f is the gradient of no single-valued function on the plane, but is the gradient of a function defined on any simply connected region not containing the origin. This result is often expressed by physicists using the highly formal expression $d^2\theta_C = \partial \times \partial\theta_C = \delta$. θ is said to be a multi-valued function on the plane, in the sense that it's value is well defined modulo $2\pi n$, n an integer, which is determined once the integration contour is selected.

Thus we may define a single-valued angle function by pre-selecting a curve for every point x in the plane, i.e. we select a curve field $C_{x,x}$, and this introduces a cut for the angle function[2]. We have chosen this construction of the angle function because it is this definition of the angle function which is simple to carry over to the lattice. Details for this construction are presented in the lattice angle function section, ahead.

3.5 Topological Properties of the Chern-Simons Partition Function

Although this article principally concerns Chern-Simons theory in the canonical formalism, we digress briefly here to describe the partition function that is defined by abelian Chern-Simons theory coupled to external currents. The partition function may be calculated as $\mathcal{Z}[j] = \exp i\Gamma[j]$,

$$\Gamma[j] = \int_{\mathcal{M}} dx \int_{\mathcal{M}} dy j_\mu(x)\epsilon_{\mu\nu\lambda}\frac{(x-y)^\nu}{(x-y)^2}j^\lambda(y) \tag{3.14}$$

We imagine that the current j is a sum of currents $j^{(i)}$ representing single particle trajectories $j = \sum_i j^{(i)}$. Then $\Gamma[j]$ may be split into a sum of pure and mixed terms $\Gamma[j] = \sum_i \Gamma[j^{(i)}] + \sum_{i\neq k}\Gamma^{(m)}[j^{(i)}, j^{(k)}]$, with

$$\Gamma^{(m)}[j,j'] = \int_{\mathcal{M}} dx \int_{\mathcal{M}} dy j_\mu(x)\epsilon_{\mu\nu\lambda}\frac{(x-y)^\nu}{(x-y)^2}j'^\lambda(y) \tag{3.15a}$$

When the trajectories are well-behaved curves, the second term is the linking number of the link represented by the trajectories i and k, while the first term is the self-linking number of the link i[12,54]. As the mixed term will be treated in the lattice sections ahead, we examine the continuum version.

On particle trajectories j and j', $\Gamma^{(m)}[j,j']$ becomes

$$\Gamma^{(m)}[j,j'] = \int_C \int_{C'} d\ell_\mu(x)\epsilon_{\mu\nu\lambda}\frac{(x-y)^\nu}{(x-y)^2}d\ell'^\lambda(y) \tag{3.15b}$$

We may invoke Stokes' Theorem to rewrite the first integral as a surface integral over the surface B' bounded by C', $\int_C \int_{B'} d\ell^\mu(x)da^\lambda(y)\delta(x-y)\delta_{\mu\lambda}$ (da is a differential element of the area of surface B'. Thus the linking number may be expressed as a local integration kernel $K(x-y)$ which counts the signed intersections of a link with a surface. This is the formulation that will be useful in treating lattice Chern-Simons theory.

3.6. Anyon Operators

3.6.1. Canonical quantization of Chern-Simons-fermion action

With this in hand, we now present the standard calculation which demonstrates (suggests, really, this is not a mathematical proof) that abelian Chern-Simons theory coupled to fermions has excitations whose statistics are anyonic. First, we must study the appropriate action, which is*

$$S_{\text{CS-fer}} = S_{\text{CS}} + S_{\text{fer}}, \quad S_{\text{fer}} = \int dt \int_{\mathcal{M}} \Psi^\dagger \dot{\Psi} - \Psi^\dagger \alpha_i (\partial_i - A_i)\Psi \tag{3.16}$$

Following the canonical procedure, we see that the momentum for Ψ is $\Psi^\dagger$ (once again the phase space and the configuration space are the same), and we are to impose the anticommutation relations $[\Psi^\dagger(x), \Psi(y)]_+ = i\delta(x-y)$. The Hamiltonian is simply minus the second term

$$H_{\text{CS-fer}} = \int_{\mathcal{M}} \Psi^\dagger \alpha_i (\partial_i - A_i)\Psi \tag{3.17}$$

A physical picture that emerges from studying this kind of system is that electrons are created into the state by the operator $\Psi^\dagger$, while they are annihilated out of it by Ψ. In addition, because $[\Psi^\dagger(x), \Psi^\dagger(y)]_+ = 0$ for all x, y, we have $(\Psi^\dagger(x))^2 = 0$, implying that no two fermions can occupy the same point (or more generally, the same state). This is called the Pauli Exclusion Principle, or the hard-core condition, and is necessary for anyons. In particular, it means that the charge density operator $\Psi^\dagger(x)\Psi(x) = \rho(x)$ has only the eigenvalues $0, 1$.**

3.6.2. Elimination of gauge variable, construction of anyon operators

The calculation we will do is a simple change of variables[10,11,14-16,19,22]. Let Ψ and $\Psi^\dagger$ be the fermion annihilation and creation operators, respectively. As before, A is the gauge field. We define the anyon annihilation operators Φ_C as (creation operators are the Hermitian conjugate of Φ_C)

$$\begin{aligned} \Phi_C(x) &= \Psi(x)U(x)K_C(x) \\ U(x) &= \exp i \int_{\mathcal{M}} f(x-z)\cdot A(z) d^2 z \\ K_C(x) &= \exp i \int_{\mathcal{M}} \theta_C(x-z) dA(z) d^2 z \end{aligned} \tag{3.18}$$

The factor U makes the combination gauge invariant (K is invariant all by itself), by pinning a magnetic flux on top of the charge created or annihilated by $\Psi^\dagger$ or Ψ. (This gives the anyon exchange phase, below, the interpretation of half of the Aharonov-Bohm phase that attaches

---

* The α_i are matrices which act on the spin degrees of freedom of Ψ. They are irrelevant here.

** Much of what I am saying is strictly true only the lattice, and only "sort of" true in the continuum, where precise statements are made about only about the smeared fields $\Psi_f = \int_{\mathcal{M}} \Psi f$. In addition, $\rho(x)$ is not diagonalizable in the continuum. The reader is asked to forgive this sloppiness, because we are only trying to establish a physical picture, and all these statements *are* true on the lattice, which is our main interest in this paper.

to a charged operator that encircles a magnetic flux). Two things result from this. First, the creation operators for particles will obey anyonic statistics

$$\begin{aligned}\Phi_C(x)\Phi_C(y) &= \exp(\frac{2i}{k}\chi)\Phi_C(y)\Phi_C(x)\\ \chi &= \theta_C(x-y) - \theta_C(y-x)\\ &= \pi \mathrm{sgn}(x_2 - y_2) + 2\pi\nu(\mathcal{C}(x-y), \mathcal{C}(y-x))\end{aligned} \tag{3.19}$$

Second, it is usually stated that all interactions (i.e. except for those in the statistics) are cancelled out of the Hamiltonian, meaning that $H_{\mathrm{CS-fer}}[\Phi]$ is the same as $H_{\mathrm{CS-fer}}[\Psi]$, with the gauge field A removed. This is not quite true, there is actually another (singular) term. The full answer is

$$H_{\mathrm{CS-fer}} = \int_{\mathcal{M}} \Phi_C \alpha_i \partial_i \Phi_C - \Phi_C \zeta \Phi_C \tag{3.20}$$

where $\zeta(x) = \int_{C(x)} B$. Were B a smooth function, we might reasonably disregard this term, arguing that the measure of the 2-cycle over which the 2-form B is integrated is zero. However, it is not a smooth function, it is an unbounded operator on a Hilbert space, and as such can easily produce nonzero (even infinite) effects. We shall see this term more clearly on the lattice where the expression is well-defined. The problem stems from a desire to use "multivalued fields", which do not have a cut or a curve field, but are instead defined on the universal cover of *QConf*. We feel that the meaning of such fields is not yet clarified, and that using the well-understood single-valued operators the existence and nature of the interactions are manifest. Nevertheless, it might reasonably be stated that "most" of the interaction has been cancelled (again, we shall see later what this means), and this fact suggests that the anyonic operators bring the Hamiltonian "closer" to diagonal form, because the system seems to be almost like a free fermion Hamiltonian, whose operators create exact Hamiltonian eigenstates. Hence, it is generally agreed that a Chern-Simons-fermion theory is really a theory of anyons.

4. Lattice Chern-Simons Theory

In this section, we finally arrive at the purpose of this talk, a review of our recent work on lattice Chern-Simons theory. Ours is not the first such effort by any means - there have been several other treatments[15,43,46-50]. Ours however, is the only one adapted to the study of knot theory.

4.1. Notation

We work in the Hamiltonian formalism where time is continuous and where space is a finite two-dimensional square lattice, denoted $\mathcal{L}^2$ with lattice spacing 1. We define the forward shift operator S_i, $i = 1, 2$ by

$$S_i f(x) = f(x + \hat{i}) \tag{4.1}$$

and its inverse is denoted by

$$S_i^{-1} : \ S_i^{-1} f(x) = f(x - \hat{i}) \tag{4.2}$$

(For convenience of notation, also define $S_0 = 1$, so that we may use the Greek index notation S_μ, $\mu = 0,1,2$). On lattices differentiation becomes differencing, and may do so on one of two possible ways – either with forward differencing

$$d_i f(x) = f(x+\hat{i}) - f(x), \quad d_i = S_i - 1 \tag{4.3}$$

or backward differencing, denoted by

$$\hat{d}_i f(x) = f(x) - f(x-\hat{i}), \quad \hat{d}_i = 1 - S_i^{-1} = S_i^{-1} d_i \tag{4.4}$$

(Again, for convenience, define $\hat{d}_0 = d_0 = \partial_0$. We will also use the notation $\dot{f} = \partial_0 f$.) Note that summation by parts on a lattice takes the form (neglecting surface terms)

$$\sum_x f(x) d_\mu g(x) = -\sum_x \hat{d}_\mu f(x)\, g(x) \tag{4.5}$$

by virtue of the lattice Leibniz rule

$$d_\mu(fg) = f d_\mu g + d_\mu f S_\mu g \tag{4.6}$$

(no sum on μ).

The lattice Fourier transform is defined by

$$f(x) = \int_{\Omega_B} \frac{d^2k}{(2\pi)^2}\, e^{-ik\cdot x} f(k) \tag{4.7}$$

where

$$\Omega_B = \{(k_1,k_2) : -\pi < k_1, k_2 \leq \pi\} \tag{4.8}$$

is the Brillouin zone. Furthermore the Fourier transform of the difference and shift operators are

$$d_i(k) = S_i(k) - 1 = e^{-ik_i} - 1 \ , \quad \hat{d}_i(k) = 1 - S_i^{-1}(k) = 1 - e^{ik_i} \tag{4.9}$$

and obey the identities $-\hat{d}_i(k) = d_i(-k) = d_i^*(k)$, $S_i^{-1}(k) = S_i(-k) = S_i^*(k)$.

Similarly, the time domain has the usual Fourier transform

$$f(t) = \int_{-\infty}^{\infty} \frac{dk_0}{2\pi}\, e^{-ik_0 t} f(k_0) \tag{4.10}$$

and derivatives are defined analogously.

Defining the lattice Laplacian Δ as $d \cdot \hat{d}$ (see comments on lattice cohomology below), we may use the Fourier expansion of lattice functions to construct the inverse Laplacian operator $-g(x)$: $d \cdot \hat{d} g(x) = -\delta(x)$ as an integral operator $(d \cdot \hat{d})^{-1} f(x) = \sum_y g(x-y) f(y)$, in analogy with the continuum (see also ref. [42]).

$$g(x) = \int_{\Omega_B} \frac{d^2p}{(2\pi)^2} \frac{1 - e^{ip\cdot x}}{\sum_i 2(1 - \cos p_i)}, \tag{4.11}$$

The spatial components $A_i(x)$ of the gauge field are real-valued functions on the links specified by the pair $[x, \hat{i}]$, and the time component $A_0(x)$ is a function on sites. Thus the field strength tensor in this formalism has two parts. The space-space part

$$F_{ij}(x) = d_i A_j(x) - d_j A_i(x) \tag{4.12}$$

is a function on plaquettes, but mapped by our use of forward differencing to a function on sites as well, by the convention of associating a plaquette to the site at its lower left corner. Further, given any vector V_i (such as A_i or d_i), we define the notation $V_i^\perp = \epsilon_{ij} V_j$, so that we may write (in 2 dimensions) the commonly occurring expression $F_{ij}\epsilon_{ij} = d_i A_i^\perp = -d_i^\perp A_i$. When written with the shift operator S, however, it is unfortunately necessary to define it as $(S^\perp)_1 = S_2$, $(S^\perp)_2 = S_1$. The space-time components

$$F_{0i}(x) = d_0 A_i(x) - d_i A_0(x) \tag{4.13}$$

are a function on links. Defining the dual field strength in the usual way,

$$(*F)_\mu(x) = \epsilon_{\mu\nu\lambda} F_{\nu\lambda}(x) \tag{4.14}$$

the Bianchi identity takes the form

$$d_\mu (*F)_\mu = 0 \tag{4.15}$$

(i.e. using forward differencing).

Thus, we may define a lattice version of cohomology, in close analogy with the continuum, by defining d with a forward difference, $*d*$ with a backward difference, and $*d$ on scalars with a backward difference. Note that here we have not defined a wedge product or Poincaré dual – this is in fact what we seek in a Lattice Chern-Simons Theory.

4.2. The Most General Gauge Invariant Lattice Chern-Simons Action

With these thoughts in mind we seek a lattice version of Chern-Simons theory as a quadratic function on the $A_i(x)$, $A_0(x)$. In analogy with the continuum theory, we restrict ourselves to expressions odd under parity and time reversal, and at most linear in time derivatives. This leads to the expression

$$\mathcal{L}_{CS} = \frac{k}{4\pi} \sum_{x,y,i} A_0(x,t) J(x-y) B(y,t) + \frac{k}{4\pi} \sum_{x,y,i,j} \dot{A}_i(x,t) K_{ij}(x-y) A_j(y,t) \tag{4.16}$$

Though we have allowed for a nonlocal J in the above formula, we shall concentrate on the case $J(x) = \delta(x)$ for now, and leave the case of general J for later in the paper.[‡] Note here also that without loss of generality, we may assume $K_{ij}(x) = -K_{ji}(-x)$, or that the Fourier transform of K is an anti-hermitian matrix. The formalism of canonical quantization implies

$$[A_i(x), A_j(y)] = \frac{2\pi i}{k} K_{ij}^{-1}(x-y) \tag{4.17}$$

[‡] δ, when written with lattice arguments, is the Kronecker-δ. If written with time arguments, it should be understood to be the Dirac-δ function.

and we further require that the functional $S_{CS} = \int dt \sum_{\mathcal{L}^2} \mathcal{L}_{CS}$ be invariant under gauge transformations $A \longrightarrow A + d\Lambda$. This gives the condition $\hat{d}_i K_{ij}(x) = d_j^{\perp}\delta(x)$, This equation may be easily solved in k-space with the expansion

$$K_{ij}(k) = K_{\parallel\parallel}\hat{d}_i d_j + K_{\perp\perp}\hat{d}_i^{\perp} d_j^{\perp} + K_{\perp\parallel}\hat{d}_i^{\perp} d_j + K_{\parallel\perp}\hat{d}_i d_j^{\perp} \tag{4.18}$$

from which we may read off the requirements

$$K^*_{\parallel\perp} = -K_{\perp\parallel}, \quad K_{\parallel\parallel} = -K^*_{\parallel\parallel}, \quad K_{\perp\perp} = -K^*_{\perp\perp}, \tag{4.19}$$

Note that $K_{\perp\perp}$, though constrained to be an imaginary function in k-space, is still a free function. We may use this freedom to subtract off the trace, leaving us with the equivalent form

$$K_{ij} = \frac{d_i d_j^{\perp} - \hat{d}_i^{\perp}\hat{d}_j}{2d \cdot \hat{d}} + K_{\perp\perp}\hat{d}_i^{\perp} d_j^{\perp}. \tag{4.20}$$

We then attempt the anyonization on the lattice in direct analogy with that done in the continuum.

4.3. Definition of Angle Function, Lattice Cut

Recalling the inverse lattice Laplacian operator $g(x)$, let $f_i(x) = \hat{d}_i^{\perp} g(x)$, which then satisfies

$$-d_i^{\perp} f_i(x) = \delta(x) \ . \tag{4.21}$$

Then the angle function is defined[42] as a contour sum over this vector field from a base point B to the endpoint x along a curve C_x as follows:

$$\theta_{C_x}(x, y) = 2\pi \sum_{\ell \in C_x} d\ell_i f_i(\ell - y) \tag{4.22}$$

C_x is a curve going from the base point B to a point x, $d\ell_i = \pm\hat{i}$ along the directed link going from z to $z \pm \hat{i}$, and we have shifted the origin of the vector field f to y. We refer to y as the "origin of the angle function". We choose the base point B so that these curves cannot wind around it, by putting it off at infinity in some direction.

θ behaves in most respects like the continuum version. However, in one (important) respect it does not. Let $\nu(C_x, C_y)$ be the signed intersection number of C_x with C_y, defined as the number of left-handed intersections minus the number of right-handed intersections (with slight deformations all curve intersections can be brought to one of these forms). The continuum angle function obeys the identity

$$\theta_{C_x}(x, y) - \theta_{C'_y}(y, x) - \pi \mathrm{sgn}(x_2 - y_2) + 2\pi\nu(C_x, C'_y) = 0 \tag{4.23}$$

(where $y = (y_1, y_2), x = (x_1, x_2)$) however, the above-defined lattice angle function satisfies

$$\theta_{C_x}(x, y) - \theta_{C'_y}(y, x) - \pi \mathrm{sgn}(x_2 - y_2) + 2\pi\nu(C_x, C'_y) = \xi(x - y) \tag{4.24}$$

with

$$\xi(x - y) = -\frac{1}{2}\left[f_1(x - y) + f_2(x - y) + f_1(x - y + \hat{2}) + f_2(x - y + \hat{1})\right] \tag{4.25}$$

ξ satisfies $\xi(x) = -\xi(-x)$.

The important properties obeyed by the angle function are unaffected by adding any curve independent nonsingular function to θ. We may take advantage of this fact to alter the definition of the angle function, if we so choose, so as to zero out the right hand side of Eq. (4.24). The appropriate modification is

$$\tilde{\theta}_{C_x}(x,y) = \theta_{C_x}(x,y) + \frac{1}{2}\xi(x-y) \tag{4.26}$$

While it turns out that this angle function is not the one we shall need for construction of anyons, it does appear later on, in a most interesting way.

$\theta(x,y)$ cannot be regarded as a function of the one variable $x-y$ on the full lattice, but it may be so regarded on any simply connected region of the lattice that does not include the origin. This is because the vector field f has the property Eq. (4.21), which prevents the dependence of definition Eq. (4.22) on the summation contour from dropping out completely. Instead the curve dependence may be expressed as

$$\theta_C(x,y) - \theta_{C'}(x,y) = 2\pi\omega(CC'^{-1},y) \tag{4.27}$$

where $\omega(CC'^{-1},y)$ is the winding number of the closed curve CC'^{-1} around the point y. θ may be made into a function of a single variable by by fixing this curve dependence once and for all, in a way that respects the proposed translation invariance.

We define a "field of curves" C, which assigns to every pair of points x, y a curve $C(x,y)$ running from the base point B to endpoint x. In order that it allows for translation invariance of θ, we impose that $C(x,y) = C(x+z, y+z)$, i.e. $C(x,y) = C(x-y)$. If we use these curves as the summation contours in the definition of θ, then $\theta_C(x,y) = \theta_C(x-y)$, i.e. it is now a single-valued function, and has a "cut". This is normally thought of as a line of discontinuity in a function – on a lattice this clearly has no meaning. Instead, we define the cut as follows. Given two neighboring points x and $x+\hat{i}$, and the curves chosen to run to them, C_x and $C_{x+\hat{i}}$, define the closed curve $\Delta_i C_x = C_x L_{x,\hat{i}} C^{-1}_{x+\hat{i}}$ ($L_{x,\hat{i}}$ is the link based at x pointing in the $\hat{i}$ direction, and may be considered as a curve running from x to $x+\hat{i}$) as the curve that comes in from the base point along C_x, continues along the link running from x to $x+\hat{i}$, and returns to the base point along the curve $C_{x+\hat{i}}$. If this curve winds around the center point of the angle function, then the link from x to $x+\hat{i}$ cuts perpendicularly across the cut of the angle function. Thus the cut of the angle function does not run along links on the lattice, but between them, on a curve running from the center point of the angle function out to infinity. It is easy to see that $d_i^{\perp}\omega(\Delta_i C(x), 0) = \delta(x)$, where ω(closed curve, z) is the winding number of a closed curve around the plaquette associated with a point z. In terms of this cut, we have

$$d_i\theta_C(x-z) = 2\pi f_i(x-z) + 2\pi\omega(\Delta_i C(x-z), z). \tag{4.28}$$

For future reference, this latter piece satisfies

$$d_i\omega(\Delta_j C(x-z)) - d_j\omega(\Delta_i C(x-z)) = \epsilon_{ij}\delta(x-z) \tag{4.29}$$

or

$$d_i^{\perp,x}\omega(\Delta_i C(x-z), z) = d_i^{\hat{x}}\omega(\Delta_i^{\perp} C(x-z), z) = \delta(x-z). \tag{4.30}$$

The lattice vector function $\omega(\Delta_i^{\perp}C(x-z),z)$ may be understood as a "contour density", in that, when dotted into a vector field and summed in the variable z over the entire plane, the result is a contour sum along a lattice curve, with curve parameter z. This curve is a reflection across the origin of the lattice curve beginning at the base point B, (i.e. at infinity) and ends at the point x, running next to the cut of the angle function θ_C. The lattice curve defined by $\omega(\Delta_i^{\perp}C(x-z),z)$ we call the "dual cut" D_C. Such contours may contain, in addition to the open curve running from B to $x-z$, any number of closed loops, if such loops are contained in the cut of θ_C, by a perverse choice of C.

4.4. Lattice Anyon Operators

4.4.1 Anyon interactions in the Hamiltonian

In analogy with the continuum treatment, we write down the anyon creation and annihilation operators as

$$\Phi_C(x) = U(x)K_C(x)\Psi(x) \quad \Phi_C^{\dagger}(x) = \Psi^{\dagger}(x)K_C^{\dagger}(x)U^{\dagger}(x) \tag{4.31}$$

with

$$U(x) = \exp i\sum_z f_i^{\perp}(x-z)A_i(z) \qquad K_C(x) = \exp\frac{2i}{k}\sum_z \theta_C(x-z)(\rho(z)-\rho_0) \tag{4.32}$$

For future reference, we note at this point that because of the identity (4.27) for the angle function, under a change of curve (the curve field is a redundant variable in the anyon Hamiltonian) we have

$$\Phi_{C'}(x) = \exp i\left(\sum_z \omega(C_z C_z'^{-1},z)(\rho(z)-\rho_0)\right)\Phi_C(x) \tag{4.33}$$

Consider the Hamiltonian density

$$\begin{aligned} H &= \Psi^{\dagger}(x+\hat{\imath})e^{iA_i(x)}\Psi(x) \\ &= \Phi_C^{\dagger}(x+\hat{\imath})U(x+\hat{\imath})K_C(x+\hat{\imath})e^{iA_i(x)}K_C^{\dagger}(x)U^{\dagger}(x)\Phi_C(x) \end{aligned} \tag{4.34}$$

In the continuum formulation, we recall that with this change of variables it was possible to eliminate entirely the interaction from the Hamiltonian. We shall carry out this calculation on the lattice, and clarify the nature of the interaction of the anyonic fields, which were obscured by the singularities of the continuum theory.

Our Hamiltonian is simply

$$H = \sum_{x,\hat{\imath}} \Phi_C^{\dagger}(x+\hat{\imath})U(\Delta_i C(x))\Phi_C(x) \tag{4.35}$$

with

$$U(\Delta_i C(x)) = \exp i\left(\sum_z \omega(\Delta_i C(x-z),z)(\rho(z)-\rho_0)\right) \tag{4.36}$$

which may be expressed in terms of the charge enclosed inside a closed curve as

$$U(\Delta_i C(x)) = \exp i\left(\sum_z \omega(D_C(x+\hat{\imath})L_{x,\hat{\imath}}D_C(x)^{-1}, z)(\rho(z)-\rho_0)\right) \tag{4.37}$$

where $D_C(x)$ is the dual cut of the angle function θ_C with origin x, and thus does not change with z. Thus, the phase U has the interpretation of the charge (or flux) enclosed inside the closed curve formed by the dual cuts $D_C(x)$ and $D_C(x+\hat{\imath})$, and the link $L_{x,\hat{\imath}}$.

Because of the transformation property (4.33) under a change of curve, this type of interaction (the factor U) is unavoidable in a theory of anyons [51] described by single-valued anyon fields. It reflects the fact that the original hamiltonian is independent of C, but the new fields Φ are not. This is the price paid for the elimination of the gauge field. Note that this phase is of the form $1/k$ times an integer. When the statistics parameter $1/k$ is an odd integer, this phase is just 1, and if $1/k$ is rational, the phase may take on only a finite number of values. In such a case the theory becomes a Z_N "gauge" theory, where N is the rationality of $1/k$. Note further that the interaction term, indeterminate in the continuum calculation, may be defined unambiguously in this version of the theory, by appropriate choice of curve field C.

4.4.2. Computing the statistics relation of the anyon operators

Using the Baker-Campbell-Hausdorff formula, as well as the commutator (4.17), we may also compute the "statistics" of the anyon operators, i.e. the phase $\exp i\chi$ resulting from a change in the order of the operators

$$\Phi_C(x)\Phi_C(y) = -\Phi_C(y)\Phi_C(x)\exp i\chi$$

The one-dimensional representation of $\pi_1(QConf)$ dictates χ as that of Eq. (3.19). We report the results of this simple calculation

$$\begin{aligned}\Phi_C(x)\Phi_C(y) &= \\ &- \Phi_C(y)\Phi_C(x)\exp\frac{2i}{k}\left(\theta_C(x-y)-\theta_C(y-x)\right)\exp -\frac{8i\pi}{k}K_{\perp\perp}(x-y)\end{aligned} \tag{4.38}$$

In the continuum, we recall that the difference of angle functions that sits in the first exponential was $\pi\,\mathrm{sgn}(x_2-y_2)-2\pi\nu(C(x-y),C(y-x))$, with $\nu(C(x-y),C(y-x))$ the signed intersection number of the two curves $C(x-y)$ and $C(y-x)$, as we required. Doing the calculation on the lattice gives in addition the term $\xi(x-y)$, which we recall was an odd function. This would be a problem, but for the happy appearance of the function $K_{\perp\perp}$, a free function constrained only insofar that it also must be odd. We may arrive at precisely those commutation relations obtained in the continuum calculations by simply choosing $K_{\perp\perp}=\xi/4\pi$.

4.4.3. Anyon operators in terms of Wilson lines imply K^{-1} is common site local

The topological properties of lattice Chern-Simons theory are laid bare directly by rewriting the anyon creation operators, using Gauss' Law and a summation by parts, as a Wilson line times the fermion field, i.e.

$$\Phi_C(x) = \Psi(x)W(D_C(x)), \quad W(D_C(x)) = \exp i \sum_{\ell \in D_C(x)} d\ell_i A_i(\ell) \tag{4.39}$$

where the summation contour is the dual cut D_C of the lattice angle function that appeared in the original expression for Φ_C. Calculating the statistics of anyon operators with these expressions of course gives an equivalent result

$$\Phi_C(x)\Phi_C(y) = -\Phi_C(y)\Phi_C(x)\exp\frac{2i}{k}\left(\pi \mathrm{sgn}(x_2 - y_2) - 2\pi\nu[D_C(x-y), D_C(y-x)]\right) \tag{4.40}$$

but in this form the commutator leads to a new insight. Notice that this implies that the commutator of two contour sums of the form A over the cycles D_C counts their signed intersection number. This would imply that the commutator matrix-valued function K_{ij}^{-1} is in fact a local function, given the choice $\xi/4\pi$ for $K_{\perp\perp}$, in spite of the appearance of an inverse Laplacian operator in expression Eq. (4.20). Tedious (!) algebra shows that there lurks a Laplacian in the numerator as well, and that there is a local expression for K_{ij}^{-1}.

$$\begin{aligned} K_{ij} &= \frac{1}{2}\begin{pmatrix} S_2 - S_2^{-1} & -(-1 + S_2^{-1} + S_1 + S_2^{-1}S_1) \\ -1 + S_1^{-1} + S_2 + S_1^{-1}S_2 & S_1^{-1} - S_1 \end{pmatrix} \\ &= \begin{pmatrix} d_2 + \hat{d}_2 & -2 - 2d_1 + 2\hat{d}_2 + \hat{d}_2 d_1 \\ 2 + 2d_2 - 2\hat{d}_1 - \hat{d}_1 d_2 & -d_1 - \hat{d}_1 \end{pmatrix} \end{aligned} \tag{4.41}$$

We normally regard as local on the lattice anything which is a finite polynomial in shifts and their inverses, and this expression certainly meets that criterion. However, it satisfies an even stronger requirement, that of "Common-Site Locality". Considered as a quadratic form, mapping pairs of links to $\Re$, this function vanishes unless both links share a common site. In particular, the commutation of a link variable with its neighbors may be read off from Eq. (4.41) - all are nonzero, in fact are equal to $\pm 1/2$. The inverse of K_{ij}^{-1}, i.e. K_{ij}, is also local, though not common site local. The properties of K_{ij} make it suitable as a definition of the Poincaré dual operation.

Common site locality may be easily seen to be a necessary condition for a local intersection form, because it is simply an antisymmetric bilinear function on curves (or curve segments). In fact, a converse theorem may be proved, that a necessary and sufficient condition for a local intersection form is nondegeneracy, common site locality, and the gauge invariance condition, which we write as $\hat{d}_i = K_{ij}^{-1}d_j^{\perp}$. The proof is simple. A local intersection form is a form which vanishes on any pair of curves which touch but do not cross, and as well on any pair of disconnected curves. By adding disconnected pieces we may thus restate the above as the vanishing of the local intersection form on pairs of curves each of which is either closed, or has endpoints on the boundary of the lattice. Such curves, represented by integer valued maps on links j, satisfy the condition $*d*j = \hat{d}_i j_i = 0$, and so may be alternatively written as the boundary of a plaquette-valued function b: $j = *d*b, j_\mu = \hat{d}_\nu b_{\nu\mu}$. A summation by parts leads to the gauge invariance condition.

In fact, we may also study the uniqueness of the solutions. The function $J(x-y)$ which we introduced in (4.16), then fixed to a δ-function may be considered in the general case. We find that there are exactly four solutions to the conditions (i) and (ii), corresponding to

$J(x) = \delta(x)$, $S_1\delta(x)$, $S_2\delta(x)$, and $S_1S_2\delta(x)$. These four possibilities in turn correspond to the four natural ways to map a 2-form to a 0-form, by associating the value of the 2-form at a plaquette p to the value of the 0-form at the site on one of its corners. And this in turn corresponds to the four natural definition of the Hodge * operation on 2-forms and 0-forms.

4.5. *Lattice Algebra of Forms*

The four definitions of * define the multiplication $\phi \cdot b$ of 0-forms ϕ with 2-forms b in tandem with the 4 local intersection forms which define the wedge product, through the identity (ϕ a 0-form, A a 1-form)

$$d(\phi \cdot A) = d\phi \wedge A + \phi \cdot dA \tag{4.42}$$

which in fact establishes the gauge invariance of the action Eq. (4.16). The final element of this identity is the multiplication rule for 0-forms with 1-forms, the left hand side of the above. This, in all four cases, must be defined as $\phi \cdot A = (S_i^{\perp})^{-1}[\phi]A_i$, i.e.

$$\begin{aligned}(\phi \cdot A)_1(x) &= \phi(x - \hat{2})A_1(x) \\ (\phi \cdot A)_2(x) &= \phi(x - \hat{1})A_2(x)\end{aligned} \tag{4.43}$$

This completes the definition of the algebra of lattice forms, whose existence explains the surprising result that abelian Chern-Simons theory on the lattice is actually identical to its continuum cousin.

It should again be noted that this is not the first algebra of forms written down for the lattice. The first work in this direction was done by Becher and Joos[52], and indeed their work formed the basis for the lattice Chern-Simons action of ref. [15]. Theirs, however, is not suitable for the study of knot theory, as it does not define an intersection form with the necessary antisymmetry property.

4.6. *Lattice Chern-Simons Theory in the Presence of External Currents*

The wave function of Chern-Simons theory in the presence of external currents is the natural object to study once we have settled on a lattice Chern-Simons action[53]. The constraint on the lattice is solved in exactly the same way as in the continuum

$$\Psi_{\rm phys}[\lambda, j_\mu, t] = \exp\left(i\sum_x \lambda(x) j_0(x,t)\right)\ \tilde{\Psi}[j_\mu, t] \tag{4.44}$$

The canonical structure is more complicated, though.

$$B(x) = \frac{2\pi i}{k}\frac{\partial}{\partial\lambda(x)} \quad , \quad \lambda(x) = \frac{1}{d\cdot\hat{d}}\left(\hat{d}\cdot A - \frac{\hat{d}K^{-1}d}{2d\cdot\hat{d}}B(x)\right) \tag{4.45}$$

This additional term leads to an extra term in the Hamiltonian

$$H = \sum_x j_i(x)A_i(x) = \sum_x \left(j \times \hat{d}\frac{1}{d\cdot\hat{d}}B(x) + j\cdot d\frac{1}{d\cdot\hat{d}}\hat{d}\cdot A\right) \tag{4.46}$$

(where we have used the identity $\delta_{ij} = \epsilon_{ik}\hat{d}_k\frac{1}{d\cdot\hat{d}}\epsilon_{jl}d_l + d_i\frac{1}{d\cdot\hat{d}}\hat{d}_j$) and the Schrödinger equation, which is then solved by

$$\Psi_{\text{phys}} = \exp\left(i\sum_x \lambda(x)j_0(x,t) - \frac{2\pi i}{k}\int_{-\infty}^{t}\sum\left(j\times\hat{d}\frac{1}{d\cdot\hat{d}}j^0 + j\cdot d\frac{1}{d\cdot\hat{d}}\frac{\hat{d}K^{-1}d}{2d\cdot\hat{d}}j_0\right)\right) \tag{4.47}$$

The trajectory of a particle is a piecewise linear lattice curve consisting of instantaneous hoppings in spatial directions between lattice sites and temporal segments representing the particle at rest on a particular site. When evaluated on a current j representing two lattice trajectories $r_1(t)$ and $r_2(t)$, $j = j_1 + j_2$ the second of the three terms in Eq. (4.47) is of exactly the form of the second term in Eq. (3.11), and in analogy with that case, its mixed term computes exactly the lattice angle function $\theta(r_1(t) - r_2(t))$ defined in Eq. (4.22). † Recall that this angle function was troubled by the property Eq. (4.24). However, the third term in Eq. (4.47) computes exactly $(1/2)\xi(r_1(t) - r_2(t))$, so that the sum actually equals the improved lattice angle function $\tilde{\theta}(r_1(t) - r_2(t))$, which we recall has the much nicer property Eq. (4.23), and brings us to the conclusion that these external charged particles have precisely the anyonic statistics desired.

4.6. Requirements for Lattice Chern-Simons Theory with Discrete Time Using Local Intersection Forms

Finally, we consider the possibility of lattice Chern-Simons theory with time latticized as well as space. We approach this keeping in mind the lessons learned above. With time latticized, we must forego the canonical approach and concentrate on the partition function, which computes the linking numbers of closed trajectories. These linking numbers may be treated as intersection numbers (between a curve and a surface) by the expedient of potentiating the current (representing it as the boundary of a surface) as was done in the introduction for the continuum case. The result is that the partition function may be expressed in terms of the signed intersection number of a curve with a surface, and what is required, then is an appropriate local intersection form, (which we denote by T, corresponding to K^{-1}). Again, a necessary and sufficient condition is (i) common site locality, together with (ii) a gauge invariance condition $dT = \hat{d}$ (we represent the surface by its dual variable $a = *b$, $a_\mu = \epsilon_{\mu\nu\lambda}b_{\nu\lambda}$, and as a result we no longer have the $\perp$ as we do in the two-dimensional gauge invariance formula $\hat{d}_i K_{ij}(x) = d_j^{\perp}\delta(x)$). The simplest action suggested by the above is

$$\begin{aligned} S_{CS} &= \sum A\wedge dA \\ &= \sum A_\mu(x)T_{\mu\nu'}(x-y)\epsilon_{\nu'\nu\lambda}d_\nu A_\lambda(y) \\ &= \sum A_\mu(x)\hat{d}_\nu T_{\mu\nu'}(x-y)\epsilon_{\nu'\nu\lambda}A_\lambda(y) \\ &= \sum A_\mu(x)D_{\mu\lambda}(x-y)A_\lambda(y) \end{aligned} \tag{4.48}$$

† The pure terms are unfortunately ill-defined in this formalism, containing factors like $(\delta-\text{function})(t)(\text{step function})(t)$. In fact, to avoid such quantities even in the mixed term, it is necessary to insist that the particles hop at different times.

We may consider D as a summation kernel, and then the action S_{CS} projects out its symmetric part. We therefore seek a local intersection form T which satisfies (i) and (ii), but also makes D a symmetric summation kernel. Many T's are available which satisfy (i) and (ii), but there are none which satisfy the symmetric requirement with the definition of $\wedge$ as given above. A more general approach to $\wedge$ is required, and this is the subject of ongoing study.

5. Conclusion

We have reviewed some recent advances in abelian lattice Chern-Simons theory. It was found that, within the canonical formalism, such a theory exists, and that this theory shares ALL of the topological properties of the continuum version, without alteration. In particular, it was possible to recast the problem entirely into the form of a linear first order difference equation for a matrix kernel that is required to be local in a very strong sense, i.e. Common Site Local, which should also be an anti-symmetric kernel. The solution of this problem in 2-dimensions was shown to lead to a definition of a lattice exterior algebra with all the properties of the continuum version (except that the action of the Hodge * operator is still not completely defined). There are exactly four such solutions, corresponding precisely to the four corners of a plaquette on a square lattice. The problem of reproducing the continuum partition function was also cast into this form; however the solution of these conditions has not yet been found.

6. References

1. E. Witten, Comm. Math. Phys. 121, 351 (1989).
2. D. Eliezer and G. W. Semenoff, Ann. Phys. **217**, 66, 1992
3. M. Laidlaw and C. DeWitt, Phys. Rev. D3 (1971), 1375;
4. J. Leinaas and J. Myrheim, Nuov. Cim. 37B (1977), 1.
5. G. Goldin, R. Menikoff and D. Sharp, Phys. Rev. Lett. 22 (1981), 1664.
6. Y.-S Wu, Phys. Rev. Lett. 52 (1984), 2103.
7. F. Wilczek and A. Zee, Phys. Rev. Lett. 51 (1984), 2250.
8. I. Kogan and A. Morozov Sov. Phys. JETP 61 (1985), 1.
9. A. M. Polyakov, Mod. Phys. Lett. A3 (1988), 325
10. G. Semenoff, Phys. Rev. Lett. 61 (1988), 517.
11. G. Semenoff and P. Sodano, Nucl. Phys. B328 (1989), 753.
12. C. H. Tze and S. Nam, Annals of Physics 193, (1989), 419
13. G. Dunne, R. Jackiw and C. Trugenberger, Ann. Phys. (N.Y.) 194 (1989), 197.
14. T. Matsuyama, Phys. Lett. B228 (1989), 99.
15. J. Fröhlich and P. Marchetti, Comm. Math. Phys. 116 (1988), 127.
16. J. Fröhlich, F. Gabbiani and P. A. Marchetti, in *Physics, Geometry and Topology*, ed. H. C. Lee, Plenum, New York, 1990.
17. J. Schoenfeld, Nucl. Phys. B185 (1981), 157.
18. B. Binegar, J.Math.Phys.23, 1511 (1982).
19. S. Forte and T. Jolicoeur, Nucl. Phys. B (1990), in press.

20. R. Jackiw and V. P. Nair, Phys. Rev. D43, 1933 (1991).
21. I. Kogan and G. W. Semenoff, Nuc. Phys. B, 1991, in press.
22. S. Forte, Rev. Mod. Phys. 1991, in press.
23. Y. Ladegaillerie *Bull. Sci. Math.* **100**, 255 (1976)
24. J. S. Birman *Comm. Pur. App. Math.* **22**, 41 (1969)
25. G. P. Scott *Proc. Camb. Phil. Soc.* **68**, 605 (1970)
26. E. Artin *Ann. Math* **48**, 101 (1947)
27. T. D. Imbo, C. S. Imbo, and E. C. G. Sudarshan, *Phys. Lett.* **B234**, 103 (1990)
28. T. D. Imbo and J. March-Russell *Phys. Lett* **B252**, 84 (1990)
29. T. Einarsson *Phys. Rev. Lett* **64**, 1995, (1990)
30. T. Einarsson *Mod. Phys. Lett* **B5**, 675 (1991)
31. M. Bos and V. P. Nair, *Int. J. Mod. Phys* **A5**, 959 (1990)
32. M. Bergeron, D. Eliezer, and G. W. Semenoff, *Canonical Chern-Simons Theory and the Braid Group on a Riemann Surface* , submitted to *Mod. Phys. Lett.*
33. J. Lykken, J. Sonnenschein and N. Weiss, Phys. Rev. D42 (1990) 2161.
34. R. B. Laughlin, Science 242, 525 (1989).
35. V. Kalmeyer and R. B. Laughlin, Phys. Rev. Lett. 59, 2095 (1987).
36. A. Fetter, R. Hanna and R. Laughlin, Phys. Rev. B40, 8745 (1989)
37. James E. Hetrick, Yutaka Hosotani, Bum-Hoon Lee, Ann.Phys.209:151-215,1991.
39. Prasanta K. Panigrahi, Rashmi Ray, B.Sakita, Effective Lagrangian for a System of Non-relativistic Fermions in (2+1)-Dimensions Coupled to an Electromagnetic Field: Application to Anyonic Superconductors, City College of New York Preprint CCNY-HEP-89/22, Dec 1989. 31pp.
40. G. Semenoff and N. Weiss, Phys.Lett. B, 1990, in press.
41. G. Dunne, R. Jackiw and C. Trugenberger, Ann. Phys. (N.Y.) 194 (1989), 197.
42. M. Lüscher, Nucl. Phys. B326 (1989), 557.
43. G. W. Semenoff, in *Physics, Geometry and Topology*, H. C. Lee, Plenum, New York, 1990.
44. R. Jackiw, K.-Y. Lee and E. Weinberg, Phys. Rev. D42, 2344 (1990).
45. G. W. Semenoff, Proceedings of the 1991 Karpacz Winter School on Theoretical Physics, to be published.
46. R. Kantor and L. Susskind, 'A Lattice Model of Fractional Statistics', Stanford Preprint, 1990.
47. T. Honan, Doctoral Thesis, Univ. Maryland, 1985, unpublished
48. V.F. Müller, Z.Phys.C47:301-310,1990.
49. V.F. Müller, On the connection between Euclidean and Hamiltonian lattice field theories of vortices and anyons, Kaiserslautern U. preprint PRINT-91-0127, 1991.
50. E. Fradkin, Phys. Rev. Lett. 63 (1989), 322; Phys. Rev. B 1990, in press.
51. Y. Hatsugai, M. Kohmoto, and Y-S. Wu, Phys. Rev. Lett. 66, (1991), 659
52. H. Becher and P. Joos, Z. Phys. C15, 343 (1982).
53. D. Eliezer, G. W. Semenoff, *Intersection Forms and the Geometry of Lattice Chern-Simons Theory*, Phys. Lett. **B**, in press
54. H. Flanders, *Differential Forms with Applications to the Physical Sciences*, Dover Publications, New York, 1989

Extended Structures in Topological Quantum Field Theory

Daniel S. Freed

Department of Mathematics
University of Texas at Austin

An n dimensional quantum field theory typically deals with *partition functions* and *correlation functions* of n dimensional manifolds and quantum Hilbert spaces of $n-1$ dimensional manifolds. One of the novel ideas in *topological* field theories is to extend these notions to manifolds of dimension $n-2$ and lower. Such extensions inevitably lead to the introduction of categories. These ideas are very much "in the air". Some of the people involved are Kazhdan, Segal, Lawrence, Kapranov, Voevodsky, Crane, and Yetter. Mostly this has been considered for 3 dimensional theories, but recently such ideas have also appeared in relation to the 4 dimensional Donaldson invariants (see [Fu], for example). Our motivation comes from a detailed understanding of *classical* topological field theories, which we also extend to manifolds of codimension two and higher. In the particular case of gauge theory with finite gauge group we define extensions of the usual "path integral" for the extended classical theory [F1]. For an n-manifold this is the usual path integral, and for an $(n-1)$-manifold we recover the quantum Hilbert space. The result of this integration for an $(n-2)$-manifold is a *2-Hilbert space*.

In this note we briefly explain the consequences of this extended notion in an arbitrary 3 dimensional topological theory. The 2-Hilbert space $\mathcal{E}$ of a circle has additional structure: it is a "commutative, associative algebra with identity and involution". This must be understood in the categorical sense, since $\mathcal{E}$ is a category. In good cases $\mathcal{E}$ can be realized as the category of representations of a quantum group. Hence we explain the appearance of quantum groups in 3 dimensional field theories in terms of our extended path integral. We refer to [F1] for more details as well as for an example: gauge theory with finite gauge group.

Reshetikhin/Turaev [RT] start with a Hopf algebra (of a certain type) and from it construct a 3 dimensional field theory. Recent work of Kazhdan/Reshetikhin starts instead with a special type of category and construct the field theory from it. That category is then the 2-Hilbert space of the circle in the resulting theory. We find these algebraic data unnatural in isolation; our purpose is to explain their introduction in terms of general properties of field theory. Our point of view is that they are the "solution" to a field theory, rather than a natural starting point. In the last section

The author is supported by NSF grant DMS-8805684, a Presidential Young Investigators award DMS-9057144, and by the O'Donnell Foundation. This paper is based on a talk given in the *Special Session on Knots and Topological Quantum Field Theory* as part of the American Mathematical Society meeting in Dayton, Ohio held in October, 1992.

we discuss framed tangles, and so make more direct contact with the starting point in their work. However, except in the case of finite theories we cannot offer an alternative to their *constructions* of invariants.

The 3 dimensional Chern-Simons theory with positive dimensional gauge group [W] is only defined projectively for oriented manifolds; there are central extensions of diffeomorphism groups which appear. Witten realized these central extensions by certain framings. We briefly discuss a different topological structure—a *rigging*—which realizes these central extensions. Our discussion is loosely based on some remarks in Segal [S].

2-Hilbert Spaces

A finite dimensional Hilbert space over $\mathbb{C}$ is a set W with operations of addition, scalar multiplication, and hermitian inner product:

$$\begin{aligned} +&: W \times W \longrightarrow W \\ \cdot&: \mathbb{C} \times W \longrightarrow W \\ (\cdot,\cdot)&: W \times W \longrightarrow \mathbb{C} \end{aligned}$$

These operations satisfy the usual axioms, which we do not list here. A *2-Hilbert space* has an analogous definition, except we replace the set W by a *category* $\mathcal{W}$ and the field $\mathbb{C}$ by the category $\mathcal{V}$ of all finite dimensional Hilbert spaces.

A category differs from a set in that its "elements" (usually called "objects") may have automorphisms.[1] More generally, there are "morphisms" $V \to V'$ between objects in a category. For example, the morphisms between any two inner product spaces $V, V' \in \mathcal{V}$ are linear maps $V \to V'$. But $\mathcal{V}$ is much more than a category—it has a structure analogous to a (semi)ring structure on a set. Namely, there is an addition (direct sum), a multiplication (tensor product), an additive identity (the zero vector space), and a multiplicative identity (the ground field $\mathbb{C}$).

So a 2-Hilbert space $\mathcal{W}$ is a module over $\mathcal{V}$ with an inner product. In other words, it is a category endowed with addition, scalar multiplication, and an inner product:

$$\begin{aligned} +&: \mathcal{W} \times \mathcal{W} \longrightarrow \mathcal{W} \\ \cdot&: \mathcal{V} \times \mathcal{W} \longrightarrow \mathcal{W} \\ (\cdot,\cdot)&: \mathcal{W} \times \mathcal{W} \longrightarrow \mathcal{V} \end{aligned}$$

These maps are *functors*. Of course, $\mathcal{V}$ itself is a one dimensional 2-Hilbert space; the inner product is

$$(V_1, V_2) = V_1 \otimes \overline{V_2}.$$

[1]We abuse notation and write '$A \in C$' for an object A in a category C.

Analogous to the Hilbert space $\mathbb{C}^n$ is the n dimensional 2-Hilbert space $\mathcal{V}^n$ whose objects are n-tuples $\langle V^{(1)},\dots,V^{(n)}\rangle$ of inner product spaces. Note that the multiplicative identity is $\langle \mathbb{C},\dots,\mathbb{C}\rangle$; the inner product is

$$(\langle V_1^{(1)},\dots,V_1^{(n)}\rangle\,,\,\langle V_2^{(1)},\dots,V_2^{(n)}\rangle)=\bigoplus_{i=1}^{n} V_1^{(i)}\otimes\overline{V_2^{(i)}}.$$

Since a 2-Hilbert space is a category, there is an extra layer of structure beyond spaces and maps. Namely, there are maps (morphisms) between elements (objects) of a 2-Hilbert space, and so also maps (natural transformations) between maps (functors). Thus a familiar axiom for the hermitian inner product now asserts the existence of a preferred isometry

$$(W_1,W_2)\longrightarrow\overline{(W_2,W_1)} \tag{1}$$

for any elements W_1,W_2 in a 2-Hilbert space $\mathcal{W}$. There are new properties as well. We assume the existence of preferred maps

$$\mathbb{C}\longrightarrow(W,W)\longrightarrow\mathbb{C} \tag{2}$$

$$(W_2,W_1)\cdot W_1\longrightarrow W_2 \tag{3}$$

for $W,W_1,W_2\in\mathcal{W}$. Note from (1) that (W,W) has a preferred real structure, and we assume that (2) is compatible. The composition is a real number attached to each W, which we denote $\dim W$. The '$\cdot$' in (3) is scalar multiplication.

Here is a less trivial example of a 2-Hilbert space. Let G be a finite group and $\mathcal{W}=\text{Rep}(G)$ the category of finite dimensional unitary representations of G. Addition is direct sum, scalar multiplication is tensor product (with G acting trivially on the "scalar" vector space), and the inner product is

$$(W_1,W_2)=(W_1\otimes\overline{W_2})^G,$$

the vector space of G-invariants in $W_1\otimes\overline{W_2}$. (There is an additional operation—tensor products of representations—but we ignore it for now.) An *orthonormal basis* of $\mathcal{W}$ consists of a collection of representations $W_1,\dots,W_n$, one from each equivalence class of irreducible representations. Then for any representation W the maps in (3) give a preferred isometry

$$\bigoplus_{i=1}^{n}\frac{1}{\dim W_i}\,(W,W_i)\cdot W_i\longrightarrow W. \tag{4}$$

Here $\frac{1}{\dim W_i}(W,W_i)$ is the vector space (W,W_i) with its inner product multiplied by $\frac{1}{\dim W_i}$. This equation asserts that $\{W_i/\sqrt{\dim W_i}\}$ is an "orthonormal basis" of $\text{Rep}(G)$. If we replace the finite group G by a compact Lie group of positive dimension, then we obtain an infinite dimensional 2-Hilbert space.

Many constructions in linear algebra have analogues for 2-Hilbert spaces. These include the dual space, direct sum, and tensor product. The trace of an endomorphism of a 2-Hilbert space is a Hilbert space. For example, the "dimension" of Rep(G) is the Hilbert space $R(G)$ of equivalence classes of representations of G. (There is also a ring structure on $R(G)$ from the tensor product of representations in Rep(G).)

The reader may object that a 2-Hilbert space resembles an integral lattice in a Hilbert space more than it does a Hilbert space. Note, however, that if $V \in \mathcal{V}$ is a Hilbert space, and $\mu \in \mathbf{R}^+$ a positive real number, then $\mu V \in \mathcal{V}$ makes sense—it is the same underlying vector space V with inner product multiplied by μ. Note also that μV is isometric to V. We can extend this multiplication formally to complex numbers with nonzero phase, and then allow these scalars in 2-Hilbert spaces. From this point of view there is no natural lattice in Rep(G), since we have nothing to fix the scale of the inner product in an irreducible representation. In this regard there is an isometry (4) with W_i replaced by $\mu_i W_i$ for *any* positive scalars μ_i. The map in (4) then depends on the μ_i.

Topological Field Theories

An n dimensional topological field theory typically consists of assignments

$$\begin{aligned} Y^{n-1} &\longmapsto E(Y) \\ X^n &\longmapsto Z_X \end{aligned} \tag{5}$$

of a finite dimensional Hilbert space $E(Y)$ to a closed $(n-1)$-manifold, and a "path integral" $Z_X \in E(\partial X)$ to a compact n-manifold. In most theories the manifolds carry additional topological structure, such as an orientation. Symmetries of the manifolds which preserve the extra structure are implemented as symmetries on the corresponding objects in the field theory. Thus symmetries of Y act as unitary transformations on $E(Y)$ and symmetries of X leave Z_X invariant. Oppositely oriented manifolds map to the conjugate object. Most importantly, there is a gluing law for pasting together components of the boundary of an n-manifold. The axioms are spelled out in [A1] (cf. [FQ], for example). They capture the gross structure common to all topological theories; specific theories have more detailed structure, of course.

The consequences of these axioms in a 2 dimensional topological field theory are standard, and provide a good warmup to the 3 dimensional case. Let $E = E(S^1)$ denote the quantum Hilbert space of the circle. Up to isotopy there is only one orientation-reversing diffeomorphism of S^1, and it determines an *antilinear* map

$$c\colon E \longrightarrow E. \tag{6}$$

Note that $c^2 = \mathrm{id}$ since any orientation-preserving diffeomorphism of S^1 is isotopic to the identity. So c determines a *real structure* on E. The path integral Z_X over any surface X is real. The path integral Z_P over the pair of pants P determines a multiplication

$$\circ\colon E\otimes E \longrightarrow E,$$

and Z_{D^2} acts as the identity, where D^2 is the disk. One easily shows that the trilinear form

$$x\otimes y\otimes z \longmapsto \big(x\circ y, c(z)\big)_E, \qquad x,y,z\in E,$$

is symmetric, and that the multiplication is associative. It follows that E is semisimple. Conversely, let E be a commutative, associative algebra E with identity, compatible real structure, and compatible inner product. Then we can construct a 2 dimensional field theory whose associated algebra is E.

We now extend the assignments (5) to codimension two manifolds in an n dimensional theory. As mentioned in the introduction, we understand this extension in terms of a generalized "path integral" based on an extended classical theory. The result is that to a closed $(n-2)$-manifold S we assign a 2-Hilbert space $\mathcal{E}(S)$:

$$S^{n-2} \longmapsto \mathcal{E}(S).$$

There are "higher" notions for lower dimensional manifolds, but they will not concern us here. The theory also assigns to an $(n-1)$-manifold Y with boundary an element $E(Y)\in\mathcal{E}(S)$. We postulate a gluing law in terms of the inner product on $\mathcal{E}(S)$ for $(n-1)$-manifolds pasted together along S. We also assume that symmetries of S act on $\mathcal{E}(S)$ by unitary transformations, and further that homotopies of symmetries act as natural transformations (which respect the 2-Hilbert space structure.) In particular, this implies that $\pi_1 \operatorname{Diff}(S)$ acts as unitary automorphisms of the identity[2] on $\mathcal{E}(S)$. Compare with the standard assertion that for a closed $(n-1)$-manifold Y there is an action of $\pi_0 \operatorname{Diff}(Y)$ by unitary transformations of $E(Y)$.

Our main interest is in 3 dimensional theories of oriented manifolds. We examine the consequences of the axioms for the structure of the 2-Hilbert space $\mathcal{E}=\mathcal{E}(S^1)$, where S^1 is the standard oriented circle. (In the next section we discuss some modifications necessary in a *projective* theory, for example the Chern-Simons theory with positive dimensional compact gauge group.) First, $\pi_1 \operatorname{Diff}^+(S^1)\cong \mathbb{Z}$ and the positive generator, represented by a loop of rotations, acts as an automorphism of the identity on $\mathcal{E}$. That is, for each $W\in\mathcal{E}$ there is a morphism

$$\theta_W\colon W \longrightarrow W \tag{7}$$

which commutes with all morphisms in $\mathcal{E}$. Next, reflection induces a conjugation $c\colon \mathcal{E}\to\mathcal{E}$ as in (6), and we denote

$$c(W) = W^*. \tag{8}$$

[2]If $\mathcal{E} = \operatorname{Rep}(G)$ then a unitary automorphism of the identity is multiplication by a phase on each irreducible representation.

Since $c^2 = \mathrm{id}$ we have $(W^*)^* = W$. Then $\theta_{W^*} = \theta_W^*$ follows from an equation in $\mathrm{Diff}(S^1)$ relating rotations and reflections.

For any connected compact oriented 1-manifold S there is a unique isotopy class of orientation-preserving diffeomorphisms $S \to S^1$. In a 2 dimensional theory that suffices to identify the Hilbert spaces assigned to S and S^1 uniquely. In a 3 dimensional theory, however, different diffeomorphisms $S \to S^1$ yield different isometries $\mathcal{E}(S) \cong \mathcal{E}(S^1)$, although there is an isometry (natural transformation) between any two such isometries. The latter depends nontrivially on a choice of isotopy, due to the nontrivial $\pi_1 \mathrm{Diff}^+(S^1)$. Hence we must make a rigid convention to identify different circles. Namely, we require any circle S to appear in the complex line $\mathbb{C}$ and have its center in $\mathbb{R} \subset \mathbb{C}$. Then there is a unique composition of a translation and a dilation which identifies S with the standard circle $S^1 \subset \mathbb{C}$. The reflection which induces duality is reflection in the real axis. We also standardize surfaces diffeomorphic to a disk with a finite number of smaller disjoint disks removed. Such surfaces embed in $\mathbb{C}$ with standardized boundaries.

Now $\mathcal{E} = \mathcal{E}(S^1)$ has a structure analogous to the real commutative associative algebra structure (with compatible inner product) on $E(S^1)$ in a 2 dimensional theory. The path integral Z_P on the standardized pair of pants P determines a multiplication

$$\odot : \mathcal{E} \otimes \mathcal{E} \longrightarrow \mathcal{E}. \tag{9}$$

The "reality" statement is a preferred isometry

$$(W_1 \odot W_2)^* \cong W_1^* \odot W_2^*$$

for all $W_1, W_2 \in \mathcal{E}$. The path integral $Z_{D^2} = 1 \in \mathcal{E}$ over the standard disk $D^2 \subset \mathbb{C}$ acts as an identity for the multiplication in the sense that there are preferred isometries

$$1 \odot W \cong W \odot 1 \cong W$$

for all $W \in \mathcal{E}$. The associativity is a natural isometry

$$\varphi_{W_1,W_2,W_3} : (W_1 \odot W_2) \odot W_3 \longrightarrow W_1 \odot (W_2 \odot W_3) \tag{10}$$

which satisfies the "pentagon relation". The isometry (10) is constructed from the gluing law, and the pentagon follows from cuttings and pastings of the surface Q which is a disk with 3 interior disks removed. Also, the braiding diffeomorphism $\beta : P \to P$ of the pair of pants induces a natural isometry

$$R_{W_1,W_2} : W_1 \odot W_2 \longrightarrow W_2 \odot W_1.$$

Equations in $\mathrm{Diff}^+(Q)$ imply two "hexagon relations", while the relation

$$(\theta_{W_2} \odot \theta_{W_1}) \circ R_{W_1,W_2} = R^{-1}_{W_2,W_1} \circ \theta_{W_1 \odot W_2}$$

follows from an equation in $\mathrm{Diff}^+(P)$.

This is a taste of what may be extracted from the functoriality and the gluing laws. See [F1] for more details. I imagine there is an appropriate semisimplicity statement which can be made as well.

Category enthusiasts may prefer to regard $\mathcal{E}$ as an abelian category endowed with extra structure—monoidal structure (9), duality or rigidity (8), balancing (7), and braiding (10). If we are also given a *fiber functor*, that is a functor $\mathcal{E} \to \mathcal{V}$ which preserves the monoidal structure, then $\mathcal{E}$ can be recognized as the category of representations $\mathrm{Rep}(A)$ of a quasitriangular quasi-Hopf algebra A, also known as a quantum group. In [F1] we construct an obvious fiber functor for the Chern-Simons theory with finite gauge group, and so recover Hopf algebras which appeared previously in this context. In general, a field theory does not construct a fiber functor, and they do not exist for all theories.[3]

We remark that if $\mathcal{E} = \mathrm{Rep}(A)$ and $W \in \mathcal{E}$ is an irreducible representation, then θ_W is multiplication by $e^{2\pi i h_W}$, where h_W is the *conformal weight* corresponding to W.

Just as one can construct a 2 dimensional theory starting from an algebra E of the appropriate sort, so too one can construct a 3 dimensional theory starting from a 2-Hilbert space $\mathcal{E}$ with multiplication, braiding, etc. A precise version of this statement is contained in recent work of Kazhdan/Reshetikhin.

Central Extensions

The Chern-Simons theory introduced by Witten [W] involves certain central extensions, which he realizes in terms of "2-framings". We follow Segal [S] and instead propose that manifolds in the theory be endowed with an extra topological structure termed a "rigging". A rigging is a trivialization of a topological invariant which for a closed oriented 4-manifold W is the signature $\mathrm{Sign}(W)$. There are corresponding topological invariants of 3-, 2-, and 1-manifolds which we briefly explain below. Observe that three times the signature is the Pontrjagin number $p_1(W) \in \mathbb{Z}$. Topological invariants in lower dimensions stemming from $p_1(W)$ are more easily constructed than those stemming from $\mathrm{Sign}(W)$. The difference here is one between K-theory and cohomology. Also, there are invariants of spin manifolds which stem from the $\hat{A}$-genus $\hat{A}(W)$. In all three cases we can define

[3]For a finite σ-model into a space with n points, the 2-Hilbert space $\mathcal{E}$ is $\mathcal{V}^n$, which does not admit a functor $\mathcal{V}^n \to \mathcal{V}$ which preserves direct sums and tensor products. (Indeed, the image of the unit object $1 = \langle \mathbb{C}, \dots, \mathbb{C} \rangle$ must be one dimensional, but 1 is the sum of n "basis" elements.) This is related to the fact that that the quantum Hilbert space of S^2 is n dimensional. In Chern-Simons theories, on the other hand, the Hilbert space of S^2 is one dimensional. A related observation: The endomorphisms of the unit object $1 = \langle \mathbb{C}, \dots, \mathbb{C} \rangle$ form a ring which is not a field. I believe that an arbitrary unitary theory decomposes into a direct sum of such "irreducible" theories—the idempotents in this ring give such a decomposition—and it is possible that fiber functors always exist for irreducible theories.

riggings, though we focus here on the signature and associated invariants. Our constructions in this section are similar to constructions of Segal [S]. This material is preliminary as we cannot yet check all of the details.

A word about the abstractions which follow. *Torsors* and *gerbes* are concrete realizations of integral cohomology, somewhat analogous to the way that elements of K-theory are realized by vector bundles. For example, a family of $\mathbf{Z}$-torsors is a principal $\mathbf{Z}$ bundle, and it has a characteristic class in the first cohomology of the parameter space. Families of higher $\mathbf{Z}$-gerbes have characteristic classes in higher integral cohomology.

Our starting point is the observation in [F2] that a Dirac operator on a 4-manifold W with boundary has a topological index which lives in a $\mathbf{Z}$*-torsor* which is a topological invariant of the operator on the boundary. '$\mathbf{Z}$-torsor' is by definition 'principal homogeneous space for the integers'. In the case of the signature operator that $\mathbf{Z}$-torsor $T_{\partial W}$ has a canonical trivialization, and the signature is, of course, an integer. One can describe the $\mathbf{Z}$-torsor $T_X(g)$ of a closed oriented 3-manifold X with metric g in terms of the ξ-invariant ($\frac{1}{2}\eta$-invariant) of Atiyah/Patodi/Singer. The ξ-invariant of the metric determines the $\mathbf{Z}$-torsor $T_X(g)$ of real numbers x which satisfy $e^{2\pi i x} = e^{2\pi i \xi}$. The $\mathbf{Z}$-torsor T_X is the space of sections of this bundle of $\mathbf{Z}$-torsors over the space of metrics; it is a topological invariant of X.

Next, the exponentiated ξ-invariant of a Riemannian 3-manifold X with boundary is meant to live in the determinant circle C_Y of the boundary $Y = \partial X$. (We now omit the metric from the notation.) This is mentioned in [S] and is currently under investigation with Dai. The determinant circle is the set of elements of unit norm in the determinant line with respect to the Quillen metric. It is a torsor for the group $\mathbf{T}$ of unit size complex numbers. Consider the collection of all $\mathbf{R}$-torsors $\widetilde{C_Y}$ which cover the $\mathbf{T}$-torsor C_Y, i.e., the collection of all covering maps $\widetilde{C_Y} \to C_Y$ compatible with the $\mathbf{R}$ and $\mathbf{T}$ actions. The set of morphisms between any two such is a $\mathbf{Z}$-torsor, and the collection $\mathcal{G}_Y$ of all these covering maps is an example of a $\mathbf{Z}$*-gerbe*. As before, we eliminate the choice of metric by working with smooth families over the space of metrics. This construction works for any closed oriented 2-manifold Y. The exponentiated ξ-invariant of a compact oriented 3-manifold is a point in $C_{\partial X}$, and using this point we can construct a particular cover $T_X \in \mathcal{G}_{\partial X}$.

The corresponding topological invariant of a closed oriented 1-manifold S is more complicated to describe. Briefly, for each metric on S there is a self-adjoint signature operator, which is essentially two copies of the operator $i\frac{d}{dx}$ on functions. It has discrete real spectrum extending to both ∞ and $-\infty$. Let $A \subset \mathbf{R}$ be the complement of the spectrum. Then for $a, b \in A$ there is a finite dimensional Hilbert space of eigenvectors with eigenvalue between a and b. Let $C_{a,b}$ be the determinant circle of this space. These circles fit together to form a flat circle bundle over $A \times A$. Now consider a flat circle bundle $C \to A$ together with consistent isomorphisms $C_a \otimes C_{a,b} \to C_b$. The collection of all such is a $\mathbf{T}$-gerbe. Then the collection of liftings of this $\mathbf{T}$-gerbe to $\mathbf{R}$-gerbes is a "2-gerbe" over $\mathbf{Z}$. Finally, we factor out the metric to obtain a topological invariant $\mathfrak{G}_S$.

We summarize this discussion in Table 1. Each entry is a topological invariant, which means it is functorial under orientation-preserving diffeomorphisms. The invariants also obey gluing laws,

dim	closed manifold	compact manifold with boundary
4	$\mathrm{Sign}(W) \in \mathbf{Z}$	$\mathrm{Sign}(W) \in T_{\partial X}$
3	$\mathbf{Z}$-torsor T_X (lifts of $e^{2\pi i \xi}$)	$T_X \in \mathcal{G}_{\partial X}$
2	$\mathbf{Z}$-gerbe $\mathcal{G}_Y$ (covers of determinant circle)	$\mathcal{G}_Y \in \mathfrak{G}_{\partial Y}$
1	2 -gerbe $\mathfrak{G}_S$ (covers of "determinant $\mathbf{T}$-gerbe")	

Table 1: Topological invariants of oriented manifolds

and they "change sign" when the orientation is reversed. As mentioned, the $\mathbf{Z}$-torsor T_X has a natural trivialization $T_X \cong \mathbf{Z}$ when X is closed.

A *rigging* of an oriented manifold is a trivialization of the topological invariant in Table 1. For 4-manifolds this is meaningless, or it demands that we only consider 4-manifolds with vanishing signature. For a closed 3-manifold X a rigging is a choice of an element in T_X. Recalling that T_X has a natural trivialization, this amounts to choosing an integer. The existence of a canonical element in T_X corresponds to Atiyah's canonical 2-framing [A2]. For a closed 2-manifold Y a rigging is a choice of cover of the determinant circle bundle over the space of metrics. I believe that our definition of a rigging of a 1-manifold differs from Segal's. Diffeomorphisms of rigged manifolds are required to preserve the rigging. Thus the group of diffeomorphisms $\mathrm{Diff}^+_{\mathrm{rig}}(Y)$ of a closed oriented rigged 2-manifold Y is a central extension by $\mathbf{Z}$ of the group of orientation-preserving diffeomorphisms $\mathrm{Diff}^+(Y)$. The fundamental group of rigged diffeomorphisms $\pi_1 \mathrm{Diff}^+_{\mathrm{rig}}(S)$ of a rigged oriented closed 1-manifold S is a central extension by $\mathbf{Z}$ of $\pi_1 \mathrm{Diff}^+(S)$. Note that the boundary of any oriented manifold is rigged. Also, riggings glue together.

A 3 dimensional (unitary) field theory of rigged oriented manifolds has the structure described in the previous section, but modified to account for the riggings. The example we have in mind here is Chern-Simons theory with positive dimensional compact gauge group. There are homomorphisms $\mathbf{Z} \to \mathbf{T}$ induced by: (i) the change in the path integral over a closed 3-manifold X under change of rigging; (ii) the action of the kernel of $\pi_0 \mathrm{Diff}^+_{\mathrm{rig}}(Y) \to \pi_0 \mathrm{Diff}^+(Y)$ on $E(Y)$ for a closed rigged 2-manifold Y; and (iii) the action of the kernel of $\pi_1 \mathrm{Diff}^+_{\mathrm{rig}}(S) \to \pi_1 \mathrm{Diff}^+(S)$ on $\mathcal{E}(S)$ (by automorphisms of the identity) for a closed rigged 1-manifold S. I believe that these homomorphisms are universal in a theory—that is, independent of X, Y, S—and that the generator maps to $e^{2\pi i c/24}$ where c is the *central charge* of the theory. Other computations must now take into account the riggings as well. For example, I believe that it is no longer necessarily true that the square of the conjugation (8) is the identity,[4] but I cannot yet see this in terms of riggings.

Invariants of Framed Tangles

We briefly explain how a 3 dimensional topological field theory, extended as previously discussed, yields invariants of framed tangles. This is meant to make contact with the Kazhdan/Reshetikhin work.[5] See [RT] for a precise definition of framed tangles. For simplicity we ignore riggings in this section.

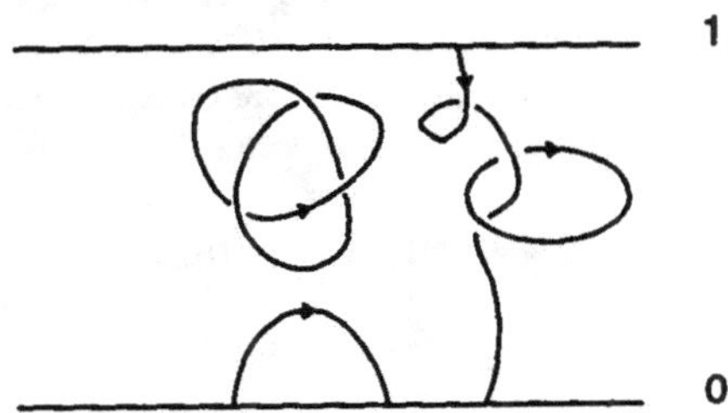

Figure 1: A framed tangle

A framed tangle D is represented by a diagram like Figure 1, where b strands intersect the line labeled 0 and t strands intersect the line labeled 1. View the lines as copies of $\mathbb{C}$ and view the whole picture as embedded in $\mathbb{R}^3$. Extend the picture to $[0,1] \times S^2$ by adding a point at ∞ to each plane and an extra strand $[0,1] \times \{\infty\}$. Now cut out tubular neighborhoods of each strand to obtain a 3-manifold X. The boundary of X is

$$\begin{aligned} \partial X = \left(\{1\} \times S^2 - (t+1)D^2\right) \cup -\left(\{0\} \times S^2 - (b+1)D^2\right) \cup \text{(annulus at } \infty) \\ \cup \bigcup_{\pi_0(D)} \text{(annulus or torus)}. \end{aligned}$$

Using the framings we can identify, up to isotopy, the annuli and tori with the standard annulus $[0,1] \times S^1$ and the standard torus $S^1 \times S^1$. Now the quantum invariant Z_X is an element of the Hilbert space $E(\partial X)$. We reinterpret it according to the decomposition induced by (4).

Applying the gluing law to (9) we find that $E(\{1\} \times (S^2 - (t+1)D^2))$ is the $(t-1)$-fold tensor product

$$\odot\colon \mathcal{E} \otimes \cdots \otimes \mathcal{E} \longrightarrow \mathcal{E}. \tag{11}$$

[4]In Hopf algebra terms this corresponds to asserting that the square of the antipode is not necessarily the identity.
[5]as explained to me by Reshetikhin, who I warmly thank.

We must choose a specific order for the multiplications since multiplication is not associative (cf. (10)). For the annuli and tori we have

$$E([0,1]\times S^1)\cong(\mathrm{id}\colon \mathcal{E}\to\mathcal{E})$$
$$E(S^1\times S^1)=V\cong \mathrm{Trace}(\mathrm{id}\colon \mathcal{E}\to\mathcal{E}).$$

This last equation defines V. Now we glue the annuli to the ends $\{0\}\times S^2$ and $\{1\}\times S^2$. When we glue along a circle we obtain an inner product in $\mathcal{E}$. We only have to be careful about orientations. We geometrically pass between the two orientations of S^1 via reflection, and so in the field theory via duality, according to (8).

This abstract description can be made somewhat more concrete. Notice that by (3) we can convert the inner products from the gluing into maps. Write

$$\pi_0(D)=\pi_0(D)_{\mathrm{annuli}}\cup\pi_0(D)_{\mathrm{tori}}.$$

Define maps (functors)

$$F_i\colon \underbrace{\mathcal{E}\otimes\cdots\otimes\mathcal{E}}_{|\pi_0(D)_{\mathrm{annuli}}|}\longrightarrow\mathcal{E},\qquad i=0,1,$$

by

$$F_i(A_1\otimes\cdots\otimes A_n)=A^{\pm}_{i_1}\odot\ldots\odot A^{\pm}_{i_n},$$

where i_j is the position of the boundary of the jth annulus among the circles in $\{1\}\times S^2$, the sign is chosen according to the orientation (a minus sign is the dual), and the product on the right hand side is (11). Then the vector space obtained by gluing the annuli to the ends $\{0\}\times S^2$ and $\{1\}\times S^2$ maps to the vector space of natural transformations from F_0 to F_1. To include the tori we simply take the tensor product with $V^{\otimes|\pi_0(D)_{\mathrm{tori}}|}$. So, finally, the partition function Z_X can be seen as an element of

$$\mathrm{NatTrans}(F_0,F_1)\otimes V^{\otimes|\pi_0(D)_{\mathrm{tori}}|}.$$

This element is an invariant of the framed link.

This is almost the description of Kazhdan/Reshetikhin, except that they realize V as the space of natural automorphisms of the trivial functor $W\mapsto 1$. This is correct if $\mathrm{End}(1)\cong\mathbb{C}$.

As a simple example we see that the partition function of the tangle pictured in Figure 2 is a natural transformation $W\odot W^*\mapsto 1$. Similarly, we find a natural transformation $1\to W\odot W^*$. This fills a gap in [F1,§5].

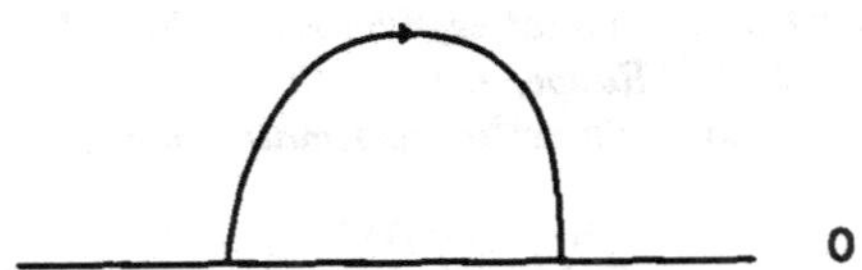

Figure 2: The natural transformation $W \odot W^* \to 1$

References

[A1] M. F. Atiyah, *Topological quantum field theory*, Publ. Math. Inst. Hautes Etudes Sci. (Paris) **68** (1989), 175–186.

[A2] M. F. Atiyah, *On framings of 3-manifolds*, Topology **29** (1990), 1–7.

[F1] D. S. Freed, *Higher algebraic structures and quantization*, Commun. Math. Phys. (to appear).

[F2] D. S. Freed, *A gluing law for the index of Dirac operators* (to appear in the 60th birthday volume dedicated to Richard Palais).

[FQ] D. S. Freed, F. Quinn, *Chern-Simons theory with finite gauge group*, Commun. Math. Phys. (to appear).

[Fu] K. Fukaya, *Floer homology for 3-manifolds with boundary* (University of Tokyo preprint, 1993).

[RT] N. Reshetikhin, V. G. Turaev, *Invariants of 3-manifolds via link polynomials and quantum groups*, Invent. math. **103** (1991), 547–597.

[S] G. Segal, *The definition of conformal field theory* (preprint).

[W] E. Witten, *Quantum field theory and the Jones polynomial*, Commun. Math. Phys. **121** (1989), 351–399.

Department of Mathematics, University of Texas, Austin, TX 78712

E-mail address: dafr@math.utexas.edu

A Method for Computing the Arf invariants of links

Patrick Gilmer
Department of Mathematics, Louisiana State University
Baton Rouge, LA, 70803, U.S.A.
e-mail: gilmer@marais.math.lsu.edu

ABSTRACT

We will describe a method for calculating the Arf invariant of a link from a link diagram. This method is related to the Gordon-Litherland method of calculating signatures from the Goeritz form. We also relate it to a state sum formula of Jones for certain evaluations of the Jones polynomial.

With a view to applications to real algebraic curves[1], we have shown how to define and calculate Arf invariants for links in rational homology 3- spheres[2,3]. Here we give a simple description of this method as applied to links in the 3-sphere.

First we must discuss the Brown invariant of a proper quadratic function on a finite dimensional $\mathbb{Z}_2$ inner product space. Let W be a $\mathbb{Z}_2$ vector space equipped with a possibly singular symmetric bilinear form $<\,,\,>$ which we will call a inner product. A quadratic function φ on a W is a map $\varphi : W \to \mathbb{Z}_4$ such that

$$\varphi(u+v) = \varphi(u) + \varphi(v) + \jmath < u, v >$$

where $\jmath$ is the monomorphism $\mathbb{Z}_2 \to \mathbb{Z}_4$.

Define the radical R to be $\{u \in W | < u, v >= 0 \quad \forall v \in W\}$. A quadratic function is called nonsingular if its radical is trivial. A quadratic function is called proper if it vanishes on the radical. In this case there is an induced nonsingular quadratic form $\bar{\varphi}$ defined on W/R.

The Brown invariant of a proper quadratic function $\beta(\varphi) \in \mathbb{Z}/8\mathbb{Z}$ is defined by

$$e^{\frac{2\pi i \beta(\varphi)}{8}} = \frac{1}{\sqrt{2}^{\left(\dim(W)+\dim(R)\right)}} \sum_{v \in W} i^{\varphi(v)}$$

This formula is due to Brown[4] for nonsingular forms. Its extension to singular proper forms is due to Kharlamov and Viro[5]. One has that $\beta(\varphi) = \beta(\bar{\varphi})$. If q is not proper, let r denote a nonzero element of R such that $\varphi(r) = 2$; then $i^{\varphi(v)} + i^{\varphi(v+r)}$ is zero. It follows that, if q is not proper, the right hand side of the above equation yields zero. In this case we say β is undefined, and write $\beta(\varphi) = U$.

We can describe or picture a quadratic function as a $\mathbb{Z}_4$ weighted graph. Let G be a finite graph with at most one edge joining any two vertices and no loops. We say two vertices which are joined by an edge are adjacent. Assign to each vertex v a weight $\varphi(v)$ in $\mathbb{Z}_4$. Let W be the $\mathbb{Z}_2$ vector space generated by $V(G)$, the vertices of G. For all $v,\ v' \in V(G)$, define $\langle v, v\rangle = \varphi(v) \bmod 2$ and $\langle v, v'\rangle$ to be one or zero according to whether v and v' are adjacent or not. For any subset $S \in V(G)$, define

$$\varphi(\sum_{v \in S} v) = \sum_{v \in S} \varphi(v) + \jmath(\text{ the number of edges with both endpoints in S}).$$

By *Lemma*(5.4)[3], this defines a quadratic function on W .

Moreover, any quadratic form arises in this way. Given a quadratic form φ on a vector space W and a basis v_i for W, we form a graph G with vertices v_i. Two vertices v_i and v_j are joined by an edge if and only if $\langle v_i, v_j' \rangle$ is nonzero. One then weights each vertex v_i with $\varphi(v_i)$.

The Brown invariant is additive under direct sum. The best way to calculate β is make use of additivity for direct sums by viewing the quadratic function as isomorphic to the direct sum of simple building block forms. Our building block forms and their Brown invariants are given below.

$$\beta(\overset{1}{\bullet}) = 1 \qquad \beta(\overset{-1}{\bullet}) = -1$$

$$\beta(\overset{0}{\bullet}) = 0 \qquad \beta(\overset{2}{\bullet}) = U$$

$$\beta(\overset{0}{\bullet}\!\!-\!\!\!-\!\!\overset{?}{\bullet}) = 0 \qquad \beta(\overset{2}{\bullet}\!\!-\!\!\!-\!\!\overset{2}{\bullet}) = 4$$

This isomorphism to a sum of building blocks is achieved by changing the basis. This can easily be done via moves on our graph:

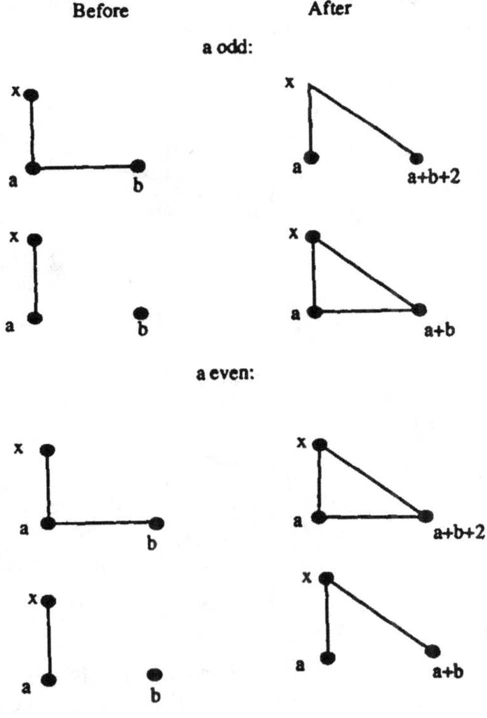

Here the vertex weighted a is being added to the vertex weighted b, and the vertex weighted x represents any other vertex which has an edge joining it to the vertex labelled a. After each move one reduces modulo two the number edges joining any two vertices. Here is an algorithm which reduces any graph via these moves to a disjoint union of the above basic building blocks. If any vertex has an odd weighing (we call this an odd vertex), we can add it to each vertex which is adjacent to it. In this way, we may split off any odd vertices. The non-building-block components of the resulting graph will have only even weightings. Our further moves below preserve this property. If we have a vertex v with an even weighting which is adjacent to only one other vertex v' (we call this a "hanging even vertex"), we may add v to every other vertex which is adjacent to v'. In this way we split off a barbell with two vertices. If some non-building-block component of the graph has no hanging even vertices, then each vertex is joined to at least two other vertices. Pick a vertex v. If we add one vertex adjacent to v to another vertex adjacent to v, then v will have one fewer adjacent vertices. In fact the new graph is obtained from the original one by deleting an edge which meets v. Continuing in this way, we may make v a hanging even vertex. Then we can split off a barbell as above and continue in this manner.

Here is a worked example. We draw an arrow to indicated which vertex is to be added to which other vertex. If there is no edge joining these vertices, we draw one in grey. The Brown invariant of this form is five.

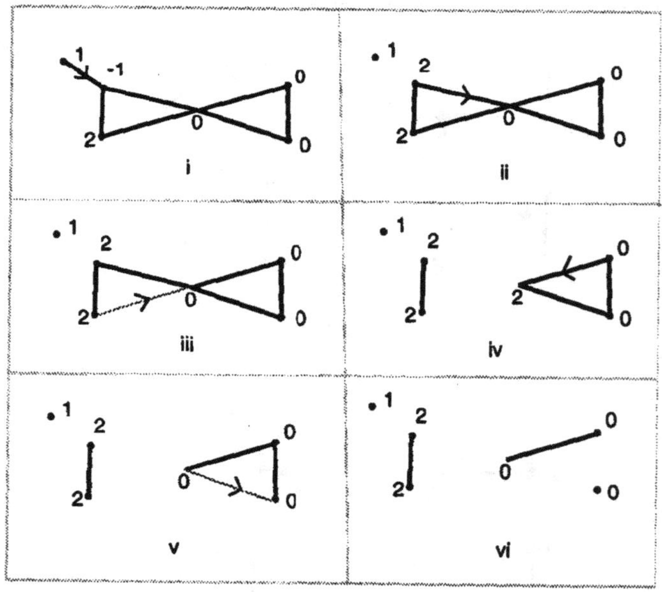

Let L be an oriented link in S^3. Let F be a possibly non-orientable, possibly disconnected spanning surface for L. Gordon and Litherland showed [6] that any two spanning surfaces for the same link can be connected by a sequence of moves of the following three types:

isotopy

adding or removing a small twisted band

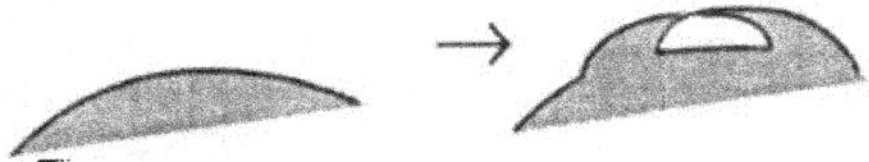

adding or removing a hollow handle

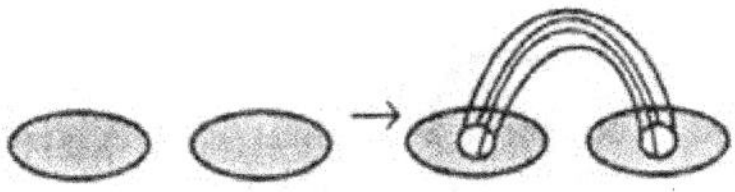

The path described by the core of a hollow handle can be any path whose interior misses the original spanning surface. Any invariant of a spanning surface unchanged by these moves is a link invariant.

Let $\hat{K}_i$ be is a parallel copy of K_i on F oriented in the same direction as K_i. Let $\hat{L}$ be the oriented link whose components consist of the $\hat{K}_i$. Then define $\mu(F) = Lk(L, \hat{L})/2$. It is easy to see that $\mu(F)$ is an integer.

One may define a quadratic function φ_F on $H_1(F, \mathbb{Z}_2)$ equipped with the intersection form as follows [3]. Given $x \in H_1(F, \mathbb{Z}_2)$, pick an embedded curve J on F representing x. $\varphi_F(x)$ is simply the number of positive half twists on a neighborhood $\nu(J)$ of J in F. More precisely, one picks an orientation for J and computes the linking number of J with the boundary of $\nu(J)$ oriented parallel to J. Pinkall used this same function to study regular isotopy of closed immersed surfaces in the 3-sphere[7]. φ_F is proper if and only if the linking number of any component of L with the union of all the other components of L is even. In that case, L is called proper.

One defines

$$Arf(L) \equiv \beta(\varphi_F) - \mu(F) \in \mathbb{Z}/8\mathbb{Z}.$$

It is well defined as it is invariant under the three moves above. Since one can always choose an orientable F, $Arf(L) \equiv 0 (\text{mod} 4)$. Thus one can define

$$R(L) \equiv Arf(L)/4 \pmod 2.$$

If F is orientable, the definition of $R(L)$ is easily seen to agree with one of Robertello's original definitions[8] of the Arf invariant.

Suppose we are given a connected regular diagram of L in S^2. All the regions of the diagram are disks which we can color alternating black and white. The black regions are the shadow of a spanning surface for L, say F. We draw a decorated graph in S^2 as follows[9,10]. We place a vertex in each white region. For each crossing we draw an edge e passing through the crossing joining the vertices of the white regions on either side. If these white regions are the same, we have a loop. We assign a sign to each edge according to the following rule. If the edge crosses the shadow of a band of F with a positive (negative) half twist we label it with a plus (minus) sign. We write $\eta(e) = +1$ or -1 according to this sign. Thus, as is well-known, a signed graph in S^2 encodes an unoriented link in S^3. Actually, we have chosen the opposite of the usual sign convention, as we plan to calculate our invariants using the spanning surface F which consists of a disk for each shaded region and a twisted band for each edge. In point of fact, using the usual sign convention would not effect our answer, since a link and its mirror image have the same Arf invariant. Thus we make this choice for expository purposes. We illustrate this procedure with the figure eight knot:

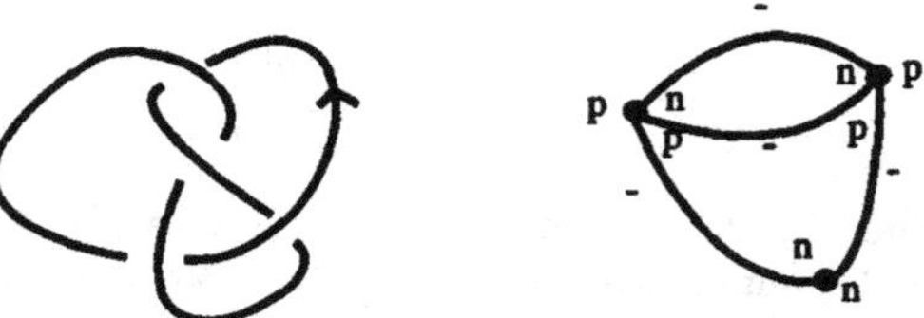

To encode the orientation of L, we label the regions around each vertex separated by the edges meeting that vertex by p or n depending on whether the oriented link is traveling counterclockwise or clockwise around that vertex in that region. Note also that a single p or n will propagate itself around a graph much like an arrow in a traditional diagram. In the case of a knot a single n or p will force a complete labeling. The rule of propagation is that if p (or n) appears at one end of an edge then it must appear on the opposite side and opposite end of this edge. We say an edge is of Type II if the regions adjacent to its end point are both labeled with the same symbol n or p. Like:

Let $G, E(G), V(G)$ denote this graph, its edges and its vertices. Gordon and Litherland showed [6]

$$\mu(F) = \sum_{\substack{e \in E(G) \\ \text{Type II}}} \eta(e).$$

Here $\eta(e)$ is plus or minus one according to whether the edge is labeled plus or minus. Note also that the boundary of each white region gives an element of $H_1(F, \mathbb{Z}_2)$. These elements generate $H_1(F, \mathbb{Z}_2)$ and are subject to the relation that their sum is zero. This actually gives a presentation of $H_1(F, \mathbb{Z}_2)$. This follows from Alexander duality in the plane. We may obtain a graph G' which encodes a quadratic form with the same Brown invariant from the signed graph which encodes the knot by weighting each vertex with the number of edges which meet the vertex counted with sign (ignoring loops) and then reducing the number of edges joining any two vertices modulo two. Below we complete the calculation of the Arf invariant of the figure eight knot. Since $\mu = -2$ and $\beta = 2$, we conclude that the Arf invariant is four, and the Robertello invariant is minus one.

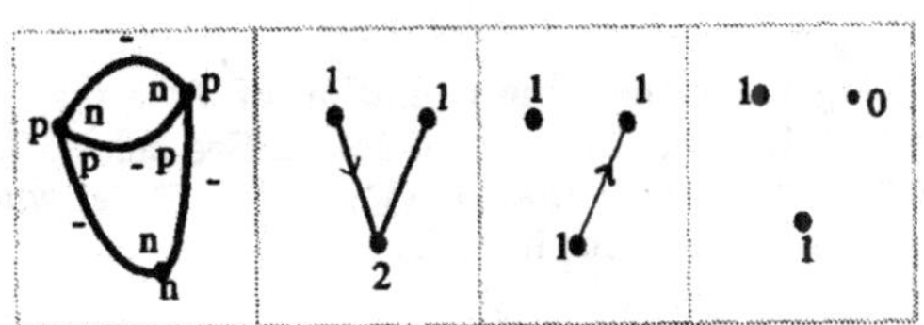

It is interesting that this algorithm also discovers if the link is not proper. In that case, we will obtain a building block consisting of a single vertex weighted two. Now consider Conway's 11-crossing Alexander polynomial one knot. Below we show the associated decorated graph. One calculates $\mu = 0$. The Brown invariant of the associated G' is easily calculated to be zero.[3] So this knot has a trivial Arf invariant, as it should since it has a trivial Alexander polynomial.

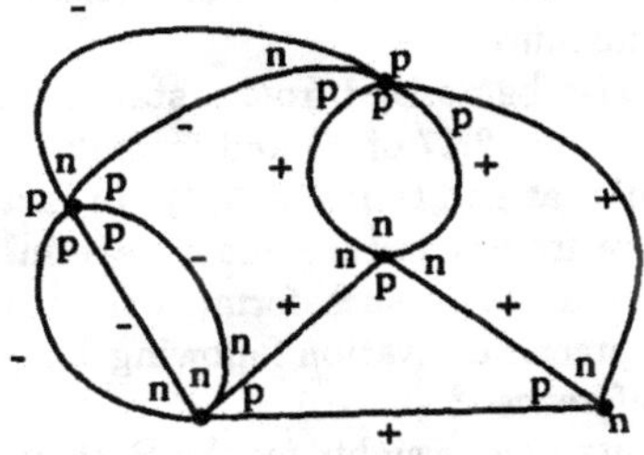

Consider now the 11-crossing knot encoded by the following decorated graph. $\mu = 1$. The graph which encodes the quadratic form is our earlier worked example with Brown invariant five. Thus the Arf invariant of this knot is non-trivial.

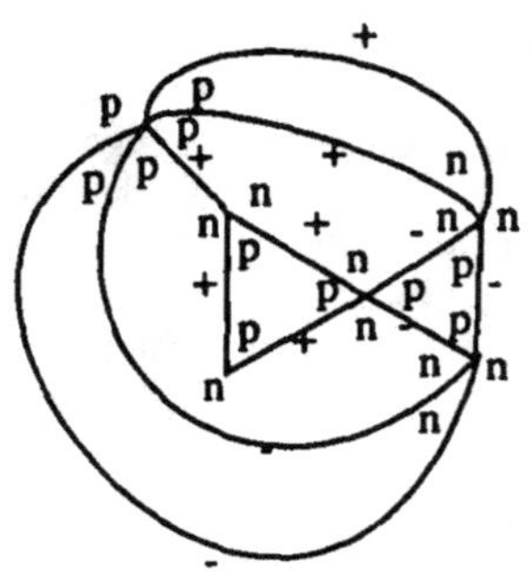

Finally, by applying the Gauss sum definition of β we may obtain a state sum formula for $-1^{R(L)}$ which is essentially due to Jones. See below. A state is defined to be a subset S of $V(G) = V(G')$. Note that $\varphi^{G'}(\sum_{v\in S} v)) = \sum \eta(e)$ where the sum is over edges of G with exactly one endpoint in S. Thus

$$i^{[\varphi^{G'}(\sum_{v\in S} v)]} = \prod_{e\in E(G)} W(e,s)$$

where

$$W(e,s) = \begin{cases} i^{\eta(e)} & \text{if exactly one endpoint of } e \text{ is in } S \\ 1 & \text{otherwise} \end{cases}$$

Loops are considered to have two identical end points. So $W(\text{loop}, s)$ is always one.

Note the dimension of the radical of $\varphi^{G'}$ is the number of components of the link $\beta_0(L)$. So, applying the Gauss sum formula, we have:

$$(-1)^{R(L)} = \frac{\omega^{-\mu(G)}}{\sqrt{2}^{\left(|V(G)| + \beta_0(L)\right)}} \sum_{S\subsetneq V(G)} \prod_{e\in E(G)} W(e,S)$$

where $\omega = \frac{1+i}{2} = \sqrt{i}$ and $|V(G)|$ denotes the number of vertices in G. Moreover the right hand side is zero if $R(L)$ is undefined.

The above formula may also be derived from a state sum approach to the Jones polynomial due to Jones (Example 2.17 of [10]) and formula of Murakami's [11,12] which evaluates the Jones polynomial at i in terms of $R(L)$. In fact, we learned that there was a formula like that above from a lecture Jones gave in Austin in Spring 1988, and then realized that we could obtain such formula from our approach to the Arf invariant. However to give a precise derivation following Jones, we must correct some minor misprints/errors in Reference [10].

First we note that the Boltzmann weights for the Potts model with n states given in 2.17[10] do not satisfy the "star-triangle relation" Equation 2.5 [10], even for $n = 2$. However, a multiple of these weights do:

$$w(a,b) = \begin{cases} -d^{-3}, & \text{if } a = b; \\ d, & \text{if } a \neq b. \end{cases}$$

where d is any solution of $d^2 + d^{-2} = -\sqrt{n}$. Actually these weightings (up to some ambiguity in the choice of solutions of $2+d^4+d^{-4} = n$) actually appear in a handwritten manuscript for Reference [10]. Next if we change the definition of the unoriented link invariant on bottom of page 321 of [10] by changing the exponent of $1/\sqrt{n}$ to $V+1$ instead of $V-1$, we obtain an invariant which is one on the unknot. The resulting oriented link invariant 2.13[10] can then be shown to satisfy the skein axiom for the Jones polynomial evaluated at $t = d^4$, where we choose $\sqrt{t} = d^2$. The Jones polynomial evaluated at i is then given by a state sum with the above weights evaluated at $d^2 = -\omega$. Then it not hard to derive the state sum for $R(L)$ given above.

Acknowlegment This research was supported by a grant from the Louisiana Education Quality Support Fund

References

1. P. Gilmer, Real Algebraic Curves and Link Cobordism, *Pacific J. Math.* **153** (1992) 31-69.
2. P. Gilmer, Link invariants in rational homology three-spheres, *Abstracts of the A.M.S.* **8** (1987)180.
3. P. Gilmer, Link Cobordism in Rational Homology Spheres, to appear in Jour. of Knot Th. and its Ramf.
4. E. Brown, Generalizations of the Kervaire Invariant, *Ann. of Math.* **95** (1972) 368-383
5. V.M Kharlamov and O.Ya.Viro, Extensions of the Gudkov-Rohlin congruence, in *Topology and Geometry-Rohlin Seminar*, ed. O. Ya. Viro, Springer-Verlag, New York, Lecture Notes in Math. **1346**, (1988) 357-406
6. McA. Gordon and R. A. Litherland, On the signature of a link, *Invent. Math.* **47** (1978) 53-69.
7. U.Pinkall, Regular Homotopy classes of immersed surfaces *Topology* 24 (1985) 421-434
8. R. Robertello, An invariant of knot cobordism, *Comm. Pure and Appl. Math.* **18** (1965) 543-555
9. L. H. Kaufmann, New Invariants in the theory of knots, *Amer. Math. Monthly* **95** (1988),195-242.
10. V. F. R. Jones, On knot invariants related to some statistical mechanics models, *Pacific J. Math.* **137** (1989) 311-334.
11. H. Murakami, A recursive calculation of the Arf invariant of a link, *J. Math. Soc. Japan* **38** (1986) 335-338.
12. W. B. R. Lickorish and K. C. Millett, Some evaluations of link polynomials, *Comment. Math. Helv.* **61** (1986) 349- 359.

DIFFERENTIAL EQUATIONS FOR LOOP EXPECTATIONS IN QUANTUM GAUGE THEORIES

CHRISTOPHER KING

Department of Mathematics
Northeastern University
Boston MA 02115

December, 1992

ABSTRACT.
We present a method for computations of Wilson loop expectations in quantum gauge theories. The main idea is to find a differential equation of evolution for the expectation of the operator obtained by cutting the Wilson lines at a fixed time. On a compact Riemannian manifold, a Morse function plays the role of the time coordinate. This viewpoint is explored for three dimensional Chern-Simons theory and for two dimensional Yang-Mills theory. In both cases, the equation of evolution is first order and linear, and the infinitesimal braid relations of knot theory aid in its solution. It is also shown how the partition function of Yang-Mills theory on a compact two dimensional manifold may be obtained by this method, and some algebraic structures reminiscent of tangles are displayed.

1. Introduction.

The purpose of this note is to describe a method for the calculation of Wilson loop expectations in two different quantum gauge theories, namely three dimensional Chern-Simons theory and two dimensional Yang-Mills theory. The former is a topological quantum field theory, and the latter has been called an "area preserving quantum field theory" [W1]. Both theories have an infinite dimensional symmetry group (in addition to the usual gauge invariance) and both are close to being exactly solvable. The descriptions of these theories given below borrow heavily from Witten's work [W1,W2,W3]. Earlier versions of the ideas in this paper appeared in [FK1,FK2,GKS].

2. Chern-Simons theory.

We begin with three dimensional Chern-Simons theory. Let G be a compact, semi-simple Lie group, with an Ad-invariant inner product on its Lie algebra $\mathfrak{g}$. We will use $TrA^{\dagger}B$ to denote the inner product of two elements A, B of $\mathfrak{g}$. Let M be an oriented compact three manifold without boundary. Let A be a connection on a principal G-bundle over M. Then the Chern-Simons action is

$$S[A] = \frac{k}{4\pi}\int_M Tr(A \wedge dA + \frac{2}{3} A \wedge A \wedge A) \tag{1}$$

The second term $Tr(A \wedge A \wedge A)$ can be rewritten as $\frac{1}{2}\epsilon^{ijk} Tr(A_i[A_j, A_k])$. A gauge transformation is a bundle automorphism which preserves the fiber, and so it is given by a map $g : M \to G$. The connection A transforms inhomogeneously under the gauge transformation; however the Chern-Simons action is almost invariant. It changes only by the addition of a multiple of the winding number of g, which is the integer specifying the element of $\pi_3(G)$ to which g belongs [W3]. By choosing the normalization of the inner product on $\mathfrak{g}$, we can arrange so that for all $k \in \mathbf{Z}$, $S[A]$ changes by the addition of an integer multiple of 2π under any gauge transformation. This implies invariance of the expression $exp(iS[A])$. The partition function of Chern- Simons theory on M is now defined as

$$Z = \int DA \; exp(iS[A]) \tag{2}$$

This formal expression is the starting point for Witten's analysis of Chern-Simons theory. The topological invariance of the theory is manifested in this formal expression by the invariance of the action $S[A]$ under diffeomorphisms of M. We are particularly interested in Witten's conclusions concerning Wilson loop expectations. In order to explain this, it is convenient to introduce a Riemannian metric on M and to consider a Morse function $f : M \to \mathbf{R}$ (this idea was also introduced in [FK2]). The function f has isolated non-degenerate critical points. Let $[a, b]$ be an interval in which f has no critical values. Standard Morse theory [H] shows that $f^{-1}([a, b])$ is diffeomorphic to $f^{-1}(a) \times [a, b]$. The preimage $f^{-1}(a)$ is a disjoint union of compact two manifolds. Let us pick one component Σ of $f^{-1}(a)$, and consider the corresponding three manifold $N = \Sigma \times [a, b]$. Using the invariance under diffeomorphisms, we can pull back the Chern-Simons theory onto N. For convenience we will assume that Σ is contractible in M, and that $||\nabla f|| = 1$ on $f^{-1}([a, b])$, so that f is a normal coordinate for the family of surfaces $\{f^{-1}(c)\}, c \in [a, b]$.

One of Witten's main observations was that the physical state space $\mathcal{H}$ of the quantum Chern-Simons theory on a two manifold like Σ is finite dimensional. Witten presents two descriptions of this vector space, both of which require introducing a complex structure on Σ. First, $\mathcal{H}$ is the space of conformal blocks of a certain rational conformal field theory on Σ; this vector space is described also by Segal [S]. Secondly, $\mathcal{H}$ is the space of global holomorphic sections of a complex line bundle on the moduli space of flat connections on Σ. The vector space $\mathcal{H}$ depends on both the gauge group G and the integer k appearing in the Chern-Simons action.

Witten also describes the physical state space $\mathcal{H}_n$ of Chern-Simons theory on a surface with n marked points, corresponding to the intersection of the surface with Wilson loops in the three manifold. We review this description below; for simplicity we restrict to the case $\Sigma = S^2$. By deforming the Wilson loops if necessary, we can assume that their preimages on N are given by n parametrised curves $(\gamma_1, \ldots, \gamma_n)$ satisfying the following properties:

$$\begin{aligned} &i)\ \gamma_i : [a, b] \to N, \qquad \gamma_i(t) = (z_i(t), t) \qquad 1 \le i \le n \\ &ii)\ |z_i'(t)| < \infty \\ &iii)\ z_i(t) \neq z_j(t), \qquad 1 \le i < j \le n \end{aligned} \tag{3}$$

The main point is that each curve γ_i is the graph of a differentiable function $z_i(t)$ over the interval $[a,b]$, with values in Σ, and that these curves do not intersect in N. Let us denote by $U_i(t)$ the parallel transport along γ_i, which in the case of a smooth connection A on N is the solution of the differential equation

$$\frac{dU_i}{dt} = < A(\gamma_i(t)), \gamma_i'(t) > U_i(t), \qquad U_i(a) = e \tag{4}$$

If it exists, the solution $U_i(t)$ is in G for all $t \in [a,b]$. Actually, a typical connection in the Chern-Simons theory is too singular to allow such an equation; we will ignore this problem, bearing in mind that the partition function itself does not have a rigorous definition (yet).

Associated with each Wilson loop γ_i there is a unitary representation $\rho_i : G \to GL(V_i)$, where V_i is a finite dimensional vector space. We let W_n denote the subspace of $V_1 \otimes \cdots \otimes V_n$ which is invariant under the diagonal action of G ("charge conservation" implies that W_n is non empty [W3]). For large k, Witten states that W_n is the physical state space $\mathcal{H}_n$; for other values of k, $\mathcal{H}_n$ may be a proper subspace of W_n. $\mathcal{H}_n$ may also be identified with the vector space of n-point correlation functions of primary fields in the appropriate conformal field theory.

Now consider the following quantity, for all $t \in [a,b]$:

$$\phi(t) = < \rho_1(U_1(t)) \otimes \cdots \otimes \rho_n(U_n(t)) > \tag{5}$$

The "expectation" on the right hand side is taken with respect to the Chern-Simons theory on the three manifold $\Sigma \times [a,t]$. Gauge invariance of the Chern-Simons action implies that $\phi(t)$ preserves the subspace W_n; Witten observed that in fact $\phi(t)$ preserves $\mathcal{H}_n$. When $\mathcal{H}_n$ is identified with the space of n-point correlation functions in a conformal field theory, $\phi(t)$ can be viewed as the monodromy matrix arising from analytic continuation of the correlation functions along the paths $\{z_i(t)\}$. This implies that $\phi(t)$ satisfies the differential equation describing this analytic continuation, namely the Knizhnik-Zamolodchikov equations [KZ]. In the simplest case where $G = SU(N)$, these are

$$\frac{d\phi}{dt} = -\frac{1}{k+N} \sum_{1\leq i<j\leq n} \frac{\Omega_{ij}}{z_i - z_j}(z_i' - z_j')\phi \tag{6}$$

The operators Ω_{ij} are given as follows: let $\{T_a\}$ $(a = 1,\ldots,r)$ be an orthonormal basis of the Lie algebra $\mathfrak{g}$. Then

$$\Omega_{ij} = \sum_{a=1}^{r} 1 \otimes \cdots \otimes \rho_i(T_a) \otimes \cdots \otimes \rho_j(T_a) \otimes \cdots \otimes 1 \tag{7}$$

where the two non-identity factors occur in the i and j places.

An alternative derivation of these equations for Chern-Simons theory on $C \times R$ is presented in [FK1]. The topological invariance of Chern-Simons theory on $S^2 \times [a,b]$ follows directly from this equation. It is locally integrable, which implies that there is a (multi-valued) solution $\phi(z_1,\ldots,z_n)$. This integrability is a consequence of the infinitesimal braid relations:

$$[\Omega_{ij}, \Omega_{kl}] = [\Omega_{ij}, \Omega_{ik} + \Omega_{jk}] = 0, \qquad \textit{for all } i \neq j \neq k \neq l \tag{8}$$

Therefore the solution $\phi(b)$ is invariant under ambient isotopies on N which preserve the splitting $\Sigma \times [a,b]$ and which are the identity on $\Sigma \times \{a\}$ and $\Sigma \times \{b\}$.

3. Two-dimensional Yang-Mills theory.

We now turn to our other model, namely two dimensional (Euclidean) Yang-Mills theory. Computations in this theory are most easily performed using the lattice formulation, since this involves only finite dimensional integrals. Nevertheless it is interesting to consider it from the point of view adopted in the previous section on Chern-Simons theory. The following description is taken from [W1].

Let Σ be a compact two dimensional Riemannian manifold, without boundary. Let A be a connection on a principal G-bundle E over Σ. The curvature $F = dA + A \wedge A$ is a $\mathfrak{g}$-valued 2-form on Σ, and the Yang-Mills action on Σ is

$$S_{YM}[A] = \frac{1}{4e^2}\int_\Sigma Tr(F \wedge *F)\, d\mu \tag{9}$$

where e is the coupling constant and μ is the Riemannian measure on Σ. This action is invariant under all gauge transformations of the bundle E, and all area-preserving diffeomorphisms of the surface Σ. The Yang-Mills partition function is defined as

$$Z_{YM} = \int exp(S_{YM}[A])\, DA \tag{10}$$

The Yang-Mills measure implied in this definition has been constructed in several different ways [F, Sen, W1]. As Witten [W1] points out, the lattice gauge theory approach produces a measure which sums over the contributions of all bundles on Σ. A direct continuum construction has been used to define the measure for each bundle separately [Sen].

We will analyse the model as follows. Let $f : \Sigma \to R$ be a Morse function, and consider again an interval [a,b] in which f has no critical values. Each component of $f^{-1}([a,b])$ is a cylinder, which is diffeomorphic to the manifold $S^1 \times [a,b]$. Let Λ be one such component. Again the Yang-Mills theory can be pulled back to Λ. We will assume again that $||\nabla f|| = 1$ on $f^{-1}([a,b])$. Suppose also that the Wilson loops on Σ are pulled back to n curves $\{\sigma_i(t)\}$ on Λ, where they appear as the graphs of differentiable S^1-valued functions over the interval [a,b]. The situation is now quite analogous to the previous discussion of Chern-Simons theory, but in one dimension lower. Again we would like to compute the quantity

$$\psi(t) = < \rho_1(U_1(t)) \otimes \cdots \otimes \rho_n(U_n(t)) >,$$

where the expectation is taken with respect to the Yang-Mills measure on $S^1 \times [a,t]$.

Since Yang-Mills theory is not topologically invariant (except in the case where the coupling vanishes), the operator $\psi(t)$ will depend on the curves $\sigma_i(s)$ for all $a \le s \le t$. However the invariance under area preserving diffeomorphisms simplifies this dependence greatly. Our goal is to write down a differential equation satisfied by $\psi(t)$, which will be the analog of the Knizhnik-Zamolodchikov equations. We will first consider a simpler limiting case in which S^1 is replaced by R. We can introduce coordinates (y,t) in $R \times [a,b]$, so the curves are given as $\{\sigma_i(t) = (y_i(t),t)\}$. As in Chern-Simons theory, the operator $\psi(t)$ acts on $V_1 \otimes \cdots \otimes V_n$, and it leaves invariant the "physical Hilbert space" W_n. In [GKS], Yang-Mills theory on R^2 (with the

Euclidean metric) was analysed using complete axial gauge, and the equation of evolution of $\psi(t)$ was found (where it was called Bralic's equation). We quote the result below, with some changes of notation. For convenience we assume that the curves do not cross between a and b; they are labelled so that $y_i(t) < y_j(t)$ for all $i < j$.

$$\begin{aligned} \frac{d\psi}{dt} &= \psi(t)\, H(t) \\ H(t) &= -\frac{e^2}{2} \sum_{1 \le i < j \le n} |y_i(t) - y_j(t)|\, \Omega_{ij} \end{aligned} \tag{11}$$

Like the Knizhnik-Zamolodchikov equations in Chern-Simons theory, these equations imply that the solution is invariant under area-preserving diffeomorphisms. This invariance is most easily understood using the following "local commutativity" result [GKS]. For any vector $\chi \in W_n$, and any $s, t \in [a, b]$,

$$[H(s), H(t)]\chi = 0 \tag{12}$$

This implies that the solution is given by

$$\psi(b) = \psi(a)\, exp[\int_a^b H(s)\, ds], \tag{13}$$

which may be rewritten in terms of the areas of the strips between the curves $\{\sigma_i\}$. This implies invariance under area-preserving diffeomorphisms which are the identity at $t = a$ and $t = b$. The local commutativity theorem is proved using the infinitesimal braid relations. If the curves cross between a and b, the solutions of the Bralic equation for each non-crossing interval may be composed to give the operator $\psi(b)$. This procedure is explained in detail in [GKS].

For the sake of completeness we also present the corresponding equation on the cylinder $S^1 \times [a, b]$. In contrast to the Chern-Simons theory, the state space is infinite dimensional; in addition to the representation spaces $\{V_i\}$ associated with the Wilson loops, there is an additional factor $L^2(G)$ associated with the gauge field on the circle S^1. The gauge group G acts on the space $L^2(G)$ by the adjoint action $f(h) \mapsto f(g^{-1}hg)$, and G then acts on $V_1 \otimes \cdots \otimes V_n \otimes L^2(G)$ by the diagonal action. The physical state space K_n is the invariant subspace under this action. For example, K_0 is the space of central functions on G.

In order to write down the equation for this case, we introduce some additional notation. We label the marked points on the circle as $\{P_1, \ldots, P_n\}$, where P_i and P_{i+1} are adjacent for all $i = 1, \ldots, n-1$. It will be assumed again that this order does not change over the interval $[a, b]$, so the curves do not cross. We associate the factor $L^2(G)$ with the interval I between P_n and P_1. There is an induced Riemannian metric on the circle, coming from the metric on Σ; using this, we define $l_{ij}(t)$ to be the distance between P_i and P_j, measured without passing through I. We also define $l_{0,i}(t) = l_{1,i}(t)$ and $l_{i,n+1}(t) = l_{i,n}(t)$ and $l_i(t) = l_{i,i+1}(t)$ for $1 \le i \le n-1$, and $\kappa(t)$ to be the length of I. Next we introduce differential operators on $L^2(G)$, given by the usual left and right invariant vector fields on G, for $a = 1, \ldots, r$:

$$(X_a^L f)(g) = \frac{d}{dt} f(g e^{tT_a})|_{t=0}$$

$$(X_a^R f)(g) = \frac{d}{dt} f(e^{-tT_a} g)|_{t=0}$$

This leads to the following operators on $V_1 \otimes \cdots \otimes V_n \otimes L^2(G)$:

$$\begin{aligned} T_a^{(i)} &= 1 \otimes \cdots \otimes \rho_i(T_a) \otimes \cdots \otimes 1 \qquad 1 \leq i \leq n,\ a = 1, \ldots, r \\ T_a^{(0)} &= 1 \otimes \cdots \otimes 1 \otimes X_a^R \\ T_a^{(n+1)} &= 1 \otimes \cdots \otimes 1 \otimes X_a^L \end{aligned}$$

For all $0 \leq i < j \leq n+1$ we define

$$\tilde{\Omega}_{ij} = \sum_{a=1}^{r} T_a^{(i)} T_a^{(j)}$$

This is an extension of the operators previously defined in (7). Finally we define

$$\Delta_G = \sum_{a=1}^{r} (T_a^{(0)})^2$$

Using the lattice definition of Yang-Mills theory [W1], it is not difficult to derive a differential equation for $\psi(t)$. The equation is

$$\begin{aligned} \frac{d\psi}{dt} &= \psi(t) \tilde{H}(t) \\ \tilde{H}(t) &= -\frac{e^2}{2} \sum_{0 \leq i < j \leq n+1} l_{ij}(t) \tilde{\Omega}_{ij} + \frac{e^2}{2} \kappa(t) \Delta_G \end{aligned} \tag{14}$$

We recover the previous equation on $R \times [a, b]$ by restricting to the invariant subspace of $V_1 \otimes \cdots \otimes V_n \otimes C$, corresponding to constant functions in $L^2(G)$. Note also that for the case $n = 0$, $\tilde{H}(t)$ is proportional to the scalar Laplace-Beltrami operator on G, acting on the subspace of central functions in $L^2(G)$.

Once again there is a local commutativity result which greatly simplifies the solution of this equation. For each $0 \leq i \leq n+1$, and $1 \leq a \leq r$, we define

$$D_a^{(i)} = \sum_{0 \leq j \leq i} T_a^{(j)}$$

$$E_a^{(i)} = \sum_{i+1 \leq j \leq n+1} T_a^{(j)}$$

Then any state χ in the physical state space K_n (and also in $D(\Delta_G)$, the domain of Δ_G) satisfies

$$D_a^{(n+1)} \chi = 0$$

We introduce the operators

$$h_i = \sum_{a=1}^{r} D_a^{(i)} E_a^{(i)},$$

in terms of which $\tilde{H}(t)$ may be rewritten as

$$\tilde{H}(t) = -\frac{e^2}{2} \sum_{1 \le i \le n-1} l_i(t) h_i + \frac{e^2}{2} \kappa(t) \Delta_G \tag{15}$$

A straightforward calculation shows that (assuming $i < j$)

$$[h_i, h_j] = \sum_{a,b,c=1}^{r} f_{ab}^c E_b^{(j)} D_c^{(i)} D_a^{(n+1)} \tag{16}$$

where $\{f_{ab}^c\}$ are the structure constants of $\mathfrak{g}$, given by

$$[T_a, T_b] = \sum_c f_{ab}^c T_c$$

These identities imply that for any χ in $K_n \bigcap D(\Delta_G^2)$, and any s, t in $[a, b]$ (assuming the curves do not cross between a and b)

$$[\tilde{H}(s), \tilde{H}(t)]\chi = 0 \tag{17}$$

This implies again that the solution of (14) is given by the exponential of the integral of $\tilde{H}(s)$.

4. An algebraic view of YM_2.

To conclude this discussion of two-dimensional Yang-Mills theory, we make some observations about the calculation of the partition function Z_{YM} when the coupling e^2 is zero. The theory is now topologically invariant. Let us consider the Morse function $f : \Sigma \rightarrow R$ as the analog of the time coordinate for the quantum theory on Σ. If there are no critical values in [a,b], then f behaves more or less like the usual "time" for each of the (cylindrical) components of $f^{-1}([a, b])$. By considering simultaneously the m components of $f^{-1}([a, b])$, we obtain an operator $\Psi(t)$ on the m-fold tensor product $K_0 \otimes \cdots \otimes K_0$ (there are no Wilson lines, so n=0 on each component). The Bralic equation shows that $\Psi(t)$ is constant on [a,b]. Therefore the only changes occur at the critical points of f. By our assumptions on f, these are isolated and non-degenerate. So the number of components in $f^{-1}(t)$ changes by ± 1 as each critical value is passed. Therefore it is natural to associate with each critical point an operator from $K_0^{\otimes m}$ to $K_0^{\otimes m+1}$, or vice versa. The partition function Z_{YM} will then be obtained by composing these operators, in the order given by the corresponding critical values of f. The result will be an operator from C to C, since m=0 outside the interval $f(\Sigma)$. This viewpoint is similar to the definition of tangle invariants in [RT] (see [FK2] for a discussion of other topological field theories).

It is quite straightforward to compute the operators corresponding to each critical point, using results from lattice gauge theory [W1]. It is convenient to introduce the orthonormal basis of K_0 consisting of irreducible characters $\{\chi_\alpha\}$, and to let d_α denote the dimension of the representation α. We will also need to consider a space of distributions $\tilde{K}_0$, containing the δ-function at the identity in G.

There are four kinds of critical points of f; one each of index 0 and 2, and two of index 1. At a critical point of index 0, f has a local minimum, and the surface is locally a disk. The corresponding operator $\eta : C \to \tilde{K}_0$ is given by

$$\eta(\lambda) = \lambda \sum_\alpha d_\alpha \chi_\alpha \qquad (\lambda \in C)$$

At a critical point of index 2, where f has a local maximum, the operator $\epsilon : K_0 \to C$ is the linear extension of the operator given by

$$\epsilon(\chi_\alpha) = d_\alpha$$

At a critical point of index 1, the surface is locally a three-holed sphere. There are two cases, depending on whether $f^{-1}(t)$ is connected for values less than or greater than the critical value. In the former case, the operator $\Delta : K_0 \to K_0 \otimes K_0$ is given by

$$\Delta(\chi_\alpha) = \frac{1}{d_\alpha} \chi_\alpha \otimes \chi_\alpha$$

In the latter case, the operator $m : K_0 \otimes K_0 \to K_0$ is given by

$$m(\chi_\alpha \otimes \chi_\beta) = \frac{1}{d_\alpha} \chi_\alpha \delta_{\alpha\beta}$$

The value of Z_{YM} may be obtained by composing these operators in the manner described. The task is made especially simple by the fact that the operators m, Δ are diagonal with respect to the basis $\{\chi_\alpha\}$, and also that the only interaction between different components of $f^{-1}(t)$ occurs at the critical points of f. The result is

$$Z_{YM} = \sum_\alpha d_\alpha^{n_0 - n_1 + n_2} \tag{18}$$

where n_i is the number of critical points of f of index i. Since the Euler number of Σ is $n_0 - n_1 + n_2$, this reproduces the result in [W1], and shows that Z_{YM} is a topological invariant.

It should be noted that in the course of this calculation, it is necessary to consider informal vectors outside the Hilbert space K_0. A similar computation can be performed for Yang-Mills theory at non-zero coupling, in which case the corresponding operators can be regularized.

It may be of interest to notice that the operators $\eta, \epsilon, m, \Delta$ present the structures of an algebra and a coalgebra on the space of central functions on G. In fact, $\eta(1) = \delta_e$ is the δ-function at the identity; $\epsilon(f) = f(e)$ is the evaluation at e; and $m(f_1, f_2) = f_1 * f_2$ is the group convolution. Furthermore, the topological invariance of Z_{YM} is manifested by algebraic relations between these operators, reminiscent of the tangle relations in [RT]. For example, suppose Σ is embedded in R^3 and f is the height function. A vertical cylindrical section of Σ may be deformed by inserting an additional local maximum and minimum (and two additional saddle points). The invariance of Z_{YM} corresponds to the identity

$$(1 \otimes \epsilon)(1 \otimes m)(\Delta \otimes 1)(\eta \otimes 1) = 1 \tag{19}$$

Unlike in the tangle case [RT], the maps Δ, ϵ are not algebra maps, so they do not provide a bialgebra structure. In fact the surfaces corresponding to the maps $\Delta \circ m$ and $m \circ \Delta$ are not topologically equivalent, so there is no reason to expect a bialgebra structure.

References

[F] D. Fine, *Quantum Yang-Mills on a Riemann Surface*, Commun. Math. Phys. **140** (1991), 321-338.

[FK1] J. Frohlich and C. King, *The Chern-Simons Theory and Knot Polynomials*, Commun. Math. Phys. **126** (1989), 167-199.

[FK2] J. Frohlich and C. King, *Two-Dimensional Conformal Field Theory and Three-Dimensional Topology*, Int. Jour. Mod. Phys. A **4** (1989), 5321-5399.

[GKS] L. Gross, C. King and A. Sengupta, *Two dimensional Yang-Mills theory via stochastic differential equations*, Annals of Physics **194** (1989), 65-112.

[H] M. Hirsch, *Differential Topology*, Springer-Verlag, 1976.

[KZ] V.G. Knizhnik and A.B. Zamolodchikov, *Current algebra and Wess-Zumino models in two dimensions*, Nucl. Phys. B **247** (1984), 83-103.

[RT] N. Reshetikhin and V. Turaev, *Invariants of three manifolds via link polynomials and quantum groups*, Invent. Math. **103** (1991), 547-597.

[S] G. Segal, *Two-dimensional conformal field theories and modular functors*, IXth International Conference on Mathematical Physics (B. Simon, A. Truman and I.M. Davies, ed.), Adam Hilger, Bristol, 1989.

[Sen] A. Sengupta, *Quantum Gauge Theory on Compact Surfaces*, preprint (1992).

[W1] E. Witten, *On Quantum Gauge Theories in Two Dimensions*, Commun. Math. Phys. **141** (1991), 153-209.

[W2] E. Witten, *Two Dimensional Gauge Theories Revisited*, preprint (1992).

[W3] E. Witten, *Quantum Field Theory and the Jones Polynomial*, Commun. Math. Phys. **121** (1989), 351-399.

Triangulations, Categories and Extended Topological Field Theories [1] [2]

R.J. Lawrence[3]

Department of Mathematics
Harvard University
Cambridge, Massachusetts

Abstract. The concept of a topological field theory is extended to encompass structures associated with manifolds of codimension >1. When all the manifolds involved are considered triangulated, it is seen that such structures may be constructed from a finite quantity of data, most conveniently viewed as associated with polyhedra and their decompositions. The special cases of 2 and 3 dimensions are briefly considered, the relations with structures of higher categories, algebras and vector spaces, becoming clear. A more detailed account is currently in preparation.

1: Introduction

A d-dimensional topological field theory (TFT), associates a vector space $Z(\Sigma)$ to a closed $(d-1)$-dimensional manifold Σ, and a vector $Z(M) \in Z(\partial M)$ to a d-dimensional manifold M. In many cases the allowed manifolds M and Σ may be restricted in some way, or may be supplied with extra data, for example, a framing or triangulation. We shall assume that all manifolds considered are orientable, and supplied with an orientation, unless otherwise stated. The structure, Z, is constrained by the requirement that it satisfy certain axioms. Their precise form varies amongst authors, but the following general set are by now fairly standard [A].

$\mathfrak{A}_1$ NATURALITY An isomorphism of manifolds $\alpha: \Sigma_1^{d-1} \xrightarrow{\sim} \Sigma_2^{d-1}$ induces an isomorphism $\Sigma(\alpha): Z(\Sigma_1) \xrightarrow{\sim} Z(\Sigma_2)$ of the corresponding vector spaces, with $Z(\beta \circ \alpha) = Z(\beta) \circ Z(\alpha)$ for any suitable $\beta: \Sigma_2^{d-1} \xrightarrow{\sim} \Sigma_3^{d-1}$.

$\mathfrak{A}_2$ MULTIPLICATIVITY $Z(\Sigma_1 \coprod \Sigma_2) = Z(\Sigma_1) \otimes Z(\Sigma_2)$, where $\coprod$ denotes disjoint union.

$\mathfrak{A}_3$ VACUUM $Z(\emptyset) = K$, the base field.

$\mathfrak{A}_4$ DUALITY If Σ^{d-1} is a closed manifold, there is a natural isomorphism $Z(\Sigma^*) = Z(\Sigma)^*$ where Σ^* denotes the manifold Σ endowed with the opposite orientation.

$\mathfrak{A}_5$ NATURALITY An isomorphism $\alpha: M_1^d \xrightarrow{\sim} M_2^d$ induces an equality between the vectors $Z(M_1)$ and $Z(M_2)$ in the vector spaces $Z(\partial M_1)$ and $Z(\partial M_2)$, by the isomorphism $Z(\alpha|_{\partial M_1})$ of $\mathfrak{A}_1$.

---

[1] This paper was presented at the AMS Meeting #876, held in Dayton, Ohio, on Oct. 31 1992.

[2] This paper is supported in part by NSF Grant No. 9013738.

[3] The author is a Junior Fellow of the Society of Fellows.

$\mathfrak{A}_6$ ASSOCIATIVITY If Σ_1, Σ_2 and Σ_3 are $(d-1)$-dimensional manifolds and Y and Y' are cobordisms between Σ_1 and Σ_2, Σ_2 and Σ_3, respectively, then

$$Z(Y \cup Y') = Z(Y) \circ Z(Y') \in Z(\Sigma_1) \otimes Z(\Sigma_3)^*$$

where $Z(Y)$ and $Z(Y')$ are viewed as elements of $Z(\Sigma_1) \otimes Z(\Sigma_2)^*$ and $Z(\Sigma_2) \otimes Z(\Sigma_3)^*$, respectively, and here $\circ$ denotes the natural contraction map.

There are two other axioms that are sometimes included, viz., that of *completeness* and that variously called ***duality*** or ***conjugation*** at the level of d-dimensional manifolds. These, however, do not alter the main structure involved, the latter only being meaningful when $Z(\Sigma)$ is endowed with a $*$-structure, e.g. if it is a Hilbert space.

Suppose that M^d is a closed manifold. Then, by $\mathfrak{A}_3$, $Z(M) \in K$ is the associated invariant. It may be computed from a splitting of M into M_1 and M_2 with common boundary Σ^{d-1} as follows. Here $\partial M_1 = \Sigma$ and $\partial M_2 = \Sigma^*$, say. Then M_1 and M_2 determine vectors, $Z(M_1) \in Z(\Sigma)$ and $Z(M_2) \in Z(\Sigma)^*$. The natural pairing on these two vectors gives $Z(M)$.

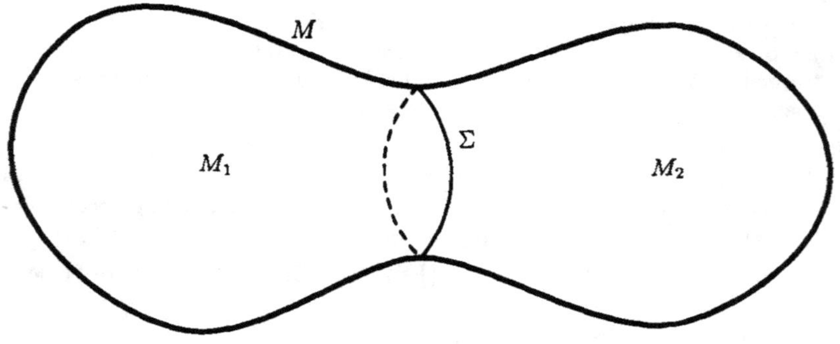

Figure 1

Suppose now that Σ, a closed $(d-1)$-dimensional manifold, is split into two parts, Σ_1 and Σ_2 with common boundary C^{d-2}. We would like to associate to Σ_1 and Σ_2 elements of an appropriate structure and its dual, dependent on C, in such a way that their natural pairing gives back $Z(\Sigma)$. Since $Z(\Sigma)$ is a vector space, the appropriate structure to be associated with C is a higher generalisation — a 2-vector space. In a similar way, closed $(d-r)$-dimensional manifolds will give rise, in an extended topological field theory (ETFT), to r-vector spaces.

In §2, the notion of an r-vector space, extending Kapranov and Voevodsky's notion of a 2-vector space, will be briefly introduced. Appropriate axioms for an ETFT are given in §3, and they are seen to embody the natural rules of composition (gluing laws) of manifolds and, therefore, an ETFT can be viewed as a functor from a universal object embodying such rules, to an object based on r-vector spaces. The precise formulation of these concepts are to appear in [L 1].

In this paper the basic ideas will be presented and illustrated in the case of dimension two. It will be seen that the formulation of invariants obtained is that of a generalised state model construction on a 'blow-up' of a triangulation of the manifold. The data involved in the state model is associated with a finite collection of suitably labelled elemental polyhedra, and it must satisfy relations geometrically expressed as the equivalences of distinct decompositions of a polyhedron into elemental forms. This is discussed in detail in §4 where $d = 2$. In §5 some examples are given; in particular the Euler characteristic and the invariant of Turaev–Viro. Further comments on the extensibility of the constructions are made in §6.

2. Higher vector spaces

Suppose K is a field. A finite dimensional vector space V over K is specified, up to isomorphism, by its dimension, a non-negative integer, n. Given two vector spaces, V and W, over K, it makes no sense to ask whether V and W are *equal*, only whether they are *isomorphic*. In particular, a linear transformation, $V \to W$, is given by an $m \times n$ matrix over K, where $m = \dim W$. From such considerations it becomes natural to consider the category of coordinatised vector spaces, $\mathcal{C}_1$, whose objects are such vector spaces and whose morphisms are linear transformations.

The structure of a higher vector space is meant to be like that of a vector space, but where K has been replaced by a category, or even a higher category. For 2-vector spaces, this notion was introduced in [KV], there being also introduced three different 2-category structures of such 2-vector spaces, each with a different degree of coordinatisation. A fully coordinatised 2-vector space is specified, up to isomorphism, by its dimension. If V is a 2-vector space of dimension n, and $\langle e_i \rangle$ is a basis for V, the *elements* of V should be thought of as formal linear combinations,

$$\sum_{i=1}^{n} \lambda_i e_i \,,$$

where $\lambda_i \in \mathcal{C}_1$ are vector spaces. That is, an element of V is equivalently an n-tuple of vector spaces. If W is a 2-vector space of dimension m, then a map $T\colon V \to W$ is given by an $m \times n$ matrix T_{ij} of elements of $\mathcal{C}_1$. The action of T is given by,

$$T(\sum \lambda_j e_j) = \sum \mu_i f_i \,,$$

where,

$$\mu_i = \bigoplus (T_{ij} \otimes \lambda_j)$$

in which $\oplus$ and $\otimes$ have their usual meanings as on vector spaces. Two such maps S and $T: V \to W$ cannot be directly compared, just as vector spaces cannot be equated. Rather there may exist a natural transformation U from S to T, and it is specified by a family,

$$U_{ij}: S_{ij} \longrightarrow T_{ij} ,$$

of linear transformations, i.e., an array of matrices. Note that the dimensions of any two entries in the array need not be in any way related. This whole structure forms a 2-category in the sense of [MoSe], objects being 2-vector spaces, 1-morphisms being maps between them and 2-morphisms being natural transformations.

In general, an $(r+1)$-vector space should be thought of as a linear space over $\mathcal{C}_r$, the r-category of r-vector spaces. Up to equivalence, an r-vector space is specified by its dimension. There is a special object in $\mathcal{C}_r$, namely the r-vector space, $\mathbf{1}_r$, of dimension 1. Whenever V is an r-vector space, the space of linear maps $V \to \mathbf{1}_r$, thought of as maps between linear spaces over $\mathcal{V}_{r-1}$, form another r-vector space, denoted by V^*. There are clearly also notions of direct sum, and tensor product, defined in similar ways to those for the usual vector spaces. Finally, as can be seen from the case of 2-vector spaces above, the higher morphism structures involve multiple nested indices.

3. Extended Topological Field Theories

The axioms for a topological field theory, as given in §1, may be embodied in a statement of the form, Z is a functor from the category $\mathcal{M}_1$ to the category $\mathcal{V}_1$ of vector spaces, preserving suitable additional structures. Here and throughout this section the top dimension, d, of the theory will be assumed fixed and therefore all dependencies upon d will be omitted. The category $\mathcal{M}_1$ has objects and morphisms given by,

$$\mathrm{Obj}_{\mathcal{M}_1} = \{\text{closed } (d-1)\text{-dim. oriented manifolds up to isomorphism}\}$$

$$\mathrm{Morph}_{\mathcal{M}_1}(\Sigma_1, \Sigma_2) = \left\{ \begin{array}{l} d\text{-dimensional oriented manifolds } M \\ \text{with } \partial M \simeq \Sigma_1^* \coprod \Sigma_2, \text{ up to isomorphism} \end{array} \right\}$$

with additional special structures,

(i) $\emptyset \in \mathrm{Obj}_{\mathcal{M}_1}$, the empty object;

(ii) $\coprod$, disjoint union, providing a monoidal structure on $\mathcal{M}_1$;

(iii) $*$, the operation of reversing orientation, which gives a contravariant functor $\mathcal{M}_1 \longrightarrow \mathcal{M}_1$.

The category $\mathcal{V}_1$ consists of vector spaces, that is,

$$\mathrm{Obj}_{\mathcal{V}_1} = \mathbf{Z}^+$$
$$\mathrm{Morph}_{\mathcal{V}_1}(m, n) = \{\alpha: [m] \times [n] \longrightarrow K\}$$

where K is the chosen base field, and $[m] = \{1,2,\ldots,m\}$ as in §2. The additional extra structures corresponding to those for $\mathcal{M}_1$ are,

(i) $1 \in \mathrm{Obj}_{\mathcal{V}_1}$, the vector space K;

(ii) $\otimes$, the operation of tensor product, which acts as multiplication on objects;

(iii) $*$, the duality operation, which on objects takes $n \longmapsto n$, and on morphisms takes $\alpha: [m] \times [n] \to K$ to $\alpha \circ P: [n] \times [m] \to K$, where P is the map permuting the first two factors.

With the above definitions of tensor category structures, the axioms $\mathfrak{A}_1$–$\mathfrak{A}_6$ of §1 are now precisely embodied in the functorial nature of Z. The category $\mathcal{M}_1$, with the above extra structure, embodies the gluing rules of manifolds of dimension d, and of closed $(d-1)$-dimensional manifolds.

An extended TFT (ETFT) is a structure similar to a TFT in which gluing rules of manifolds of all codimensions up to d are embodied. An s-ETFT contains only gluing rules of manifolds with codimensions $\leq s$, so that a 1-ETFT is just a TFT, while a d-ETFT is just an ETFT. Thus an s-ETFT is a functor,

$$\mathcal{M}_s \longrightarrow \mathcal{V}_s\,,$$

where $\mathcal{M}_s$ and $\mathcal{V}_s$ are suitable structures, both not unlike that of an s-category; $\mathcal{M}_s$ is a structure in which the r-morphisms label $(d-s+r)$-dimensional manifolds, $r \in [s]$, and the objects label closed $(d-s)$-dimensional manifolds. In $\mathcal{V}_s$, the r-morphisms are elements of $(s-r+1)$-vector spaces, as defined in §2, while the objects are s-vector spaces.

A d-dimensional s-ETFT associates to each closed $(d-r)$-dimensional manifold, M with $0 \leq r \leq s$, an r-vector space $\overline{Z}_r(M)$; and to each $(d-r)$-dimensional manifold, M with $0 \leq r < s$, an element $Z_r(M)$ of the $(r+1)$-vector space $\overline{Z}_{r+1}(\partial M)$; and does this in such a way that certain axioms are satisfied. As for the case of a TFT, the manifolds involved may be restricted in some way, or they may be endowed with extra structures. The properties which Z and $\overline{Z}$ must satisfy include the following.

$\mathfrak{E}_1$ NATURALITY An isomorphism, α, of closed manifolds, Σ_1^{d-r} and Σ_2^{d-r}, induces an isomorphism of the corresponding r-vector spaces, under which the elements corresponding to isomorphic manifolds M_1 and M_2 of dimension $d-r+1$ whose boundaries are Σ_i, are identified, whenever α is the restriction of the isomorphism to the boundary.

$\mathfrak{E}_2$ VACUUM $\overline{Z}_r(\emptyset) = \mathbb{1}_r$, while if M is a closed $(d-r)$-dimensional manifold, $\overline{Z}_r(M) \in \mathbb{1}_{r+1}$ may be identified with $Z_r(M) \in \mathcal{V}_r$.

$\mathfrak{E}_3$ DUALITY If Σ^{d-r} is a closed manifold $(1 \leq r \leq s)$, there is a pairing $\overline{Z}_r(\Sigma) \otimes \overline{Z}_r(\Sigma^*) \longrightarrow \mathbb{1}_r$, where Σ^* denotes Σ endowed with the opposite orientation.

$\mathfrak{C}_4$ MULTIPLICATIVITY $\overline{Z}_r(\Sigma_1 \coprod \Sigma_2) = \overline{Z}_r(\Sigma_1) \otimes \overline{Z}_r(\Sigma_2)$.

$\mathfrak{C}_5$ ASSOCIATIVITY

The most important property is $\mathfrak{C}_5$, the analogue of the associativity axiom (axiom $\mathfrak{A}_6$ of §1) for TFT's. This is a generalised gluing law, by which whenever a manifold M is decomposed into a union of other manifolds, M_i, with gluing taking place along boundaries (possibly only along part-boundaries), there is a procedure by which $Z(M)$ may be obtained naturally from $\{Z(M_i)\}$ by composing only structures already associated in the theory to the M_i, their boundaries, common part-boundaries,..., up to codimension s incidence properties. However, the exact statement of this property, in the general setting, is complex. The case of 2-ETFT with $d = 2$ is illustrated in the next section.

4. 2-ETFT

In this section the structure of a 2-ETFT is investigated. The basic objects are d-, $(d-1)$-, and $(d-2)$-dimensional manifolds, the $(d-2)$-dimensional manifolds involved all being closed. There are gluing operations of the same type as those arising in 1-ETFT. Thus, if $\partial M_1 = \Sigma$ and $\partial M_2 = \Sigma^*$ where M_1 and M_2 are $(r-1)$-dimensional manifolds while Σ is a closed r-dimensional manifold, then there is a manifold M, obtained by gluing M_1 and M_2 along Σ, where $r = d$ or $d-1$. There is also the slightly more general gluing along a (still closed) part-boundary, as in $\mathfrak{A}_6$ of §1. However, in 2-ETFT, there is an additional type of gluing of d-dimensional manifolds M_i $(1 \leq i \leq n-1)$, for which $\partial M_i = \Sigma_{i+1} \cup \Sigma_i^*$, where Σ_i are $(d-1)$-dimensional manifolds with common boundary C, $(1 \leq i \leq n)$. Here C is a closed $(d-2)$-dimensional manifold. As in the case of the first type of gluing operation, there is a slightly more general form in which the codimension 2 manifold, C, along which the gluing takes place, is only a (still closed) part boundary of Σ_i. It may be observed that the types of gluing laws appearing in s-ETFT are independent of d, so that to study 2-ETFT, all the structures needed appear when $d = 2$.

We shall assume that all manifolds considered are endowed with a triangulation and an orientation. In this context, the gluing laws are best observed by transforming a triangulated k-dimensional manifold, $(M, \mathcal{T})$, to its 'blow-up' $(M, \mathcal{T}^*)$, in which each top-dimensional cell in $\mathcal{T}^*$ corresponds to a simplex or sub-simplex in $\mathcal{T}$ (see Figure 2).

When M is closed the vertices of $\mathcal{T}^*$ are labelled by flags of simplices in $\mathcal{T}$, $X_0 \subset \cdots \subset X_k$, while faces in $\mathcal{T}^*$ are labelled by incomplete flags in $\mathcal{T}$. When M is not closed, $\mathcal{T}^*$ is defined as for the closed case, except that $\mathcal{T}$ is altered by adjoining a 'virtual' top-dimensional simplex with the incidence property that it contains all simplices in ∂M. In this case, there is also another decomposition for which the vertices are in 1–1 correspondence with actual (complete) flags of simplices in $\mathcal{T}$. As geometric decompositions, $\mathcal{T}^*$ can be obtained from $\mathcal{T}_*$ by adjoining a cylinder on $(\partial\mathcal{T})^*$, where $\partial\mathcal{T}$ is the triangulation of ∂M induced from $\mathcal{T}$ by restriction.

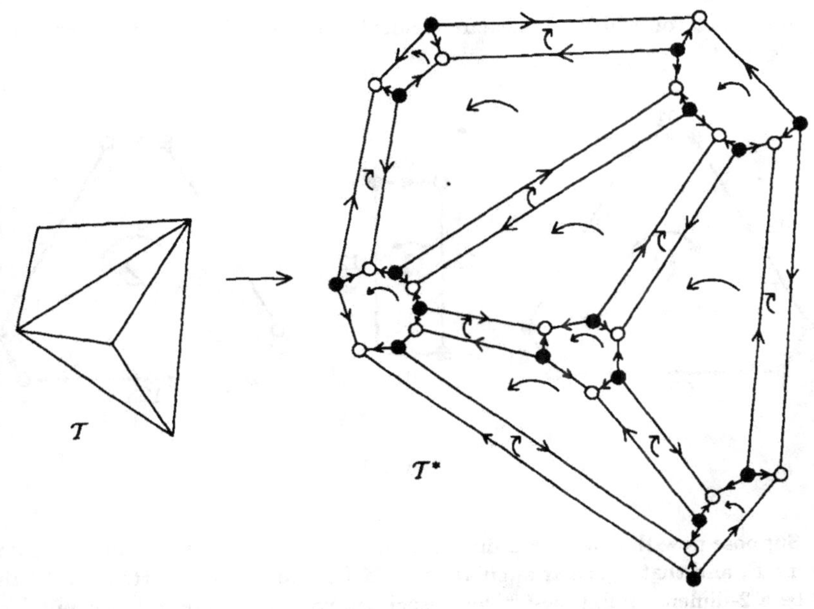

Figure 2

By definition, every cell, C, of $\mathcal{T}^*$ may be labelled by a subset, I, of $\{0,1,\ldots,k\}$ specifying the dimensions of the simplices in the corresponding partial flag. In such a case, the dimension of C is $k+1-|I|$, and I is called the type of C. Let $i = \max(I)$. Given I, the geometric form of C is completely determined by the link of X_i in $\mathcal{T}$.

From an orientation on M it is possible to canonically define orientations on all cells and sub-cells in $\mathcal{T}_*$. For $k = 2$ the result is illustrated in Figure 2, in which the orientations on r-dimensional cells are depicted by arrows, for $r = 1$ and 2, and filled and open circles, for $r = 0$. The general definitions of the orientations on the cells and subcells of $\mathcal{T}^*$ and $\mathcal{T}_*$ may be found in [L 1].

For $k = 2$, there are 7 possible non-empty subsets, I, of $\{0,1,2\}$, the dimension of an associated cell in $\mathcal{T}^*$ being $3 - |I|$. For ease of notation, the type, I, of a cell will be denoted by the string of its elements, in ascending order, and without separators. There are two kinds of type-012 cell, namely points with positive or negative orientation. There is one kind of cell of each of the types 01, 02 and 12, each of these being intervals bounded by two type-012 cells, with opposite orientations. The cells in $\mathcal{T}^*$ of type 0,1 and 2 are 2-dimensional, there being one kind of each of the last two types. For any $n \geq 2$, there is a $2n$-gon of type 0 in $\mathcal{T}^*$, associated to a point in $\mathcal{T}$ on which n edges are incident, and we say such a cell is of type 0^n. See Figure 3.

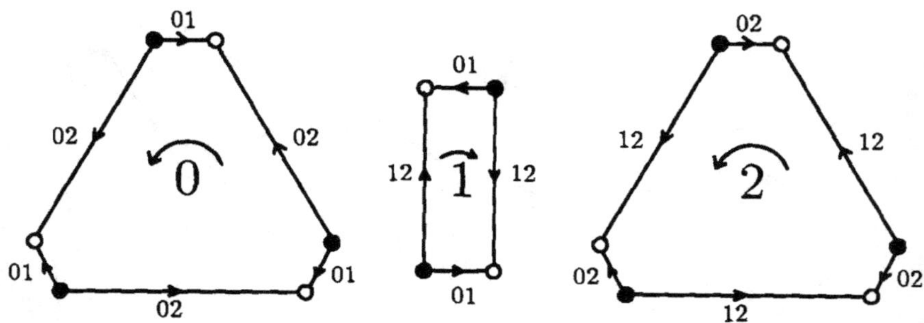

Figure 3

Suppose now that M is a 2-dimensional manifold whose boundary C contains a point P, and that $\mathcal{T}$ is a triangulation of M for which P is a vertex. In $\mathcal{T}^*$ there will be a 2-dimensional, type-0^m cell associated with P, where m is the number of 1-simplices in M containing P. In $\mathcal{T}_*$ there will be similarly associated a truncated type-0 cell with $2(m-1)$ edges, illustrated for $m = 3$ in Figure 4(i); such a cell will be said to have type 0_m. Any cell in $\mathcal{T}_*$ corresponding to a point in $\mathcal{T}$ will be of type 0^m or of type 0_m, for some m, and can be decomposed into the two types, 0_3 and 0^2 of Figure 4, to be called $0a$ and $0b$, respectively. We now investigate the effect of a gluing operation upon the associated blow-ups.

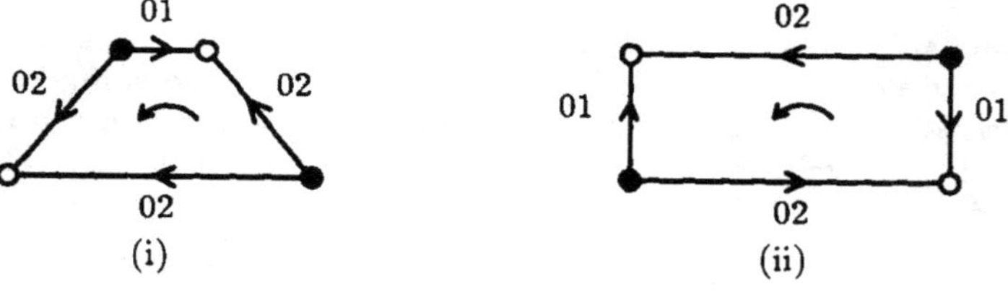

Figure 4

Suppose that N is a 2-dimensional manifold with boundary C^*, and triangulation $\mathcal{U}$ compatible with $\mathcal{T}$ on C. Let n be the number of 1-simplices in N containing P. Then $\mathcal{T}_*$ and $\mathcal{U}_*$ each contain a cell associated with P and of types 0_m and 0_n, respectively. The cell associated with P in the blow-up $(\mathcal{T} \cup \mathcal{U})_*$ will have type 0^{m+n-2}. This cell may be obtained by gluing together the two associated cells in $\mathcal{T}_*$ and $\mathcal{U}_*$, with a cell of type $0b$ between these two cells.

Suppose instead that N is such that its boundary C' contains P and shares a partial boundary C_1^* with C^* while $P \in \partial C_1$. Then the cell in $(\mathcal{T} \cup \mathcal{U})_*$ associated with P will be of type 0_{m+n-1}. This new cell may be obtained by simply gluing together the two cells associated with P in $\mathcal{T}_*$ and $\mathcal{U}_*$, with a single cell of type $0a$, between these two cells.

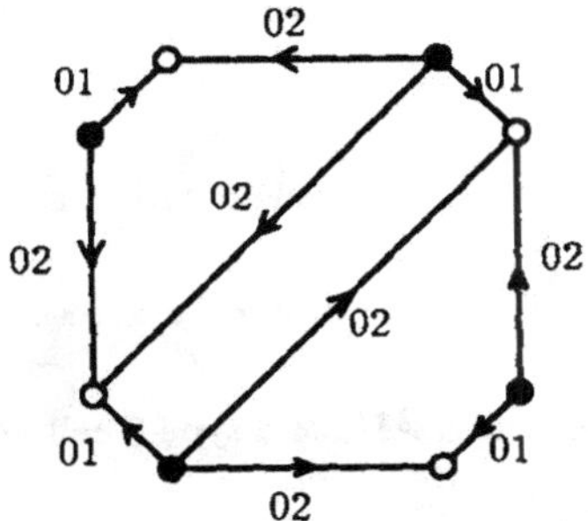

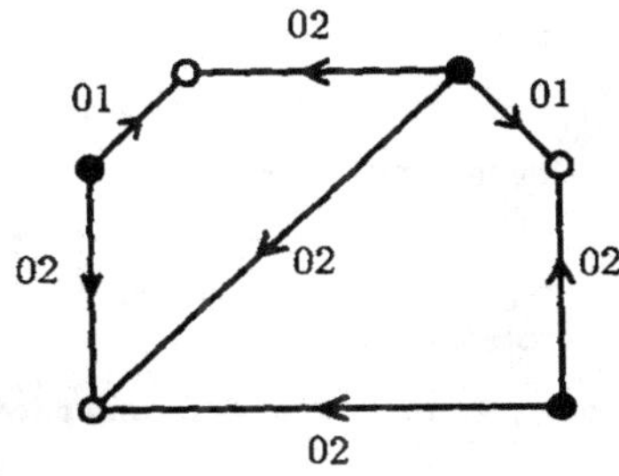

Figure 5

For the case $m = n = 3$, the two local gluing rules just discussed are illustrated in Figure 5. The gluing rules in a 2-ETFT with $d = 2$, enable the invariant associated with M in the theory, to be obtained from those associated with the elemental triangles in $\mathcal{T}$, using gluing operations along the edges and vertices in $\mathcal{T}$. By changing our viewpoint from the consideration of $\mathcal{T}$ to that of $\mathcal{T}^*$, the structures associated with elemental triangles and the gluing operations are seen to be on an equal footing, both being geometrically described by top-dimensional cells of $\mathcal{T}^*$, the gluing operation coming from those cells of types $0a$, $0b$ and 1. To specify the weights for a 2-ETFT with $k = 2$, it thus suffices to give those on the cells of Figures 4(i) and 4(ii), along with the two right hand cells of Figure 3. What we will call a *polyhedral data* for a 2-ETFT consists of $\mathfrak{D}_1$ to $\mathfrak{D}_3$ below.

- $\mathfrak{D}_1$ Two sets I_+ and I_-, to be considered as the allowed sets of labels on type-012 cells with positive and negative orientations, respectively.
- $\mathfrak{D}_2$ Sets $I_{ij}(a,b)$ for each $(a,b) \in I_+ \times I_-$ and $\{i,j\} \subset \{0,1,2\}$, the allowed labels on type-ij cells whose boundary has already been labelled by a and b.
- $\mathfrak{D}_3$ Weights $w_\lambda(\{a_v\},\{\alpha_e\}) \in K$ for $\lambda \in \{0a, 0b, 1, 2\}$ whenever $\{a_v\}$ is an

allowed vertex labelling and $\{\alpha_e\}$ is a compatible allowed edge labelling of a type-λ cell.

Thus, in $\mathfrak{D}_3$, $a_v \in I_+ \cup I_-$ for each vertex v of a type-λ cell, while, for each edge e, α_e is an element of a suitable set $I_{ij}(a_{v_1}, a_{v_2})$. In order for the w_{0a} and w_{0b} data to determine, via contraction on internal edges, a well-defined weight on an arbitrary cell of type 0, it is necessary for,

$\mathfrak{T}_1$ w_{0b} to remain unchanged when the vertex and edge labels are changed by performing a rotation through π;

$\mathfrak{T}_2$ the combination in Figure 6(i) to be invariant under rotations through $\pm{}^{2\pi}/3$;

$\mathfrak{T}_3$ the identity in Figure 6(ii) to be satisfied.

In order for w_1 and w_2 to determine well-defined weights on cells of type 1 and type 2, it is necessary for,

$\mathfrak{T}_4$ w_1 to remain unchanged when all labels are permuted as a result of a rotation through π;

$\mathfrak{T}_5$ w_2 to be unchanged under rotation through $\pm{}^{2\pi}/3$ of a type-2 cell and its labels

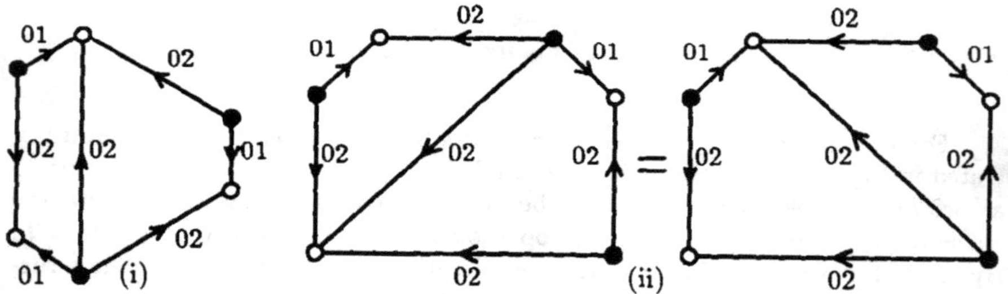

Figure 6

The purely combinatorial data above may be alternatively expressed in terms of vector spaces and 2-vector spaces in the following way. Let n_+ and n_- be $|I_+|$ and $|I_-|$, respectively. Consider two 2-vector spaces, V_+ and V_- of dimensions n_+ and n_-, respectively. The data $\mathfrak{D}_2$ may be viewed as three pairings,

$$V_+ \otimes V_- \longrightarrow \mathbb{1}_2$$

under which the basis vector $e_a \otimes f_b$ maps to the element of $\mathbb{1}_2$, given by the vector space whose basis is labelled by $I_{ij}(a, b)$. Denote these vector spaces $V_{ij}(a, b)$. The weights in $\mathfrak{D}_3$ may now be considered as maps between the vector spaces V_{ij}. Indeed, the relative orientations of the boundary edges on a cell may be used to determine which vector spaces appear in the tensor product giving the domain, and which appear in the image space. For example, the weights associated with the two right hand cells of Figure 3 may be viewed as maps,

$$V_{12}(a',b) \otimes V_{12}(a,b') \longrightarrow V_{01}(a,b) \otimes V_{01}(a',b')$$
$$V_{02}(b,a') \otimes V_{02}(c,b') \otimes V_{02}(a,c') \longrightarrow V_{12}(a,a') \otimes V_{12}(b,b') \otimes V_{12}(c,c')$$

or equivalently as tensors with each index appropriately raised or lowered. The vertex labels may appear an arbitrary number of times in any allowed composition of weights. However, edge labels, those coming from bases for $V_{ij}(a,b)$, may only appear twice, once as upper indices and once as lower indices.

Starting from a polyhedral data, one can attempt to construct a 2-ETFT, that is, to define the appropriate operations Z and $\overline{Z}$ of §3. To do this, first note that, to any triangulated M^2 with boundary, there is associated a contracted product of weights which gives rise to an element of the vector space,

$$\Big(\bigotimes \check{V}_{12}(a_.,a_.)\Big) \otimes \Big(\bigotimes V_{02}(a_.,a_.)\Big) \tag{4.1}$$

for each assignment of labels a to vertices of $\partial(\mathcal{T}_*) \simeq (\partial\mathcal{T})^*$; here $\check{}$ denotes the dual. The tensor products above are over all edges of types 12 or 02 in $\partial(\mathcal{T}_*)$, or equivalently of types 1 or 0 in $(\partial\mathcal{T})^*$. The direct sum of the vector spaces (4.1) over all vertex labels $a_.$ is the vector space $\overline{Z}(\partial M,\partial\mathcal{T})$ to be associated with ∂M, while the direct sum of the vectors in (4.1) associated with M, defines a vector $Z(M,\mathcal{T}) \in \overline{Z}(\partial M,\partial\mathcal{T})$. Indeed, given any closed 1-dimensional triangulated manifold, $(\Sigma,\mathcal{U})$, the construction above provides a family of vector spaces, depending upon the local vertex labels, and this structure can be contracted to give the two important parts,

(i) the direct sum of these vector spaces, $\overline{Z}(\Sigma,\mathcal{U})$;

(ii) a pairing $\overline{Z}(\Sigma,\mathcal{U}) \otimes \overline{Z}(\Sigma^*,\mathcal{U}) \longrightarrow K$, via the contraction of a product of tensors with types $0b$ and 1 along common type-01 edges.

To a not necessarily closed manifold Σ^1 similar constructions associate a family of vector spaces dependent upon the vertex labels on the boundary points. This family of vector spaces defines an element $Z(\Sigma,\mathcal{U})$ of the 2-vector space denoted $\overline{Z}(\partial\Sigma,\partial\mathcal{U})$, obtained as a tensor product of V^+ and V^- spaces, one for each point in $\partial\Sigma$. The gluing operation between the families associated with two such 1-dimensional manifolds sharing a common boundary point, is given by the vector spaces associated with type-02 edges. Our conclusions are summarised below.

Proposition 1 *Given a polyhedral data for 2-ETFT, an associated 2-ETFT on 2-dimensional, oriented, triangulated manifolds may be constructed, so long as the data satisfies $\mathfrak{T}_1$–$\mathfrak{T}_5$ above.*

However, in order that the theory extends to one on all oriented 2-dimensional manifolds, without chosen triangulations, it is necessary that the data satisfy, additionally, the relations given by Figures 7 and 8.

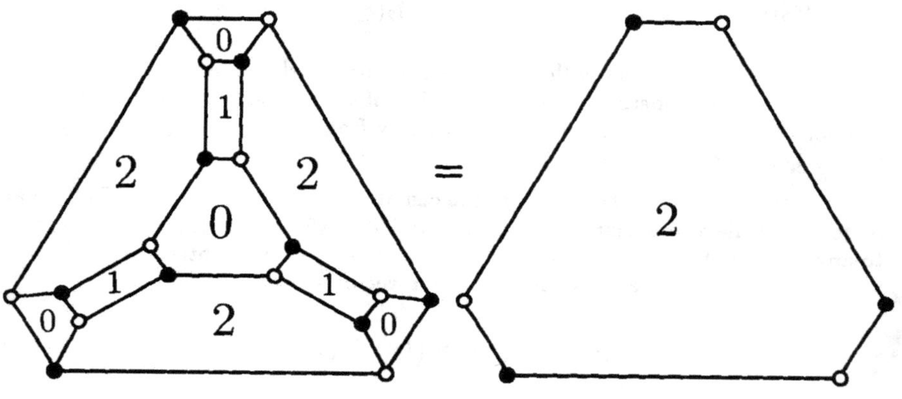

Figure 7

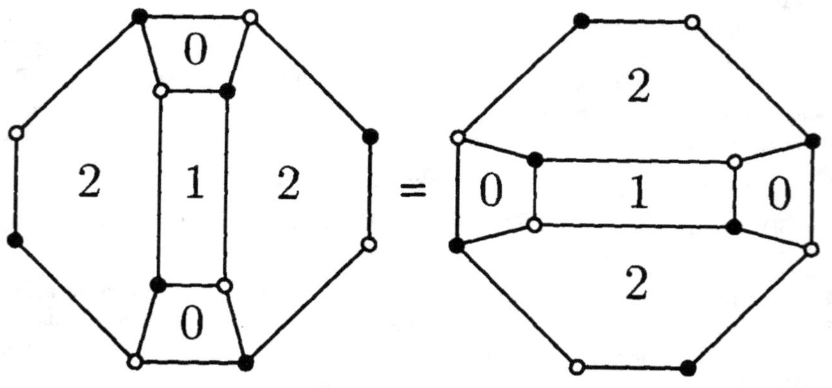

Figure 8

To see this, note first that any two triangulations of a closed manifold, M^2, may be obtained from each other by a sequence of moves under each of which only a local piece of the triangulation is affected, via a change as illustrated in either of the two diagrams of Figure 9. Two such triangulations, $\mathcal{T}_1$ and $\mathcal{T}_2$, representing the same manifold, will give rise to identical invariants, $Z(M,\mathcal{T}_1)$ and $Z(M,\mathcal{T}_2)$, since the local moves on triangulations in Figure 9 give rise to the local changes in the blow-up illustrated in Figures 7 and 8. For manifolds with boundary, two different triangulations matching on ∂M may be transformed into each other by the same types of move, so that $Z(M,\mathcal{T})$ is a vector in $\overline{Z}(\partial M,\partial\mathcal{T})$, independent of the choice of triangulation, $\mathcal{T}$ of M, once the restricted triangulation on ∂M is fixed. Different subdivisions $\partial\mathcal{T}$ of ∂M may give rise to different vector spaces.

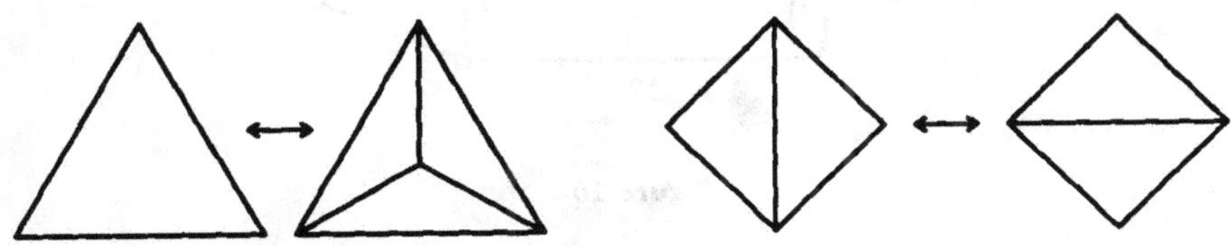

Figure 9

Suppose that $\mathcal{U}_1$ and $\mathcal{U}_2$ are two triangulations of a manifold Σ^1. These give rise in general to different families of vector spaces $[Z(\Sigma,\mathcal{U}_i)]^a$, $i=1,2$, indexed by an allowed labelling a of $\partial\Sigma$. However, a change between two such subdivisions may be accomplished via moves in which an interval is subdivided into two, or conversely, an internal vertex is removed. These moves on triangulations $\mathcal{U}$ of Σ translate into moves on $\mathcal{U}_*$, and for each such move there is a natural transformation between the corresponding vector spaces. This is given by contractions of the tensors in Figures 3 and 4. Figure 10 is such an example in the case of the addition of an internal vertex. By this procedure, for each sequence, μ, of local moves bringing $\mathcal{U}_1$ to $\mathcal{U}_2$, a map $\theta^a_\mu(\Sigma;\mathcal{U}_1,\mathcal{U}_2)\colon \big(Z(\Sigma,\mathcal{U}_1)\big)^a \longrightarrow \big(Z(\Sigma,\mathcal{U}_2)\big)^a$ is defined. It turns out that the identity in Figure 7 along with $\mathfrak{T}_1$–$\mathfrak{T}_5$ ensure that θ^a_μ is independent of μ. Now we define $Z(\Sigma)^a$ to be the inverse limit of $\big(Z(\Sigma,\mathcal{U})\big)^a$ over triangulations $\mathcal{U}$ of Σ.

Proposition 2 *Given a polyhedral data for 2-ETFT satisfying conditions $\mathfrak{T}_1$–$\mathfrak{T}_5$ above, along with the equalities of tensor contractions depicted in Figures 7 and 8, an associated 2-ETFT on 2-dimensional oriented manifolds may be constructed.*

5. Examples

In this section some examples of ETFT's will be given.

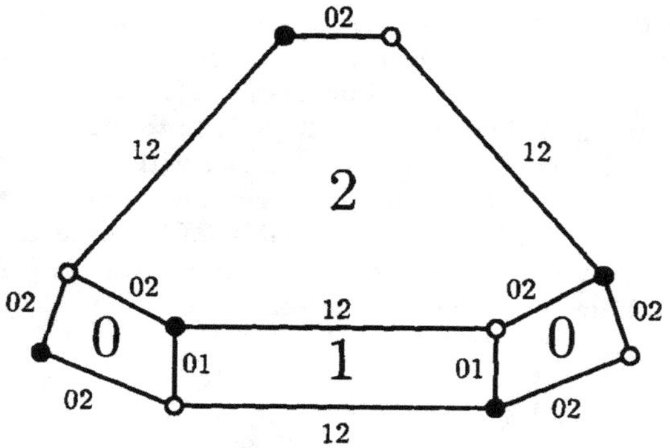

Figure 10

Example 1 Consider a polyhedral data for a 2-ETFT, that is $\mathfrak{D}_1$, $\mathfrak{D}_2$ and $\mathfrak{D}_3$. Suppose that in $\mathfrak{D}_1$, $|I_+| = |I_-| = 1$, so that $\mathfrak{D}_2$ supplies three sets, I_{01}, I_{02} and I_{12}. Suppose that $|I_{01}| = |I_{02}| = 1$. Then $\mathfrak{D}_3$ gives rise to weights,

$$w_{0a} \text{ and } w_{0b} \in K$$
$$w_1 : I_{12} \times I_{12} \longrightarrow K$$
$$w_2 : I_{12} \times I_{12} \times I_{12} \longrightarrow K$$

Conditions $\mathfrak{T}_1$, $\mathfrak{T}_2$ and $\mathfrak{T}_3$ are automatically satisfied, while $\mathfrak{T}_4$ and $\mathfrak{T}_5$ require w_1 to be symmetric and w_2 to be invariant under cyclic permutation of the three indices. Put $I = I_{12}$. The identities depicted in Figures 7 and 8 give rise to,

$$\sum_{\lambda,\mu} w_2(a_1,a_2,\lambda)w_2(\mu,a_3,a_4)w_1(\lambda,\mu) = \sum_{\lambda,\mu} w_2(a_4,a_1,\lambda)w_2(\mu,a_2,a_3)w_1(\lambda,\mu) \tag{5.1}$$

$$w_{0a}^{-4} w_{0b}^{-1} w_2(a_1,a_2,a_3) = \sum_{\lambda_i,\mu_i} w_2(a_1,\nu_1,\mu_2)w_2(a_2,\lambda_1,\nu_2)w_2(a_3,\mu_1,\lambda_2) \\ w_1(\lambda_1,\lambda_2)w_1(\mu_1,\mu_2)w_1(\nu_1,\nu_2) \tag{5.2}$$

for all $a_i \in I$. Whenever w_1 and w_2 satisfy (5.1) and (5.2), a 2-dimensional ETFT is obtained.

A specific solution of (5.1) and (5.2) is obtained from any finite group G, by setting $I = G$ and

$$w_1(g,h) = c_1\delta(gh = 1)$$
$$w_2(g,h,k) = c_2\delta(ghk = 1)$$

where (5.1) is automatically satisfied and (5.2) gives the constraint,

$$c_1^3 c_2^2 w_{0a}^4 w_{0b} |G| = 1 \,. \tag{5.3}$$

For a closed triangulated 2-dimensional manifold $(M, \mathcal{T})$, the associated invariant is,

$$Z(M) = c_1^{n_1} c_2^{n_2} |G|^{n_1 - n_2 + c} \prod_v (w_{0a}^{d_v - 2} w_{0b})$$

where n_i is the number of i-simplices in $\mathcal{T}$, and c is the number of components of M. Here d_v denotes the degree of the vertex v, the product being over all vertices v of $\mathcal{T}$. By (5.3), since $3n_2 = 2n_1$, $\sum(d_v - 2) = 2n_1 - 2n_0$ and $\chi = n_0 - n_1 + n_2$, thus,

$$Z(M) = (w_{0a}^{-2} w_{0b})^{\chi} |G|^c$$

giving rise to the only two global invariants present, namely χ and c.

For an arbitrary solution of (5.1) and (5.2), let V be a vector space with basis indexed by I. Then w_1 gives a symmetric pairing $V \otimes V \to K$, and using it to identify V^* with V, it is seen that w_2 gives maps $m: V \otimes V \to V$ and $\Delta: V \to V \otimes V$. By (5.1), m is associative and Δ is coassociative. It is therefore not surprising to see that solutions of (5.1) and (5.2) are indexed by algebraic structures. The general solution is indexed by ambialgebras, c.f. [Q].

Example 2 The detailed analysis of §4 can be similarly carried out for 3-ETFT's. A polyhedral data for 3-ETFT will associate a class of allowed labellings to 0-, 1- and 2-dimensional cells in a blow-up $\mathcal{T}_*$ of an arbitrary 3-dimensional triangulation, while allowing weights to be associated to already labeled 3-dimensional cells. Just as in §4, there are an infinite number of cells of types 0 and 1, but they can be decomposed into a finite number of elemental cells. For the purposes of defining a specific 3-ETFT, it suffices to define all the allowed cell labellings and weights on *all* the possible cell types, rather than just for the elemental ones.

There are two forms of type-0123 cell, namely, points with either of two orientations. Associate to both cells the 3-vector space $\mathbb{1}_3$, that is, only one allowed label exists on points. The 1-dimensional cells are of type-012, -013, -023 and -123, to which there are associated sets giving the allowed labels, or equivalently 2-vector spaces. Suppose that the first three sets are of order 1, the last being I. The 2-dimensional cells come in 6 types, there being only one geometric form to each type, except type-01. Suppose that there is only one allowed label on all type-01, -02 and -03 cells. Two elements of I suffice to label the vertices and edges of either type-12 or type-13 cells, three being necessary for type-23 cells. The allowed sets of labels on these cells will be denoted by X_{ij}, Y_{ij} and A_{ijk}, respectively, $(i, j, k \in I)$. Both this and the next example use the data as given so far.

For a finite group G use $I = G$, while X_{ij}, Y_{ij} and A_{ijk} are sets of order $\delta(ij = 1)$, $\delta(ij = 1)$ and $\delta(ijk = 1)$, respectively. The polyhedral data is completed by specifying weights on all type-0, -1, -2 and -3 cells. In the case when these weights are all independent of the allowed labelling of the vertices, edges and faces, the invariant obtained of a closed 3-manifold, M, depends only on the number of components, the Euler characteristic of M, and $|\operatorname{Hom}(\pi_1(M), G)|$. A more general system of weights can be defined in terms of a choice of 3-cocycle, and the resulting 3-ETFT gives rise to TFT's of the form investigated in [FQ].

Example 3 This example differs from the last, in that here I is chosen to index a set of irreducible representations of a quantum group, A. In the case $A = U_q\mathfrak{sl}_2$, at a root of unity $q = \exp(\pi i/r)$, I is chosen to be $\{0, 1/2, \ldots, (r-2)/2\}$, labelling the generators of the semi-simple part of the category of representations. The sets X_{ij} and Y_{ij} are both chosen to be of order 1 precisely when i and j represent dual representations, being of order 0 otherwise. The set A_{ijk} is chosen to specify the multiplicity of the trivial representation in the tensor product of the representations labelled by i, j and k.

The associated weights to be placed on labelled 3-dimensional cells of types 0,1,2 and 3 are now all given in terms of structure constants of A. In particular, the geometric form of a type-3 cell is that of a tetrakaidekahedron, which has twenty-four vertices, eight hexagonal and six square faces. A complete labelling of the vertices, edges and faces of such a cell is given by six elements of I, and four multiplicity labels. The weight to be associated with this cell is now a generalised quantum $6j$-symbol. For the case of $U_q\mathfrak{sl}_2$, these symbols were investigated in [KR]. The conditions on the weights, which in the case of a 3-ETFT are equivalent to the constraints of Figures 7 and 8 in the case of a 2-ETFT, are satisfied due to relations amongst the quantum $6j$-symbols, known as the orthogonality and the Elliot-Biedenharn relations.

As for any ETFT derived from polyhedral data, the associated (scalar) invariant of closed, top-dimensional manifolds has the form of a sum, over allowed labellings, of a product of local weights. Thus,

$$Z(M) = \sum_{\sigma} \prod_{C \subset T^*} w_C(\sigma|_C), \tag{5.4}$$

where T is a triangulation of M, σ ranges over allowed labellings of the cells and subcells of T^* of codimension at least one, and $w_C(\tau)$ is the weight associated with cell C when labelled according to τ. The product is over all top-dimensional cells in T^*. For top-dimensional manifolds with boundary, the contraction of weights given by the right hand side of (5.4) provides a tensor, indexed by the boundary labelling $\sigma|_{\partial M}$.

The sum (5.4), in the case of this example, is that found in [TV] for $U_q\mathfrak{sl}_2$.

6. Further remarks

In §4 it was seen in detail how the combinatorics of triangulations of manifolds of dimension 2 translates directly into rules for a structure, which was termed a polyhedral data for 2-ETFT, of weights and allowed labellings for the cells in blow-ups of arbitrary such triangulations. From such data, a 2-dimensional ETFT could be constructed, and it was apparent that such data is indexed by algebraic structures of a suitable type. The same combinatorics may be interpreted as giving the basic elements and axioms of a d-dimensional 2-ETFT, for any $d > 1$.

The translation from the structure which we have defined as a polyhedral data for 2-ETFT, to that of a d-dimensional 2-ETFT, is accomplished by introducing an additional dependence into the weights and sets of allowed labels on different cell types, upon a choice of suitable manifolds. Thus for any particular cell, C, in the blow-up of a 2-dimensional triangulation, $\mathcal{T}$, the dependence introduced will be upon $(d-2)$-, $(d-1)$- and d-dimensional manifolds, one manifold being given for each 0-, 1- and 2-simplex in $\mathcal{T}$ related to C. The manifolds will satisfy boundary constraints given by the incidence relations existing amongst the associated simplices in $\mathcal{T}$. Those structures attached to cells C, whose type is other than 2, 12 or 012, are traditionally thought of as gluing rules.

In an arbitrary dimension d, s-ETFT's can similarly be defined along with polyhedral data for s-ETFT's. Just as in the case of $s = 2$, the number of cell types in a blow-up of an s-dimensional triangulation is infinite, but they can be decomposed into a finite number of elemental cells. Thus, a polyhedral data may be given by specifying a (finite) set of allowed labellings and a finite number of weights (tensors). The analogue of Proposition 1 will then hold. The analogue of Proposition 2 relies on there being a finite simple set of generators for moves on (singular) s-dimensional triangulations of a manifold ([M], see also [P]). Hence there is a family of s-ETFT's coming from polyhedral data, indexed by finite collections of tensors constrained by a finite number of polynomial relations, arising from invariance under local moves on triangulations. Since for $s = 2$, a polyhedral data is equivalent to an ambialgebra, it is not unreasonable to also view such a polyhedral data for $s > 2$, as an *algebraic structure.* These structures will not be algebras in the usual sense, and will possess many, not necessarily binary, operations, rather than just a multiplication; similar structures were studied in [L 2] and [L 3]. The reader is referred to [L 1] for more details.

It may be remarked that the polynomial relations to be satisfied by a polyhedral data are not unlike a generalisation of the Moore-Seiberg polynomial relations [MoSe] of rational CFT. The form of the invariant for s-dimensional manifolds will always be that of (5.4). It may be thought of as a discretised version of the functional integrals arising in other approaches, see [W] and [F].

References

[A] M.F. Atiyah, 'Topological quantum field theories', *Publ. Math. Inst. Hautes Etudes Sci., Paris* **68** (1989) p.175–186.

[F] D.S. Freed, 'Higher Algebraic Structures and Quantization', *Preprint* (December 1992).

[FQ] D.S. Freed, F. Quinn, 'Chern-Simons theory with finite gauge group', *Commun. Math. Phys.* (to appear).

[KV] M.M. Kapranov, V.A. Voevodsky, '2-categories and Zamolodchikov's tetrahedra equations', *Preprint* (1992).

[KR] A.N. Kirillov, N.Yu. Reshetikhin, 'Representations of the algebra $U_q(\mathfrak{sl}(2))$, q-orthogonal polynomials and invariants of links', *Infinite dimensional Lie algebras and groups* World Scientific (1988) p.285–342.

[L 1] R.J. Lawrence, 'Extended topological field theories from an algebraic perspective', *In preparation*.

[L 2] R.J. Lawrence, 'Algebras and triangle relations', *Harvard preprint* (1991).

[L 3] R.J. Lawrence, 'On algebras and triangle relations', *Proc. 2nd. Int. Conf. on Topological and Geometric Methods in Field Theory* World Scientific (1992) p.429–447.

[MaSc] Yu.I. Manin, V.V. Schechtman, 'Arrangements of Hyperplanes, Higher Braid Groups and Higher Bruhat Orders', *Adv. Studies in Pure Maths.* **17** (1989) p.289–308.

[M] S.V. Matveev, 'Transformations of special spines and the Zeeman conjecture', *Math. USSR Izvestia* **31** (1988) p.423–434.

[MoSe] G. Moore, N. Seiberg, 'Lectures on RCFT', *Physics, Geometry and Topology* Plenum Press (1990) p.263–361.

[P] U. Pachner, 'Konstruktion methoden und das kombinatorische Homömorphiseproblem für Triangulationen kompakter semilinearen Mannig faltigkeiten', *Abh. Math. Sem. Univ. Hamburg* **57** (1987) p.69–86.

[Q] F. Quinn, 'Lectures on axiomatic topological field theory', *preprint* (August 1992).

[TV] V.G. Turaev, O.Y. Viro, 'State sum invariant of 3-manifolds and quantum 6j-symbols', *Topology* **31** (1992) p.865–902.

[W] E. Witten, 'Quantum field theory and the Jones polynomial', *Commun. Math. Phys.* **121** (1989) p.351–399.

A GEOMETRIC CONSTRUCTION OF THE FIRST PONTRYAGIN CLASS

JEAN-LUC BRYLINSKI
Math. Dept., The Pennsylvania State University
University Park, PA. 16802, USA
and
DENNIS McLAUGHLIN
Math. Dept., Princeton University
Princeton, N. J. 08544, USA

A new obstruction theory for principal bundles is developed, which leads to an integer-valued formula for the first Pontryagin class of a bundle with compact structure group. A geometric representative of this class is given, in terms of a glueing problem for gerbes. The upshot is that there is a natural sheaf of bicategories over the base manifold. Analogous constructions are discussed for finite groups, leading to a proof of the reciprocity theorem of Segal and Witten.

Introduction

In the past few years, there has been a renewal of interest in degree 4 characteristic classes. The inspiration has come mainly from physics.

For a compact simple Lie group G, the inverse transgression in the universal bundle $H^4(BG) \to H^3(G)$ gives the natural correspondence between 3-dimensional Chern-Simons gauge theory and 2-dimensional Wess-Zumino-Witten theories [13].

If G is connected, there are related results concerning the transgression for the free loop space $H^4(BG) \to H^3(LBG)$. The image of this map corresponds to those central extensions of the loop group LG which have the reciprocity property [24].

Perhaps the most relevant development from our point of view has been the discovery that the first Pontryagin class p_1 is the obstruction to defining the spinor bundle on loop space [20] [22] [26]. This means that p_1 plays a role in string theory analogous to that played by the second Stiefel-Whitney class w_2 in point particle physics.

In the case of the tangent bundle of a smooth manifold, there is a well-known formula for an explicit cocycle representing w_2 [9], as well as formulas for p_1 [15] [23]. In this paper, for any principal bundle over any space, we construct geometrically a degree 4 integral Čech cocycle representing the first Pontryagin class. The basic data in this "formula for p_1" are the transition cocycles themselves and tetrahedra in the group which have them as vertices. Our approach to p_1 may be viewed as a generalization of the theory of line bundles with connections, due to A. Weil [25] and Kostant [21].

One proof of the formula involves the lifting of the **Z**-valued Čech cocycle to a cocycle with values in a smooth version of the Deligne complex of sheaves, using a connection on the principal bundle. However, this is not the way the formula was found. We were thinking about the geometrical meaning of p_1 in connection with the theory of gerbes and gauge theory, and we found a canonical geometric object corresponding to p_1, which recasts classical obstruction theory in the language of categories. We then realized that, using the holonomy of a gerbe around the boundary of a tetrahedron, we could write down an explicit formula for a representative of p_1. The geometric construction appears in fact to have a deeper meaning, as the origin of the Chern-Simons topological quantum field theory of Witten. For compact Lie groups, this is still a conjecture, presented in [8], but for a finite group G and a class in $H^3(G, \mathbf{C}^*)$, an analog of the conjecture can be established, as is explained in §3.

1. Statement of the main result

Let G be a connected compact almost simple Lie group, with $\pi_1(G) = \mathbf{Z}/N \cdot \mathbf{Z}$, a finite number. Then $H_2(G, \mathbf{Z}) = 0$ and $H_3(G, \mathbf{Z}) = \mathbf{Z}$. Suppose that $P \to M$ is a principal G-bundle over M. We define $p_1 \in H^4(M, \mathbf{Z})$ to be the class obtained by transgression of N times the generator of $H^3(G, \mathbf{Z})$.

Choose an open covering of M by contractible open sets U_i, indexed by the set I, such that all non-empty intersections of these open sets are contractible. Choose sections $s_i : U_i \to P$, and let g_{ij} be the associated transition cocycles. So g_{ij} is the continuous function from $U_i \cap U_j$ to G, characterized by $s_j = s_i \cdot g_{ij}$.

For $y \in U_i \cap U_j$, choose a path $\gamma_{ij}(y, t)$ in G, from the identity to $g_{ij}(y)$. We require that γ_{ij} be a continuous function of $(y, t) \in (U_i \cap U_j) \times [0, 1]$, and a smooth function of t. For $y \in U_i \cap U_j \cap U_k$, denote by $\gamma_{ijk}(y)$ the loop given by the composition $\gamma_{ij} * g_{ij}\gamma_{jk} * \gamma_{ik}^{-1}$ (Figure 1).

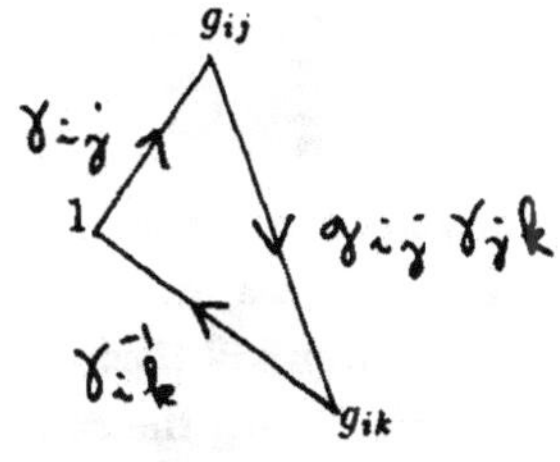

Figure 1

The loop γ_{ijk} may not be a boundary, but $N\cdot\gamma_{ijk}$ will bound a 2-simplex σ_{ijk}. Again one chooses $\sigma_{ijk}(y,x)$ to be a continuous function $(U_i\cap U_j\cap U_k)\times\Delta_2\to G$, and a smooth function on the 2-simplex Δ_2. For $y\in U_i\cap U_j\cap U_k\cap U_l$, the linear combination $g_{ij}\sigma_{jkl}-\sigma_{ikl}+\sigma_{ijl}-\sigma_{ijk}$ is a singular cycle. We think of this as a singular 2-cycle on $Map(U_i\cap U_j\cap U_k\cap U_l,G)$; it is then the boundary of a 3-chain $T_{ijkl}(y)$, which we symbolically draw as a tetrahedron in Figure 2 (note that if $N=1$, we may choose T_{ijkl} to consist of just one 3-simplex).

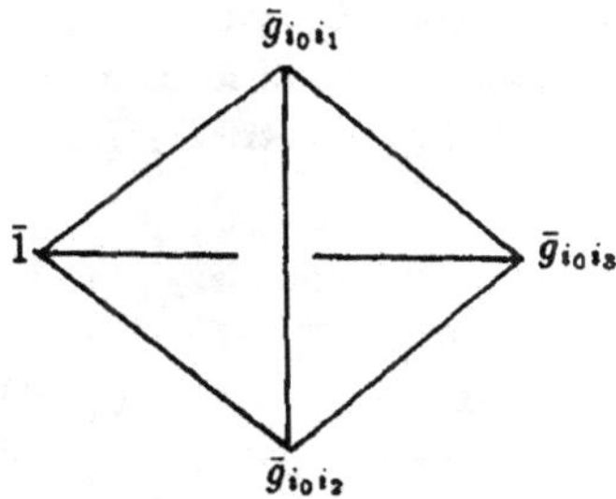

Figure 2

Finally, for $y\in U_i\cap U_j\cap U_k\cap U_l$, let V_{ijklm} be the 3-cycle

$$T_{ijkl}-T_{ijkm}+T_{ijlm}-T_{iklm}+g_{ij}T_{jklm}.$$

Theorem 1. *Let ν be the closed, bi-invariant integral 3-form on G whose cohomology class generates $H^3(G,\mathbf{Z})$. Define*

$$\beta_{ijklm}(y)=\int_{V_{ijklm}(y)}\nu.$$

Then β is a $\mathbf{Z}$-valued Čech cocycle of degree 4 which represents $-p_1$.

Remarks.

1. *The formula is combinatorial in nature, since if M is the underlying topological space of a simplicial complex, we may take a covering by the stars of the vertices. A Čech cocycle for this cover will then give a simplicial cocycle. It would be interesting to relate our approach with the combinatorial formula for p_1 of the tangent bundle of a smooth manifold, due to Gabrielov, I. Gel'fand, Losik, and to MacPherson [15] [23]. The formula for p_1 may be generalized to all Chern classes and Pontryagin classes. We also point out other formulas for characteristic classes, due to Gel'fand and MacPherson [17] and to Goncharov [19].*

2. *Volumes of tetrahedra also appear in the work of Cheeger and Simons [10] on differential characters. Their tetrahedra are geodesic tetrahedra in the fibers of a sphere bundle associated to a flat vector bundle.*

2. Holonomy of gerbes and the proof of the formula

There is available a direct proof of the formula [6], which however does not shed light on its geometric significance. We will focus here on obstruction theory for a principal bundle. First we recall the geometrical meaning of low degree cohomology with $\mathbf{Z}$ coefficients.

For X a paracompact space, the boundary map in the exponential exact sequence induces an isomorphism: $H^p(X, \underline{\mathbf{T}}) \xrightarrow{\sim} H^{p+1}(X, \mathbf{Z})$. It is well-known that, for X a smooth manifold, the sheaf cohomology group $H^1(X, \underline{\mathbf{T}})$ classifies line bundles over X. There is a similar interpretation of $H^2(X, \underline{\mathbf{T}})$ as equivalence classes of gerbes bound by the sheaf $\underline{\mathbf{T}}$ [18] [4] [5]. Such a gerbe $\mathcal{C}$ is a sheaf of categories over X, in the sense that for U open in X, there is given a category $\mathcal{C}_U$, and for $V \subset U$, there is a "restriction functor" $\mathcal{C}_U \to \mathcal{C}_V$. This functor will be denoted by $P \mapsto P_{/V}$. Each category $\mathcal{C}_U$ is to be a groupoid, which means that every morphism is invertible. It is assumed that two objects P_1 and P_2 of $\mathcal{C}_U$ are locally isomorphic, in the sense that any $x \in U$ has an open neighborhood V such that the restrictions of P_1 and P_2 to V are isomorphic. Also X is covered by open sets over which the category is not empty. However, it may (and will often) happen that the global category $\mathcal{C}_X$ is empty.

A glueing axiom must be satisfied, which allows one to obtain an object of $\mathcal{C}_X$ from an open cover (U_i), objects P_i of $\mathcal{C}_{U_i}$, and isomorphisms $(P_i)_{/U_i \cap U_j} \xrightarrow{\sim} (P_j)_{/U_i \cap U_j}$

which satisfy the natural cocycle condition. With all these conditions, one has a gerbe over X. We say that the gerbe C is bound by $\underline{\mathbf{T}}$ if the sheaf of automorphisms of any local object is isomorphic to $\underline{\mathbf{T}}$. A typical example is: C_U is the category of hermitian line bundles over U, a morphism $L_1 \to L_2$ is an isomorphism of line bundles, the restriction functors are the obvious ones. This example is indeed typical in the sense that any gerbe bound by $\underline{\mathbf{T}}$ is locally equivalent to this one.

We now turn to the question of differentiable structures. For line bundles, there is the Weil-Kostant theory of line bundles with connection and their curvature [25] [21], and the relation with so-called "smooth Deligne cohomology", due to Deligne (see [12] for the holomorphic case). We recall this briefly. The smooth Deligne complex of sheaves $\mathbf{Z}(p)^{\infty}_{\mathcal{D}}$ may be described as the complex of sheaves

$$\underline{\mathbf{T}} \xrightarrow{d \log} \sqrt{-1} \cdot \underline{A}^1_X \to \cdots \to \sqrt{-1} \cdot \underline{A}^{p-1}_X,$$

where $\underline{\mathbf{T}}$ is placed in degree 1. Given a hermitian line bundle with hermitian connection (L, ∇), one derives a class in the hypercohomology group $H^2(X, \mathbf{Z}(2)^{\infty}_{\mathcal{D}})$, which is the total cohomology of the double complex of Čech cochains with coefficients in $\mathbf{Z}(2)^{\infty}_{\mathcal{D}}$. Let (U_i) be a nice open covering of X, and let s_i be a non-vanishing section of L over U_i. Set $g_{ij} = \frac{s_i}{s_j}$ and $\alpha_i = \frac{\nabla(s_i)}{s_i}$. Then $(g_{ij}, -\alpha_i)$ is a Čech cocycle with coefficients in $\mathbf{Z}(2)^{\infty}_{\mathcal{D}}$. In this way, Deligne identifies the group of isomorphism classes of pairs (L, ∇) with $H^2(X, \mathbf{Z}(2)^{\infty}_{\mathcal{D}})$.

As regards gerbes, two levels of differentiable structures on gerbes bound by $\underline{\mathbf{T}}$ were introduced in [4], where they were baptized "connective structure" and "curving". A "connective structure" associates to each object P of C_U a sheaf $Co\ (P)$ of "connections" on P, which is a principal homogeneous space under $\sqrt{-1} \cdot \underline{A}^1_X$. A "curving" associates to each connection $\nabla \in Co\ (P)$ its curvature $K(\nabla)$, which is an honest purely imaginary 2-form on X. The conditions satisfied by these notions are explained in [4]. Note in particular that $K(\nabla + \alpha) = K(\nabla) + d\alpha$, for $\alpha \in \sqrt{-1} \cdot \underline{A}^1_X$ a 1-form on X.

The equivalence classes of gerbes on X with connective structure and curving are classified by the smooth Deligne cohomology group $H^2(X, \underline{\mathbf{T}} \xrightarrow{d \log} \sqrt{-1} \cdot \underline{A}^1_X \to \sqrt{-1} \cdot \underline{A}^2_X)$, i.e. the group $H^3(X, \mathbf{Z}(3)^{\infty}_{\mathcal{D}})$. The 3-curvature Ω of such an object is the 3-form which is equal to $dK(\nabla)$, for ∇ a connection on an object of C defined locally. It is deduced from a morphism of complexes of sheaves from $\mathbf{Z}(3)^{\infty}_{\mathcal{D}}$ to $\sqrt{-1} \cdot \underline{A}^3_X$, put in degree 3.

Now we turn to a compact Lie group G, which we assume to be simply-connected. A concrete example of a gerbe on G can be found from the path-loop fibration $PG \to G$ (see also [5] [7]). This is the universal ΩG-bundle. Locally on G, it is possible to lift the structure group to $\widetilde{\Omega G}$, an extension of ΩG by the circle $\mathbf{T}$. These local liftings are the objects of a category, in which the morphisms are constrained to induce the identity on PG. This sheaf of categories is a gerbe bound by the sheaf $\underline{\mathbf{T}}$. The corresponding obstruction in $H^2(G, \underline{\mathbf{T}}) \xrightarrow{\sim} H^3(G, \mathbf{Z})$ is the obstruction to finding the global lift. For G simply-connected, this obstruction class is a generator of $H^3(G, \mathbf{Z})$.

As G is 2-connected, one easily checks that $H^3(G, \mathbf{Z}(3)^{\infty}_{\mathcal{D}})$ is isomorphic to the group $\Lambda^3(G)$ of closed 3-forms on G with integral periods. Now consider a G-bundle $P \to M$ as in §2. To prove Theorem 1, we take the gerbe Q on G with 3-curvature $\Omega := 2\pi \cdot \sqrt{-1} \cdot \nu$, and we try to glue it all over M. To analyze the glueing problem, one may look at gerbes on P, bound by $\underline{\mathbf{T}}$, which are equipped with a connective structure "along the fibers of π", and a relative curving which assigns a relative 2-form to each connection. Such gerbes are classified by the cohomology group $H^2(P, \underline{\mathbf{T}} \xrightarrow{d\log} \sqrt{-1} \cdot \underline{A}^1_{P/M} \to \sqrt{-1} \cdot \underline{A}^2_{P/M})$, where $\underline{A}^j_{P/M}$ is the sheaf of germs of real relative j-forms.

One has an exact sequence of complexes of sheaves on P:

$$0 \to \pi^{-1}\underline{\mathbf{T}}_M \to \left(\underline{\mathbf{T}} \to \sqrt{-1} \cdot \underline{A}^1_{P/M} \to \sqrt{-1} \cdot \underline{A}^2_{P/M}\right) \to \sqrt{-1} \cdot \underline{A}^3_{P/M,cl}[-2] \to 0,$$

where $\underline{A}^3_{P/M,cl}$ denotes the sheaf of germs of closed relative 3-forms, and for $K^\bullet$ a complex, $K^\bullet[j]$ denotes the complex obtained by translating $K^\bullet$ by j steps to the left. The obstruction κ to constructing a class in $H^2(P, \underline{\mathbf{T}} \to \sqrt{-1} \cdot \underline{A}^1_{P/M} \to \sqrt{-1} \cdot \underline{A}^2_{P/M})$ mapping to $2\pi\sqrt{-1} \cdot \nu$ belongs to $H^3(M, \underline{\mathbf{T}}) \subset H^3(P, \pi^{-1}\, \underline{\mathbf{T}}_M)$, and it maps to p_1 under the exponential isomorphism $H^3(M, \underline{\mathbf{T}}) \xrightarrow{\sim} H^4(M, \mathbf{Z})$.

We can explicitly write down a cocycle representing κ. Over the open set U_i, we have: $\pi^{-1}(U_i) \xrightarrow{\sim} U_i \times G$, so from Q one obtains by pull-back a gerbe C_i on $U_i \times G$, with relative connective structure and curving. Over $U_i \cap U_j$, we may find an equivalence between the restrictions of the gerbes C_i and C_j. In fact, such an equivalence is naturally obtained from the choice of a path $\gamma_{ij}(y)$ from 1 to $g_{ij}(y)$, which depends smoothly on $y \in U_i \cap U_j$, as this path gives a path in the group of diffeomorphisms of $(U_i \cap U_j) \times G$, from Id to the diffeomorphism $(y, g) \mapsto (y, g_{ij}(y) \cdot g)$, and one has a notion of "parallel transport" for a gerbe Q along such a path of diffeomorphisms (see [5, Chapter 6]).

Over $U_i \cap U_j \cap U_k$, the composition of three equivalences gives rise to an equivalence of C_i with itself, hence to a hermitian line bundle $\mathcal{L}_{ijk}$. The choice of the two-simplex σ_{ijk} with boundary γ_{ijk} produces a section $u_{ijk}(y)$ of norm 1 of this line bundle. One verifies from general principles that over $U_i \cap U_j \cap U_k \cap U_l$, the line bundle $\mathcal{L}_{jkl} \otimes \mathcal{L}_{ikl}^{\otimes -1} \otimes \mathcal{L}_{ijl} \otimes \mathcal{L}_{ijk}^{\otimes -1}$ is canonically trivialized. Hence $\kappa_{ijkl} := \frac{u_{jkl} \cdot u_{ijl}}{u_{ikl} \cdot u_{ijk}}$ is a smooth function from $U_i \cap U_j \cap U_k \cap U_l$ to $\mathbf{T}$, which gives a $\underline{\mathbf{T}}$-valued Čech cocycle.

To compute this function κ_{ijkl}, one interprets it as the inverse of the holonomy of the gerbe Q on G around the boundary of the tetrahedron T_{ijkl}. This holonomy $H(\Sigma)$ is defined for any closed oriented surface Σ mapping to G; it is computed from an object P of the restriction of Q to Σ, with connection ∇ and curvature K. Then we have $H(\Sigma) = exp(-\iint_\Sigma K) \in \mathbf{T}$; this is easily seen to be independent of the choice of (P, ∇).

In case $\Sigma = \partial T$, an application of Stokes' theorem shows that $H(\Sigma)$ equals $exp(-\int_T \Omega)$. Hence $\kappa_{ijkl}(y) = exp(2\pi\sqrt{-1} \cdot \int_{T_{ijkl}(y)} \nu)$, which proves Theorem 1.

3. Conclusion

Recently, Breen [3] has constructed the geometric objects classified by $H^3(X, \underline{\mathbf{T}})$. He calls them 2-gerbes bound by the sheaf $\underline{\mathbf{T}}$. Such an object is a sheaf of bicategories [2], where the 1-arrows between two objects form a gerbe, and the 2-arrows between given 1-arrows form a $\underline{\mathbf{T}}$-torsor. We have actually encountered this structure in the above discussion of obstruction theory for gerbes, and we record it in the following

Theorem 3. *Assume $\pi_1(G) = 0$. Let ν be a closed left-invariant 3-form on G, with integral periods, and $\pi : P \to M$ a principal G-bundle. The bicategory over U, whose objects are gerbes bound by $\underline{\mathbf{T}}$ over $\pi^{-1}(U)$, with relative connective structure and curving such that the 3-curvature is $\Omega = 2\pi\sqrt{-1} \cdot \nu$, gives a 2-gerbe on M, bound by $\underline{\mathbf{T}}$. The cohomology class in $H^3(M, \underline{\mathbf{T}})$ defined by this 2-gerbe is the transgression of Ω in the fibration $P \to M$.*

This 2-gerbe gives a geometric interpretation of the corresponding "holonomy" gerbe on LM and of the reciprocity law it satisfies (see the introduction), which involves constructing an object of a gerbe over the space of mappings $\Sigma \to M$, for Σ a surface with boundary. Details will be forthcoming.

Bicategories are also implicit in the Chern-Simons field theory of Witten [27]. This is a 2+1-dimensional topological quantum field theory associated to a characteristic class $\alpha \in H^4(BG;\mathbf{Z})$, for G a compact Lie group. It has several layers of geometric structure which are completely unexpected from the viewpoint of classical topology. Our observation is that all this structure falls naturally into place if one represents α by a 2-gerbe $\mathcal{B}$. We will illustrate this in the case where G is a finite group [13]. To a finite group G and a class in $H^3(G,\mathbf{C}^*)$, Dijkgraf and Witten associate a TQFT in 2+1 dimensions; this has been studied also by mathematicians [14].

Our first task is to construct the vector space $V(\Sigma)$ associated to a surface Σ without boundary. For a space X, let $\mathcal{M}_X = Map(X;BG)$ denote the "moduli space" of (necessarily flat) G-bundles on X. Beilinson has given a purely cohomological construction of a flat line bundle $\mathcal{L}$ on $\mathcal{M}_\Sigma$. The procedure is to pull back $\alpha \in H^3(BG;\mathbf{C}^*) \cong H^4(BG;\mathbf{Z})$ by the evaluation $ev : \mathcal{M}_\Sigma \times \Sigma \to BG$ and then integrate over the fiber Σ. This produces a class $\int_\Sigma ev^*\alpha \in H^1(\mathcal{M}_\Sigma;\mathbf{C}^*)$ and $\mathcal{L}$ is the corresponding flat line bundle. The vector space $V(\Sigma)$ is defined to be the space of horizontal global sections of $\mathcal{L}$. The transgression $\alpha \mapsto \int_\Sigma ev^*\alpha$ is realized geometrically by mapping $\mathcal{B}$ to the $\mathbf{C}^*$-torsor $p_* ev^*\mathcal{B}$, where $p : \mathcal{M}_\Sigma \times \Sigma \to \mathcal{M}_\Sigma$ is the projection. This $\mathbf{C}^*$-torsor is described in the following statement.

Theorem 4. *Let S be the set of triples (ϕ, A, z), where ϕ is a point of $\mathcal{M}_\Sigma$, A is an object of the restriction of $ev^*\mathcal{B}$ to $p^{-1}(\phi)$ and $z \in \mathbf{C}^*$. Define an equivalence relation on S by setting*

$$(\phi, A_1, z_1) = (\phi, A_2, z_2)$$

if $A_2 \cong A_1 \otimes Q$ for Q a $\mathbf{C}^*$-gerbe on Σ and $z_2 = (\int_\Sigma [Q])z_1$, where $[Q]$ denotes the cohomology class of Q.

Then the quotient of S by this relation is a principal homogeneous space for the action of $\mathbf{C}^*$ defined by $w \cdot (\phi, A, z) = (\phi, A, wz)$. The cohomology class of this $\mathbf{C}^*$-bundle in $H^1(\mathcal{M}_\Sigma;\mathbf{C}^*)$ is exactly $\int_\Sigma ev^*\alpha$.

The assignment $\phi \mapsto [(\phi, A, z)]$ defines a typical section of $\mathcal{L}$. From this point of view it is clear that reversing the orientation of Σ changes $V(\Sigma)$ to its dual.

We now show how a 3-manifold M with boundary Σ determines a vector in $V(\Sigma)$. Set $v_M(\psi) = [(\psi, A, z)]$, where now ψ is a point of $\mathcal{M}_M$ and A is an object of $ev^*\mathcal{B}$ on $\{\phi\} \times M$. Then v_M defines a section of the pullback of $\mathcal{L}$ to $\mathcal{M}_M$. For another choice of object $A' = A \otimes Q$, where Q is a $\mathbf{C}^*$-gerbe on M, we have

$[(\psi, A', z)] = [(\psi, A, (\int_\Sigma [Q])z)]$. But the restriction map $H^2(M;\mathbf{C}^*) \to H^2(\Sigma;\mathbf{C}^*)$ is trivial, so that $\int_\Sigma [Q] = 1$. It follows that v_M is independent of the choice of object A and therefore defined globally on $\mathcal{M}_M$. Note also that it is constant (i.e. horizontal) on each component. Now for any space X with basepoint x_0, a point η of $\mathcal{M}_X$ can be represented as a homomorphism $\tilde{\eta} \in Hom(\pi_1(X, x_0), G)$. We will take the basepoint to lie in Σ and set $v_M(\tilde{\psi}) = v_M(\psi)$, for $\psi \in \mathcal{M}_M$. We now define a section w_M of $\mathcal{L}$. Let ϕ be a point of $\mathcal{M}_\Sigma$, corresponding to a homomorphism $\tilde{\phi} : \pi_1(\Sigma) \to G$. Then we put:

$$w_M(\phi) = \sum_{f \in \{Hom(\pi_1(M), G) : f|_{\pi_1(\Sigma)} = \tilde{\phi}\}} v_M(f)$$

then gives the required global section of $\mathcal{L}$ on $\mathcal{M}_M$.

The next step is to consider two oriented manifolds M_1, M_2 with $\partial M_1 = \Sigma$ and $\partial M_2 = -\Sigma$. Let M be the manifold obtained by glueing M_1 and M_2 along their common boundary. It follows easily from Van Kampen's Theorem that

$$(w_{M_1}, w_{M_2}) = \frac{1}{|G|} \sum_{\tilde{\psi} \in Hom(\pi_1(M), G)} \psi^* \alpha[M],$$

where $(\ ,\ )$ is the pairing between $V(\Sigma)$ and $V(\Sigma)^*$ and $[M]$ is the fundamental class of M. This recovers the invariant $Z(M)$ of Dijkgraaf-Witten [13].

There is another, deeper level of structure associated to α. This reflects the connection between Chern-Simons theory and conformal field theory discovered by Witten [27]. We can see this abstractly using the transgression procedure applied to 1-manifolds rather than surfaces. Consider the case of the circle S^1. As above, let $ev : \mathcal{M}_{S^1} \times S^1 \to BG$ be the evaluation map and $p : \mathcal{M}_{S^1} \times S^1 \to \mathcal{M}_{S^1}$ the projection. Pulling back α and integrating over S^1, we obtain a class $\int_{S^1} ev^*\alpha \in H^2(\mathcal{M}_{S^1};\mathbf{C}^*)$. Geometrically, this corresponds to a $\mathbf{C}^*$-gerbe $p_* ev^* \mathcal{B}$ on $\mathcal{M}_{S^1}$. Denote this gerbe by $\mathcal{C}$. The objects of $\mathcal{C}$ over the point $\phi \in \mathcal{M}_{S^1}$ are the global objects of the restriction of $ev^*\mathcal{B}$ to $p^{-1}(\phi)$. Given any two such objects A_1, A_2, there is an equivalence $A_2 = A_1 \otimes Q$ for some well-defined $\mathbf{C}^*$-gerbe Q on S^1. Then in $\mathcal{C}$ we set $Hom(A_1, A_2) = p_* ev^* Q$; here $p_* ev^* Q$ is a $\mathbf{C}^*$-torsor. This describes the fiber of $\mathcal{C}$ at ϕ.

The gerbe $\mathcal{C}$ has several remarkable properties.

Theorem 5. *Let Σ be an oriented surface whose boundary is a disjoint union of r parametrized circles $C_1, ..., C_r$ and let $b_j : \mathcal{M}_\Sigma \to \mathcal{M}_{C_j}$ be the natural restriction.*

(1) There is a canonical global object A_Σ of the pull back gerbe $\mathcal{C}_\Sigma = \bigotimes_{j=1}^{r} b_j^*\mathcal{C}$.

(2) A_Σ is invariant under $Diff^+(\Sigma)$.

(3) The object A_Σ is natural with respect to the operation of glueing surfaces along boundary circles.

This theorem is one version of the reciprocity law of Segal and Witten [24] [13]. The object A_Σ may be described pointwise as follows. For each map $\phi : \Sigma \to BG$, choose a global object A of $\phi^*\mathcal{B}$. For any other choice A', we have $A' = A \otimes Q$ where Q is a gerbe on Σ. But considered as objects of $\mathcal{C}_\Sigma$, A and A' differ by tensoring with the $\mathbf{C}^*$-torsor $\bigotimes_{j=1}^{r} b_j^*\mathcal{H}(Q)$, where $\mathcal{H}(Q) := p_* ev^* Q$ is the torsor over $\mathcal{M}_{S^1}$ described before Theorem 5. Now this $\mathbf{C}^*$-torsor is canonically trivial as $\sum_{j=1}^{r} C_j$ is homologous to zero in Σ. Therefore, we obtain a canonical global object of $\mathcal{C}_\Sigma$, which proves (1). The rest of the Theorem follows easily.

There is a compatibility between this layer of structure and the vector space V constructed above; Suppose that two surfaces Σ_1, Σ_2 are glued together along their boundary circles giving a new surface Σ without boundary. For $i = 1, 2$, let $r_i : \mathcal{M}_\Sigma \to \mathcal{M}_{\Sigma_i}$ be the natural restriction maps. Then the gerbes $r_1^*\mathcal{C}_{\Sigma_1} \otimes r_2^*\mathcal{C}_{\Sigma_2}$ and $\mathcal{C}_\Sigma$ on $\mathcal{M}_\Sigma$ are canonically equivalent. The canonical global objects $r_1^* A_{\Sigma_1} \otimes r_2^* A_{\Sigma_2}$ and A_Σ correspond to each other in this equivalence.

The next (and final) step is to apply the transgression procedure to manifolds of dimension 0 and interpret the resulting cohomology class geometrically. But this is exactly the problem of representing α by a 2-gerbe $\mathcal{B}$. Abstractly, we know that $\mathcal{B}$ can be found and that the Dijkgraaf-Witten theory can be recovered as above. A concrete description of $\mathcal{B}$ will be presented in [8].

We conclude from this discussion that 2-gerbes are the fundamental geometric objects in Chern-Simons theory. A similar observation has been made by D. Kazhdan.

Acknowledgements

Each author thanks the N.S.F for its support though research grants. The second author thanks MSRI for support during part of this research.

References

1. A. Beilinson, *Higher regulators and values of L-functions*, J. Sov. Math. **30** (1985), 2036–2070.
2. J. Bénabou, *Introduction to bicategories*, Midwest Category Seminar, Lecture Notes in Math. vol. 47, Springer-Verlag, 1967, pp. 1–77.
3. L. Breen, *Théorie de Schreier supérieure*, Ann. Sci. Ec. Norm. Sup. **25** (1992), 465–514.
4. J.-L. Brylinski, *The Kaehler geometry of the space of knots on a smooth threefold*, preprint 1990.
5. J.-L. Brylinski, *Loop Spaces, Characteristic Classes and Geometric Quantization*, Progress in Math. vol. 107, Birkhäuser.
6. J.-L. Brylinski and D. McLaughlin, *Čech cocycles for characteristic classes*, preprint 1992.
7. J.-L. Brylinski and D. McLaughlin, *The geometry of degree four characteristic classes*, preprint 1992.
8. J.-L. Brylinski and D. McLaughlin, *The geometry of degree four characteristic classes II*, (in preparation).
9. J. Cheeger, *A combinatorial formula for Stiefel-Whitney classes*, Topology of Manifolds, Markham, 1970, pp. 470–471.
10. J. Cheeger and J. Simons, *Differential characters and geometric invariants*, Lecture Notes in Math. vol. 1167, Springer-Verlag, 1985, pp. 50–80.
11. S. S. Chern and J. Simons, *Characteristic forms and geometric invariants*, Ann. of Math. **99** (1974), 48–69.
12. P. Deligne, *Le symbole modéré*, Publ. Math. IHES **73** (1991), 147–181.
13. R. Dijkgraaf and E. Witten, *Topological gauge theories and group cohomology*, Comm. Math. Phys. **129** (1990), 393–429.
14. D. Freed and F. Quinn, *Chern-Simons theory with finite gauge group*, Comm. Math. Phys. (to appear).
15. A. Gabrielov, I. M. Gel'fand and M. V. Losik, *Combinatorial calculation of characteristic classes*, Funct. Anal. Appl. **9** (1975), 48–50, 103–115, 186–202.
16. K. Gawedzki, *Topological actions in two-dimensional quantum field theories*, Nonperturbative Quantum Field Theories, ed. G't Hooft, A. Jaffe, G. Mack, P. K. Mitter, R. Stora, NATO ASI Series vol. 185, Plenum Press, 1988, pp. 101–142.
17. I. Gel'fand and R. MacPherson, *A combinatorial formula for the Pontrjagin classes*, Bull. Amer. Soc. **26, n0. 2** (1992), 304–309.
18. J. Giraud, *Cohomologie Non-Abélienne*, Ergeb. der Math. vol. 64, Springer-Verlag, 1971.
19. A. Goncharov, *Explicit construction of characteristic classes*, Adv. in Soviet Math. (in press).
20. T. P. Killingback, *World sheet anomalies and loop geometry*, Nucl. Phys. B **288** (1987), 578–588.
21. B. Kostant, *Quantization and unitary representations. Part I: Prequantization*, Lecture Notes in Math. vol. 170, Springer-Verlag, 1970, pp. 87–208.
22. D. A. McLaughlin, *Orientation and string structures on loop space*, Pac. J. Math. **155 no. 1** (1992), 143–156.
23. R. MacPherson, *The combinatorial formula of Gabrielov, Gel'fand and Losik for the first Pontryagin class*, Séminaire Bourbaki. Exposés 498-506, Lecture Notes in math. vol. 677, Springer-Verlag, 1977, pp. 105–124.

24. G. Segal, *The definition of conformal field theory*, to appear.

25. A. Weil, *Variétés Kähleriennes*, Hermann, 1958.

26. E. Witten, *The index of the Dirac operator on loop space*, Elliptic Curves and Modular Forms in Algebraic Topology, Lecture Notes in Math. vol. 1326, Springer-Verlag, 1988, pp. 161–181.

27. E. Witten, *Quantum field theory and the Jones polynomial*, Comm. Math. Phys. **121** (1989), 351–389.

The Casson Invariant for Two-fold Branched Covers of Links

David Mullins
Division of Natural Sciences
New College of USF
Sarasota, FL 34243

Abstract

This article describes a new family of link invariants, λ_p, derived from the generalized Casson invariant of their branched covers. For the case λ_2, the invariant is computable by a skein technique, described fully in Mullins[7]. Furthermore, this skein description shows the relationship between λ_2 and the derivative of the Jones polynomial evaluated at -1.

1 Introduction

For any link in S^3, there is a family of 3-manifolds associated with the link; its p-fold branched covers. Since these are invariants of the link, whenever these are rational homology spheres one can use Walker's generalization of the Casson invariant, λ, to obtain rational invariants of the link. Although the p-fold branched covers are not always rational homology 3-spheres, one obtains an infinite number of invariants of knots, since for any *prime* p, the p-fold branched cover of a knot is a $\mathbf{Z}/p\mathbf{Z}$ homology 3-sphere.

Define $\lambda_p(L) = \lambda(\Sigma^p_L)$, whenever Σ^p_L is a rational homology sphere, where Σ^p_L is the p-fold cyclic branched cover of the link L. These should be useful invariants if one can calculate them. For example, these invariants are distinct from the Alexander matrix invariants; in particular, λ_2 of the untwisted double K^2 of a knot K is easily calculated as $\pm 4\Delta''_K(1)$, while the Alexander matrix of K^2 is the same as that of the unknot. (Amy Davidow has done independent work in this area. For torus knots and doubles of knots, Davidow calculates Casson's integer invariant for r-fold branched covers, when Σ_K^r is an *integral* homology sphere, see Davidow[4].) Further, they should be useful in distinguishing knots from their mirror image, because λ_p is zero on amphicheiral knots, since the orientation reversing homeomorphism of the knot complement extends to the p-fold branched cover, and λ changes sign under change of orientation.

If one restricts oneself to two-fold branched covers of knots, then there exists a natural surgery description of the two-fold branched cover of a knot $K \subset S^3$, with the surgery tori coming from crossings in a knot diagram of K. One can theoretically calculate the Casson invariant directly from these surgery tori using the surgery description of the Casson invariant. However, there is an alternate approach which avoids calculating the second derivative of the Alexander polynomial of the surgery tori.

If L^+, L^- and L^o represent links differing only at a single crossing, then $\lambda_2(L^+)$ can be calculated from $\lambda_2(L^-)$ and the Alexander polynomial of a surgery torus. Alternately, $\lambda_2(L^+)$ can be calculated from $\lambda_2(L^o)$ and the Alexander polynomial of the same surgery torus. Then one has two different formulas for $\lambda_2(L^+)$. So, using elementary algebra, one can can calculate $\lambda_2(L^+)$ in terms of $\lambda_2(L^-)$ and $\lambda_2(L^o)$, and eliminate the need to calculate the Alexander polynomial altogether, and hence, one has a recursive formula for $\lambda_2(L^+)$ in terms of $\lambda_2(L^-)$ and $\lambda_2(L^o)$. However, to have a rigorous skein definition of λ_2, one must develop a variant of skein theory for non-zero determinant links, since the 2-fold branched cover of a link is a rational homology sphere if and only if the link has non-zero determinant. We will ignore this problem in this paper; it is done fully in Mullins[7].

It is interesting, but somewhat disappointing, that λ_2 is just a combination of the logarithmic derivative of the Jones polynomial at -1 and the signature,

$$\lambda_2(L) = \frac{-\frac{d}{dt}V_L(-1)}{6V_L(-1)} + \frac{1}{4}\sigma(L).$$

Therefore, λ_2 cannot distinguish knots any better than these two invariants. However, this does imply that the right hand side of the equation above is an *unoriented* link invariant, since the 2-fold branched cover does not depend on the orientation of the link. Moreover, by using Kauffman's bracket polynomial, Kauffman[5], in combination with Murasugi's unoriented signature, Murasugi[8], one can generalize the right hand side to a rational function *unoriented* link invariant, $\Lambda_L(t)$:

$$\Lambda_L(t) = \frac{-\frac{d}{dt}V_L(t)}{6V_L(t)} - \frac{t^{-1}}{4}\sigma(L),$$

and hence, $\lambda_2(L) = \Lambda_L(-1)$. It is possible that the other link invariants, λ_p are also related to knot polynomials, but these have proven more difficult to calculate.

Finally, It is worth noting that since λ_2 can be easily calculated, one is provided with a technique for calculating the generalized Casson invariant on a large class of three-manifolds, the two-fold branched covers of links.

2 The Casson Invariant

The original Casson invariant was defined for integral homology three-spheres.

Definition 1 (Casson) Let $K \subset M$ be a knot in an oriented, integral homology 3-sphere, M, and let $M_{1/n}$ denote $1/n$ Dehn surgery on K. Then the Casson invariant, λ, is defined inductively by:

1. $$\lambda(M_{1/n}) = \lambda(M) + n\frac{d^2}{dt^2}\Delta_K(1)$$

2. $$\lambda(S^3) = 0.$$

Note: Casson's original formula was defined as $\frac{1}{2}\lambda$.

Kevin Walker generalizes the Casson invariant to the case when M is a rational homology 3-sphere.

Definition 2 (Walker) Let $K \subset M$ be a knot in an oriented, rational homology 3-sphere, M, and let $\Delta_K(t)$ be the normalized, symmetrized Alexander polynomial for K. Let $N(K)$ be a tubular neighborhood of K in M. Let $\langle , \rangle$ be the intersection pairing on $\mathbf{H}_1(\partial N(K); \mathbf{Z})$ induced by the orientation on M. Let $m, l \in \mathbf{H}_1(\partial N(K); \mathbf{Z})$ be the meridian and longitude of K, respectively. For p a primitive element in $\mathbf{H}_1(\partial N(K); \mathbf{Z})$, let M_p denote the manifold obtained by Dehn surgery on K sending the meridian to p. Then the generalized Casson invariant, λ, is defined inductively by:

1. $$\lambda(M_b) = \frac{\langle a,b\rangle}{\langle a,l\rangle\langle b,l\rangle}\frac{d^2}{dt^2}\Delta_K(1) + \lambda(M_a) + \tau(a,b;l)$$

2. $$\lambda(S^3) = 0.$$

Here, τ is a homological invariant, and fairly easy to calculate. For a full description of Walker's generalization and τ, see Walker[11].

Using the notation of Definition 2, let $T \subset M$ be a torus in a rational homology 3-sphere, with meridian m and longitude l, and let $y_1, y_2 \in \mathbf{H}_1(\partial T)$ be distinct primitive elements with $\langle y_1, m\rangle$, $\langle y_2, m\rangle$, $\langle y_1, l\rangle$ and $\langle y_2, l\rangle$ all non-zero. Then M_{y_1} and M_{y_2} are both rational homology 3-spheres, and so by Walker's work we have:

$$\lambda(M_{y_1}) = \lambda(M) + \tau(m, y_1; l) + c_1\Delta_T''(1),$$

$$\lambda(M_{y_2}) = \lambda(M) + \tau(m, y_2; l) + c_2\Delta_T''(1),$$

$$\text{with} \qquad c_i = \frac{\langle m, y_i\rangle}{\langle m,l\rangle\langle y_i,l\rangle} \neq 0.$$

Eliminating $\Delta_T''(1)$ gives us

$$\begin{aligned}\lambda(M) = {} & \frac{c_2}{c_2-c_1}\lambda(M_{y_1}) - \frac{c_1}{c_2-c_1}\lambda(M_{y_2}) \\ & - \frac{c_2}{c_2-c_1}\tau(m,y_1;l) + \frac{c_1}{c_2-c_1}\tau(m,y_2;l). \qquad (1)\end{aligned}$$

One uses Eq. (1) to calculate the recursive formulas of λ_2 for L^+, L^-, and L^o, since the two-fold covers of L^- and L^o are obtained from the two-fold cover of L^+ by surgery on the same torus.

3 The Surgery Torus

Given a link diagram $\mathcal{D}$ for L, and a crossing X in $\mathcal{D}$, we want to find a torus $T \subset \Sigma^2_L$ such that L^+, L^- and L^o are related via surgery on T. To compute the correction function τ in Walker's surgery formula we will need to know the surgery coefficients and the longitude with respect to some homology basis for $\mathbf{H}_1(\partial T)$.

Let D be a disk in S^3 such that D contains the crossing X and $[\partial D] \sim 0$ in $\mathbf{H}_1(S^3 \setminus L)$. Since $D \cap L = \{2 \text{ points}\}$, the preimage of D in the two-fold cover, $\widetilde{D} \subset \Sigma^2_L$, is an annulus.

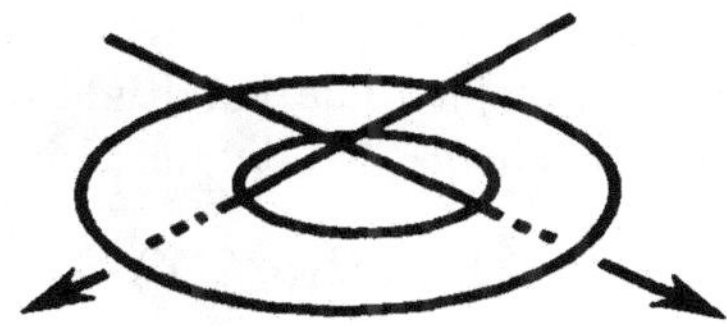

Then, if one takes a solid cylinder $D \times I$ containing the crossing, $T \stackrel{\text{def}}{=} \widetilde{D \times I}$ is a torus in Σ^2_L. $\mathbf{H}_1(\partial T)$ has a natural basis $\{x_1, x_2\}$, where x_1 is the meridian, and x_2 is one component of $\partial\widetilde{D}$, with orientation chosen such that $lk(x_1, x_2^*) = 1$, where x_2^* is the pushoff of x_2 into T.

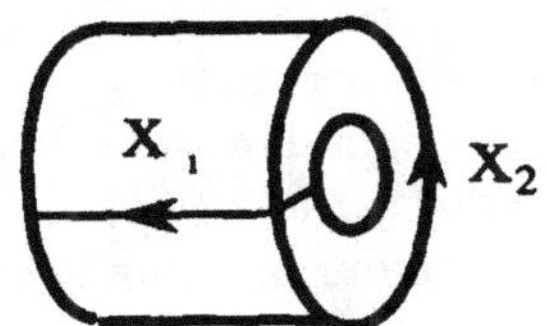

To compute the surgery coefficients one uses the fact that a meridian on the boundary of the surgery torus descends to a simple closed curve on the four-punctured sphere that bounds a disk in $S^3 \setminus L$.

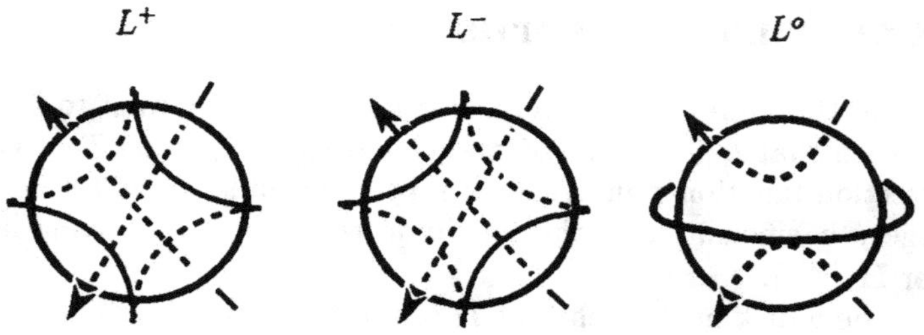

Then, by comparing the lifts of these curves one has that $\Sigma^2_{L^-}$ and $\Sigma^2_{L^o}$ are related to $\Sigma^2_{L^+}$ by $-1/2$ and 0 surgeries on $T \subset \Sigma^2_{L^+}$, respectively (in terms of the basis $\{x_1, x_2\}$). The longitude l of a solid torus T in the rational homology three-sphere Σ^2_L is by definition a generator of $ker(i_*) \cong \mathbf{Z}$, where

$$i_* : \mathbf{H}_1(\partial T; \mathbf{Z}) \to \mathbf{H}_1(\Sigma^2_L \setminus T; \mathbf{Z}).$$

Let S^+, S^-, and S' be compatible Seifert surfaces for L^+, L^-, and L^o respectively; that is, the Seifert surfaces differ only at the crossing where the links differ. Let θ_{L^+}, θ_{L^-}, and θ_{L^o} be their respective Seifert matrices. If one lets a be the determinant of $\theta_{L^o} + \theta^T_{L^o}$ and b be the determinant of $\theta_{L^+} + \theta^T_{L^+}$, then it is easily seen that in terms of the homology basis $\{x_1,x_2\}$, $ax_1 + bx_2$ is a multiple of the longitude. (One needs only view the homology of $\Sigma^2_{L^+} \setminus T$ in terms of these matrices, as well as x_1 and x_2.) a and b are not the standard determinant of links, since the signs of a and b depend on the choice of Seifert surfaces. Furthermore, $b + 2a = det(\theta_{L^-} + \theta^T_{L^-})$. Using $M = \Sigma^2_{L^+}$, $M_{y_1} = \Sigma^2_{L^-}$ and $M_{y_2} = \Sigma^2_{L^o}$, with respect to our basis we have

$$m = \binom{1}{0}, \qquad y_1 = \binom{1}{-2}, \qquad y_2 = \binom{0}{-1},$$

$$\frac{c_1}{c_2 - c_1} = \frac{2a}{b} \quad \text{and} \quad \frac{c_2}{c_2 - c_1} = \frac{b + 2a}{b}.$$

Then Eq. (1) becomes,

$$\begin{aligned}\lambda_2(L^+) =& \frac{b+2a}{b}\lambda_2(L^-) - \frac{2a}{b}\lambda_2(L^o) \\ & - \frac{b+2a}{b}\tau\left(\binom{1}{0}, \binom{1}{-2}; l\right) + \frac{2a}{b}\tau\left(\binom{1}{0}, \binom{0}{-1}; l\right).\end{aligned}$$

Evaluating τ and multiplying by b, our equation becomes

$$b\lambda_2(L^+) = (b+2a)\lambda_2(L^-) - 2a\lambda_2(L^o) + \frac{b}{6} + \frac{b+2a}{6}$$
$$+\frac{b}{4}\text{sign}(ab) - \frac{b+2a}{4}\text{sign}(a(b+2a)). \qquad (2)$$

4 The Jones Polynomial and λ_2

Throughout this section, we assume one has compatible Seifert surfaces for L^+, L^-, and L^o, with Seifert matrices θ_{L^+}, θ_{L^-}, and θ_{L^o}. Further, as before, let $b = det(\theta_{L^+} + \theta_{L^+})$, $b+2a = det(\theta_{L^-} + \theta_{L^-}^T)$, and $a = det(\theta_{L^o} + \theta_{L^o}^T)$. Since we are only considering the rational homology case, we assume b, $b+2a$, and a are nonzero. To understand Eq. (2), we note the relationship between the sign functions and the signature of the links. Since $\theta_{L^o} + \theta_{L^o}^T$ is a principal minor of both $\theta_{L^+} + \theta_{L^+}^T$ and $\theta_{L^-} + \theta_{L^-}^T$, we have the following theorem (Burde and Zieschang[2]):

Theorem 4.1

$$\sigma(L^+) - \sigma(L^o) = sign(ab) \qquad\qquad \sigma(L^-) - \sigma(L^o) = sign(a(b+2a)).$$

To understand how the Jones polynomial relates to Eq. (2), recall the skein definition of the Jones polynomial:

1. $$t^{-1}V_{L^+}(t) - tV_{L^-}(t) = (t^{1/2} - t^{-1/2})V_{L^o}(t)$$

2. $$V_{\text{unknot}}(t) = 1.$$

Then if one uses the fact that $|V_L(-1)| = |det(\theta_L + \theta_L^T)|$ together with the skein relation of the Jones polynomial at -1 (with $\sqrt{-1} = -i$), one has the following theorem:

Theorem 4.2

$$\frac{V_{L^+}(-1)}{b} = \frac{V_{L^-}(-1)}{b+2a} = \frac{-iV_{L^o}(-1)}{a} = \rho, \quad \textit{where} \quad \rho^4 = 1.$$

Then, if one multiplies Eq. (2) by ρ and uses Thm. (4.1) one obtains

$$V_{L^+}(-1)\lambda_2(L^+) = V_{L^-}(-1)\lambda_2(L^-) + 2iV_{L^o}(-1)\lambda_2(L^o) + \frac{V_{L^+}(-1)}{6} + \frac{V_{L^-}(-1)}{6} + \frac{V_{L^+}(-1)}{4}(\sigma(L^+) - \sigma(L^o)) + \frac{V_{L^-}(-1)}{4}(\sigma(L^o) - \sigma(L^-)).$$

Since $(V_{L^+}(-1) - V_{L^-}(-1))\sigma(L^o) = 2iV_{L^o}(-1)\sigma(L^o)$, if one multiplies by 6 and gathers terms which appear for all three links one has

$$\begin{aligned}&[6V_{L^+}(-1)(\lambda_2(L^+) - \tfrac{1}{4}\sigma(L^+))]\\ &\quad = [6V_{L^-}(-1)(\lambda_2(L^-) - \tfrac{1}{4}\sigma(L^-))]\\ &\quad +2i[6V_{L^o}(-1)\lambda_2(L^o) - \tfrac{1}{4}\sigma(L^o))] + V_{L^+}(-1) + V_{L^-}(-1)\end{aligned} \tag{3}$$

If one takes the skein formula for the Jones polynomial, differentiates and evaluates it at -1, one obtains

1 $\quad [-V'_{L^+}(-1)] - [-V'_{L^-}(-1)] = 2i[-V'_{L^o}(-1)] + V_{L^+}(-1) + V_{L^-}(-1)$

2. $\quad V_{\text{unknot}}(-1) = 0.$

Since $-V'_L(-1)$ and $6V_L(-1)(\lambda_2(L) - \frac{1}{4}\sigma(L))$ have the same recursive formula and are both zero on the unknot, they are equivalent. (As mentioned earlier, we have ignored the problem of avoiding links in the skein relation whose two-fold branched covers are not rational homology spheres. However, one can deal with this problem rigorously.)

References

1. S. Akbulut and J. McCarthy, *Casson's Invariant for Oriented Homology 3-Spheres: An Exposition*, Mathematical Notes (Princeton University Press, Princeton, NJ, 1990).

2. G. Burde and H. Zieschang, *Knots* (Walter de Gruyter, Inc., New York, 1985).

3. J. H. Conway, *Computational Problems in Abstract Algebra*, Pergamon Press, New York (1970), pp 329-358.

4. A. D. Davidow, *On Casson's Invariant of Branched Cyclic Covers over S^3*, Ph.D. Thesis (New York University, 1989).

5. L. Kauffman, *On Knots*, Ann. of Math. Studies, no. 115 (Princeton University Press, Princeton, NJ, 1987).

6. W. B. R. Lickorish and K. C. Millett, *Topology* **26** (1987), pp. 107-141.

7. D. Mullins "The Generalized Casson Invariant for Two-fold Branched Covers of S^3 and the Jones polynomial," *Topology* (to appear).

8. K. Murasugi, *Topology* **9** (1970), pp. 283-298.

9. J. H. Przytycki, *Canadian Journal of Math.* **41** (1989), *no.* 2, pp. 250-273.

10. D. Rolfsen, *Knots and Links* (Publish or Perish, Inc., Wilmington, DE, 1976).

11. K. Walker, *An Extension of Casson's Invariant*, Ann. of Math. Studies, no. 126 (Princeton University Press, Princeton, NJ, 1992).

D-ALGEBRAS, THE D-SIMPLEX EQUATIONS, AND MULTIDIMENSIONAL INTEGRABILITY

F.P. MICHIELSEN and F.W. NIJHOFF
Department of Mathematics and Computer Science
and Institute for Nonlinear Studies,
Clarkson University, Potsdam NY 13699-5815, USA

Recently R. Lawrence has introduced axioms for a category of d-algebras, i.e. algebras with d-ary operations. The natural associativities are defined from geometrical considerations of moves in simplex partitions. We study these algebras in connection with the tetrahedron and higher simplex equations that were proposed by Zamolodchikov and descendants a decade ago. These are the natural generalizations of the quantum Yang-Baxter equations. We discuss the connection of these algebras with the issue of integrability in multi-dimensions.

1 Introduction

The theory of integrable systems has, over the last two decades, expanded in an incredible number of directions, and the wealth of areas in the mathematical sciences which are – directly or indirectly – linked to these systems, is truly impressive. It is sufficient to focus on the theory of quantum groups and the modern developments in low-dimensional topology, to be aware of the enormous impact the study of such systems, and their ramifications, have had in mathematics, cf. e.g. [1]-[5]. From a physical point of view, most integrable situations that have been studied so far, however, deal with systems that are realized in a low number of (spatial) dimensions. In particular, integrable models described by partial differential equations deal mostly with the two-dimensional situation (one space- and one time-dimension). Truly multidimensional integrable PDE's, even nowadays, have been hardly touched upon, even on the classical level. In the statistical mechanics of exactly solvable models, the situation is slightly better: we have one famous genuinely three-dimensional integrable system, the Zamolodchikov model, [6]-[9]. The situation is hopeful, as recently generalizations of Zamolodchikov's model have been constructed, [10]. On a formal level the cornerstone to such models, namely the tetrahedron equations, can be generalized to any dimension, cf. [11], leading to the so-called d-simplex equations, cf. also [12]-[16]. However, explicit nontrivial solutions have, to our knowledge, not yet been found.

In the case of PDE's, the situation is less pronounced. Apart from systems like the ones related to the KP hierarchy, or the ones coming from the self-dual Yang-Mills equations, no systematic approach to multidimensional integrable systems exists to date. In a paper with Maillet some years ago, [17], we anticipated on possible routes towards obtaining a notion of truly multi-dimensional integrability. Inspired by the hierarchy of d-simplex equations, [14, 18], we formulated ideas that could ultimately lead to a theory of integrable classical lattice equations in any space-time dimension, [19, 20]. In [17] we speculated on the possible use of ternary (or more generally d-ary) algebras, that would be a central ingredient in the development of such a theory.

Traditionally, algebras defined with d-ary operations have not been widely studied. They form a slightly unorthodox chapter in algebra, remotely related to such exotic issues as multi-categories, [21]. In a beautiful recent paper R. Lawrence has proposed a list of axioms for ternary algebras, based on geometric considerations. Although this scheme was motivated by the simplex equations, we believe that a direct connection between the Lawrence algebras and solutions of the latter equations still remains to be clarified. In this note we will provide ideas along which ternary algebras could be connected to the tetrahedron equations. To do so we will study the deformation approach to quantum groups, which is in our view fundamental to the quantization of integrable systems.

2 3-algebras

In [22] R. Lawrence provides a scheme of axioms for ternary algebras. The natural associativities for the ternary products are formulated in that paper in terms of moves of simplex partitions. The set of axioms that she provides turns out to be quite complicated, and in our opinion can be simplified somewhat. The essential aspect is the description of natural associativities for these ternary algebras. Therefore, we modify Lawrence's prescription, and introduce the following definition of a ternary algebra.

We make the following definition for a ternary algebra:

Definition 1 (ternary algebra) *Let A be a vector space over a field k, then A has a ternary algebra structure with products $(m,\bar{m})$ if there exist two product operations:*

$$m\colon A\otimes A\otimes A\mapsto A \tag{1}$$

$$\bar{m}\colon A\otimes A\mapsto A\otimes A \tag{2}$$

satisfying the following associativity axioms:

i) $m\circ(m\otimes id\otimes id)=m\circ(id\otimes id\otimes m)\circ\sigma_{23}\circ(id\otimes id\otimes\bar{m}\otimes id)\circ\sigma_{23}$

ii) $\bar{m}\circ(m\otimes id)=(m\otimes id)\circ(id\otimes id\otimes\bar{m})\circ\sigma_{23}\circ(id\otimes id\otimes\bar{m})\circ\sigma_{23}$

ii)' $\bar{m} \circ (id \otimes m) = (id \otimes m) \circ \sigma_{23} \circ (\bar{m} \otimes id \otimes id) \circ \sigma_{23} \circ (\bar{m} \otimes id \otimes id)$

iii) $(\bar{m} \otimes id) \circ (id \otimes \bar{m}) = (id \otimes \bar{m}) \circ \sigma_{12} \circ (id \otimes \bar{m}) \circ (\bar{m} \otimes id)$

(here $\sigma_{ij}: A^{\otimes^n} \mapsto A^{\otimes^n}$ is the map which permutes the entries i,j of a tensor product.

The axioms i)-iii) provide us with the necessary ways to re-associate multiple ternary products of the type m or $\bar{m}$. One can show that the associativity axioms that are in the list of ref. [22] reduce to one of the axioms i)-iii) above. Geometrically, interpreting the action of the two products m and $\bar{m}$ in terms of labelled triangle, the axioms arise by considering different ways of moving simplex partitions into each other, cf. [22] for details.

In a more elaborate picture one would like to equip the ternary algebras defined by Def. 1, with some additional structures. A possible way of doing this is by supplying additional axioms dealing with the following situations :

i) *unital 3-algebras*, generalizing the notion of a unit with respect to the products m and $\bar{m}$. (Units with respect to m must consist of pairs of elements of the 3-algebra).

ii) *invertibility in 3-algebras*, introducing a generalized notion of invertibility with respect to m or $\bar{m}$. (Elements of a 3-algebra will have pairs of elements as their inverse).

iii) *involution*, a map $P: A \mapsto A$ such that $P^3 = id$, cf. [22].

iv) *orthogonality*, additional axioms generalizing notions of projection, [22].

v) *duality*, introducing notions of pairing in 3-algebras. (The dual of a 3-algebra with respect to m will again consist of a pair of elements).

Without going into details, we mention that all these notions, which are natural extra structures to develop further the structure of ternary algebras, need highly nontrivial ways of generalization. We have to be prepared to let go of many of our conventional ideas about any of these notions.

3 Examples of ternary algebras

Let us now provide a number of examples of ternary algebras that obey the associativity conditions of Def. 1.

The first example is that of a notion of a *3-matrix algebra*, $A = Mat^{(3)}(n;k)$, i.e. the algebra of objects which are cubic arrays (a_{ijk}), $i,j,k = 1,\dots,n$, called 3-matrices, of elements of a field k. This object A can be given a 3-algebra structure in the following way.

Example 1 *Let $(t)_{ijk}$ denote the ijk^{th}-component of t. The extensions to $A \otimes A \otimes A$ resp. $A \otimes A$ of the maps*

$$m \colon A \times A \times A \mapsto A \colon \left(m(a \otimes b \otimes c)\right)_{ijk} = \sum_{pqr} (a)_{ipq}(b)_{pjr}(c)_{qrk} \tag{3}$$

$$\bar{m} \colon A \times A \mapsto A \otimes A \colon \left(\bar{m}(a \otimes b)\right)_{ijk|pqr} = \delta_{kq}(a)_{iqp}(b)_{kjr} \tag{4}$$

endow the k-linear space A with a 3-algebra structure.

The standard basis for A consist of the elements E_{ijk} which satisfy $(E_{ijk})_{pqr} = \delta_{ip}\delta_{jq}\delta_{kr}$. In terms of this basis the 3-algebra structure on A is generated by:

$$m(E_{ipq} \otimes E_{p'jr} \otimes E_{q'r'k}) = \delta_{pp'}\delta_{qq'}\delta_{rr'} E_{ijk}$$
$$\bar{m}(E_{ijk} \otimes E_{pqr}) = \delta_{pj} E_{iqp} \otimes E_{kjr}$$

There exist a natural conjugation given by:

$$(a^{\tau})_{ijk} = (a)_{kij}$$

Related to this example is the construction of 3-algebras from associative 2-algebras.

Example 2 *Consider an (associative) 2-algebra B with automorphism t; then $A = B^{\otimes 3}$ becomes a 3-algebra if we let $(m, \bar{m})$ be generated by*

$$m((a \otimes \bar{a} \otimes \bar{\bar{a}}) \otimes (b \otimes \bar{b} \otimes \bar{\bar{b}}) \otimes (c \otimes \bar{c} \otimes \bar{\bar{c}})) = (ab) \otimes (\bar{a}c\bar{b}^t\bar{\bar{a}}\bar{\bar{c}}) \otimes (\bar{\bar{b}}\bar{c}) \tag{5}$$

and

$$\bar{m}((a \otimes \bar{a} \otimes \bar{\bar{a}}) \otimes (b \otimes \bar{b} \otimes \bar{\bar{b}})) = ((c \otimes \bar{c} \otimes \bar{\bar{c}}) \otimes (d \otimes \bar{d} \otimes \bar{\bar{d}})) \tag{6}$$
$$\text{where } c = ab, \quad \bar{\bar{b}} = \bar{\bar{c}}\bar{\bar{d}}$$
$$\bar{d} = \bar{\bar{a}}^t\bar{b}, \quad \bar{a} = \bar{c}d^t$$

this is nothing else than a 3-matrix algebra for composed objects:

$$A_{IJK} = a_{ij}\bar{a}_{i'k}\bar{\bar{a}}_{j'k'} \qquad I = (i, i'), \quad J = (j, j'), \quad K = (k, k') \tag{7}$$

A generalization of the matrix 3-algebra is given by the following example given in [22], which involves the $6j$-symbols of a quantum group, [23]. These are known to be related to the topology of 3-manifolds, cf. e.g. [24].

Example 3 *Let $\begin{vmatrix} i & j & k \\ p & q & r \end{vmatrix}$ be quantum 6j-symbols of $U_q(\mathfrak{g})$, where $\mathfrak{g}$ a simple Lie-algebra, then the maps $m \colon A \otimes A \otimes A \mapsto A$ and $\bar{m} \colon A \otimes A \mapsto A \otimes A$ generated by:*

$$m(e_{ipq} \otimes e_{p'jr} \otimes e_{q'r'k}) = \delta_{pp'}\delta_{qq'}\delta_{rr'} \begin{vmatrix} i & j & k \\ p & q & r \end{vmatrix} e_{ijk} \tag{8}$$

$$\bar{m}(e_{ijk} \otimes e_{pqr}) = \delta_{jp} \sum_s w_s^2 \begin{vmatrix} i & q & s \\ r & k & p \end{vmatrix} e_{iqs} \otimes e_{ksr} \tag{9}$$

where w_s are weight-normalistions such that

$$\sum_s w_s^2 \begin{vmatrix} i & q & s \\ r & k & p \end{vmatrix} \begin{vmatrix} s & q' & i' \\ p' & p & r \end{vmatrix} = \begin{vmatrix} q & q' & j \\ p' & p & r \end{vmatrix} \begin{vmatrix} i & j & i' \\ p' & k & p \end{vmatrix} \tag{10}$$

give the k-linear space A a 3-algebra structure.

A more general class of examples is provided by the class of *3-associative 3-algebras*, where the products m and $\bar{m}$ are defined in terms of structure constants. Thus, given a linear space V with basis $\{e^i\}_{i\in I}$, where I is some index set, and maps $G\colon V\otimes V\otimes V\mapsto V$ and $\bar{G}\colon V\otimes V\mapsto V\otimes V$ with structure constants G_s^{pqr} and $\bar{G}_{p'q'}^{pq}$ respectively relative to the given basis , then $(G,\bar{G})$ define a 3-algebra structure on V provided we have the following *pentagonal* relations for $G,\bar{G}$:

i) $G_t^{abc}G_s^{tde} = G_s^{ab't}\bar{G}_{b'd'}^{bd}G_t^{cd'e}$

ii) $G_{t'}^{abc}\bar{G}_{td''}^{t'd} = G_t^{ab'c'}\bar{G}_{c'd''}^{cd'}\bar{G}_{b'd'}^{bd}$

ii)' $\bar{G}_{a''t}^{at'}G_{t'}^{bcd} = \bar{G}_{a'b'}^{ab}\bar{G}_{a''c'}^{a'c}G_t^{b'c'd}$

iii) $\bar{G}_{a''b''}^{ab'}\bar{G}_{b'c''}^{bc} = \bar{G}_{b''c''}^{b'c'}\bar{G}_{a''c'}^{a'c}\bar{G}_{a'b'}^{ab}$

Pentagonal relations, similar to the ones given by i)-iv), are very familiar to us: they arise in connection with fusion algebras in conformal field theory (see e.g. [25]), and more recently in connection with quasi-Hopf algebras, [26]. They detect usually a departure from associativity (in 2-algebras), and hence 3-associative 3-algebras may be be considered to provide a natural context for obstructions to associativity, cf. also [27]. (Following the ideas exposed in ref. [17] these structures might be of interest in the construction of integrable systems in multi-dimensions). Let us continue along this train of thought, and investigate the emergence of such obstructions in quantum algebras.

4 Deformation Approach to Quantum Groups

Quantization of functions on a quantum group following the deformation approach, cf. e.g. [28], leads to a possible specific realization of the algebraic structures presented in the previous section. In fact, in an early paper,[29], Drinfel'd points out how to deform the commutative ring of functions on a group to a non-commutative algebra. The central object is an element $F\in U\mathfrak{g}\otimes U\mathfrak{g}$, where $\mathfrak{g}$ is the Lie algebra of the Lie group G under consideration. Realizing the universal enveloping algebra $U\mathfrak{g}$ as a polynomial algebra $\mathbb{C}[x_1,\ldots,x_n]$, where $x_1,\ldots,x_n$ are the generators of $\mathfrak{g}$, we can represent F as a polynomial $F(x,y)$ in terms of two commuting copies of the set of generators x and y. This object then defines then a new deformed product, a $*$-product, on $C^\infty(G)$. We will not give here the precise description, but what is essential is that in order for this $*$-product to be associative (it will no longer be commutative in general), we have the following condition on $F(x, y)$, namely

$$F(x+y,z)F(x,y) = F(x,y+z)F(y,z) \ . \tag{11}$$

Then, a connection with the Yang-Baxter equation is established by noting that the object

$$R(x,y) = F(y,x)^{-1}F(x,y) \ , \tag{12}$$

solves the Yang-Baxter equation. In fact, it is interesting to go through the steps of the derivation:

$$\begin{aligned}
&R(x,y)R(x,z)R(y,z) \\
&\quad = F(y,x)^{-1}F(x,y)F(z,x)^{-1}F(x,z)F(y,z)^{-1}F(y,z) \\
&\quad = F(y,x)^{-1}F(z,x+y)^{-1}F(z+x,y)F(x,z)F(z,y)^{-1}F(y,z) \\
&\quad = F(z,y)^{-1}F(z+y,x)^{-1}F(x,z+y)F(z,y)F(z,y)^{-1}F(y,z) \\
&\quad = F(z,y)^{-1}F(z+y,x)^{-1}F(x+y,z)F(x,y) \\
&\quad = R(y,z)F(y,z)^{-1}F(z+y,x)^{-1}F(x+y,z)F(y,x)R(x,y) \\
&\quad = R(y,z)F(z,x)^{-1}F(y,z+x)^{-1}F(y,x+z)F(x,z)R(x,y) \\
&\qquad\qquad = R(y,z)R(x,z)R(x,y) \ .
\end{aligned} \tag{13}$$

The steps in the derivation (13) can be represented pictorially as a series of "foldings" of the sides of a cube. Associating with x and y directions in 3-space, and representing $R(x,y)$ as a plaquette in the x, y- directions, we can think of eq. (12) as representing a decomposition of the plaquette into two triangles as follows:

$R(x,y) =$ [plaquette diagram with sides y, x, $-x$, $-y$] $= F(-x,-y)F(x,y)$

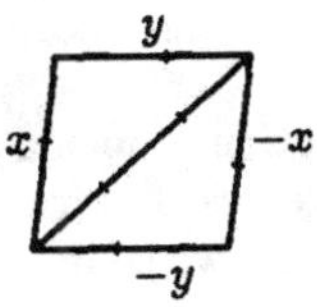

(were, in accordance with the expression in [30], we assume $F(y,x)^{-1} = F(-x,-y)$). The associativity condition (11) in this pictorial scheme is given by the following elementary move

[diagram with labels z, y, x, $x+y$] $=$ [diagram with labels $y+z$, z, y, x]

The sequence of steps leading to the Yang-Baxter equation, according to the derivation (13) can then be represented by the following sequence of elementary moves of triangles:

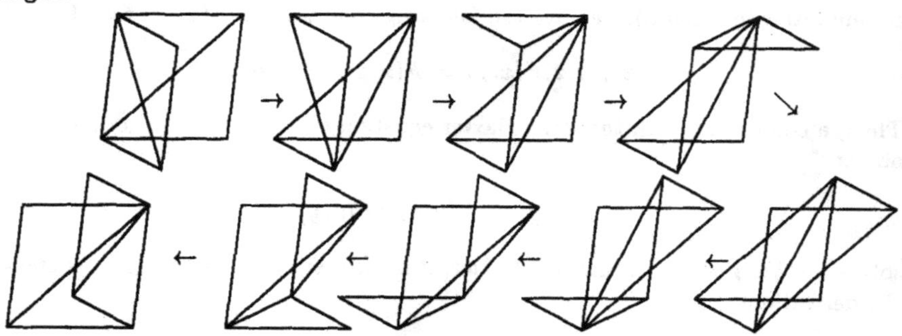

The intuitive meaning of this is that we can link associativity in binary algebras (at least in the precise picture given by the deformation quantization) to the 2-simplex equation (the Yang-Baxter equation). This gives us an idea how to generalize this, and to possibly make a connection between the 3-algebras described in the previous sections and the tetrahedron equations. As the axioms of 3-associativity are motivated purely by geometric considerations, namely in terms of simplex moves, we are led to the following fundamental question: can we generalize this picture to higher dimensions, in other words can we make a direct connection between a solution of the 3-simplex equations and structure constants of higher associativity conditions for 3-algebras? This would require the following ingredients:

i) a partition of cubes (representing the tetrahedral object) in terms of elementary tetrahedra,

ii) an elementary move representing an associativity condition of a 3-algebra, corresponding to the axioms given earlier,

iii) the verification that these elementary moves can indeed achieve the higher-dimensional folding of a configuration of cubes according to the tetrahedron equations.

It is unnecessary to say that the visualization of such a process would approach something that can only be realized in our dreams. Far from having the algebraic tools to verify the combinatorics that this requires, we could envisage the following scenario to achieve the steps i)-iii).

Having deformed, by introducing the $*$-product, the usual group-structure on $C^\infty(G)$, let us proceed with a further deformation, namely that of the relation (11), as follows

$$G(x,y,z)F(x+y,z)F(x,y) = F(x,y+z)F(y,z)\ , \tag{14}$$

introducing a new object G which plays the role of a non-trivial associator. Assuming that G commutes with F (this can be relaxed somewhat), we arrive again at the pentagonal relations

$$G(y,z,u)G(x,y+z,u)G(x,y,z) = G(x+y,z,u)G(x,y,z+u)\ , \tag{15}$$

cf. also [28]. Representing G pictorially by a tetrahedron, the pentagonal equation (15) is represented by the move:

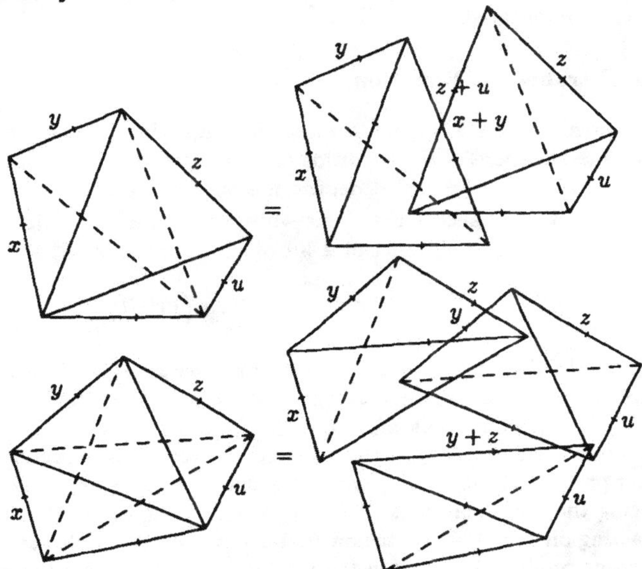

Tracing back, again, the derivation (13), now taking into account (14) instead of (11), and again assuming commutativity of G with F, we find a relation of the form

$$R(x,y)R(x,z)R(y,z) = R(y,z)R(x,z)R(x,y)R(x,y,z) , \tag{16}$$

where

$$R(x,y,z) = G(y,z,x)G(z,y,x)^{-1}G(z,x,y)G(x,z,y)^{-1}G(x,y,z)G(y,x,z)^{-1} , \tag{17}$$

which, in fact, is a form of the quasi-Yang Baxter equation, [26]. Eq. (17) corresponds to the decomposition of a cube into tetrahedra. Now this is not really what we would like to have, because the commutativity condition makes the situation too trivial. What we would like to obtain is a mechanism by which the objects $R(x,y)$ in (16) carry internal indices with which they are coupled between themselves (basically through the product in $U\mathfrak{g} \otimes U\mathfrak{g} \otimes U\mathfrak{g}$), and in addition a structure (some external indices attached to the sides of the plaquettes) by which they are coupled non-trivially to the new object $R(x,y,z)$. As will be explained in the next section, this would immediately lead to the tetrahedron equation for $R(x,y,z)$. In that case the consistency relation of (16) would be represented pictorially by the "inside-out" turning of the faces of a hypercube in dimension $d = 4$, namely by the move from the four cubes that form the sides on one side of the hypercube to the four cubes that form the remaining faces of the hypercube (we don't draw a figure, because it will be slightly too complicated to visualize).

5 The Tetrahedron Equation

As was pointed out in [18, 14, 15], the d-simplex equations, [11, 12], can be thought of as being generalizations of the condition of commutativity of matrices. In the form of [14], we have namely objects that depend not only on a single pair of matrix- indices, but on a collection of such pairs. This will give us tensorial objects $\left(S_{i_1\, i_2 \cdots i_d}^{j_1\, j_2 \cdots j_d}\right), d = 1, 2, \cdots$, and we define a product of $d+1$ of such objects by

$$\left(S^{(1)}\right)_{i_1, i_2, \cdots, i_d}^{k_1, k_2, \cdots, k_d} \left(S^{(2)}\right)_{k_1, i_{d+1}, \cdots, i_{2d-1}}^{j_1, k_{d+1}, \cdots, k_{2d-1}} \cdots \left(S^{(d+1)}\right)_{k_d, k_{2d-1}, \cdots, k_D}^{j_d, j_{2d-1}, \cdots, j_D}, \tag{18}$$

(where $D = \frac{1}{2}d(d+1)$), in which summation over repeated indices is understood. Note that the essential feature of this generalized matrix product is that every object $S^{(p)}$ is coupled once and only once to each of the other d objects by a usual matrix multiplication. Alternatively, we can regard the $\left(S^{(p)}\right)$ as being matrices acting on the tensor product $V_{\alpha_1} \otimes \cdots \otimes V_{\alpha_d}$ of d vector spaces V_{α_i}, $1 \leq \alpha_1 < \alpha_2 < \cdots < \alpha_d \leq D$. Introducing the tensor product $V^{(d)} =: V_1 \otimes V_2 \otimes \cdots \otimes V_D$ we denote by $S_{\alpha_1 \cdots \alpha_d}$ the matrix acting on $V^{(d)}$ which acts non-trivially only on the vector spaces $V_{\alpha_1}, \cdots, V_{\alpha_d}$ in the tensor product $V^{(d)}$. Then the product (18) can be written in compact form as a product of matrices acting on $V^{(d)}$ as follows:

$$S^{(1)} * S^{(2)} * \cdots * S^{(d+1)} = S^{(1)}_{1,2,\cdots,d} \cdot S^{(2)}_{1,d+1,\cdots,2d-1} \cdots S^{(d+1)}_{d,2d-1,\cdots,D}. \tag{19}$$

The d-simplex equation is the condition of commutativity of a d-fold product as in (18), i.e.

$$S^{(1)} * S^{(2)} * \cdots * S^{(d+1)} = S^{(d+1)} * \cdots * S^{(2)} * S^{(1)}. \tag{20}$$

Let us write down explicitely the first few members of the hierarchy of d-simplex equations, writing $A = S^{(1)}, B = S^{(2)}, C = S^{(3)}$, etc.

d=1: $A_i^k B_k^j = B_i^k A_k^j$
(Matrix Commutativity)

d=2: $A_{i_1 i_2}^{k_1 k_2} B_{k_1 i_3}^{j_1 k_3} C_{k_2 k_3}^{j_2 j_3} = C_{i_2 i_3}^{k_2 k_3} B_{i_1 k_3}^{k_1 j_3} A_{k_1 k_2}^{j_1 j_2}$
(Yang-Baxter Equations)

d=3: $A_{i_1 i_2 i_3}^{k_1 k_2 k_3} B_{k_1 i_4 i_5}^{j_1 k_4 k_5} C_{k_2 k_4 i_6}^{j_2 j_4 k_6} D_{k_3 k_5 k_6}^{j_3 j_5 j_6} = D_{i_3 i_5 i_6}^{k_3 k_5 k_6} C_{i_2 i_4 k_6}^{k_2 k_4 j_6} B_{i_1 k_4 k_5}^{k_1 j_4 j_5} A_{k_1 k_2 k_3}^{j_1 j_2 j_3}$
(Tetrahedron Equations)

d=4: $A^{k_1k_2k_3k_4}_{i_1i_2i_3i_4} B^{j_1k_5k_6k_7}_{k_1i_5i_6i_7} C^{j_2j_5k_8k_9}_{k_2k_5i_8i_9} D^{j_3j_6j_8k_{10}}_{k_3k_6k_8i_{10}} E^{j_4j_7j_9j_{10}}_{k_4k_7k_9k_{10}} =$

$$= E^{k_4k_7k_9k_{10}}_{i_4i_7i_9i_{10}} D^{k_3k_6k_8j_{10}}_{i_3i_6i_8k_{10}} C^{k_2k_5j_8j_9}_{i_2i_5k_8k_9} B^{k_1j_5j_6j_7}_{i_1k_5k_6k_7} A^{j_1j_2j_3j_4}_{k_1k_2k_3k_4}.$$

(Bazhanov-Stroganov Equations)

In [18, 15] we described a scheme describing ways to build the hierarchy in a recursive fashion, by introducing an obstruction mechanism. This scheme served as a motivation to formulate a notion of multi-dimensional classical integrability on lattices, [19, 20]. Thus, in order to generate the $(d+1)$-simplex equation from the d-simplex equation for any $d \geq 1$, we proceed as follows. Let us consider $d+1$ collections of objects ${}_{\alpha_k}S^{(k)}$ defined in a similar way as the objects $S^{(k)}$ we used in eq.(19) but now depending on an additional (vector) index α_k. However, we will not suppose that these operators do verify the d–simplex equation (20) for any values of the indices α_k, but rather that there exists a tensor $R^{\beta_1\cdots\beta_{d+1}}_{\alpha_1\cdots\alpha_{d+1}}$ (depending on these additional vector indices α_k) that describes an obstruction to eq.(20) for the family of operators ${}_{\alpha_k}S^{(k)}$ in the following way :

$$R^{\beta_1\cdots\beta_{d+1}}_{\alpha_1\cdots\alpha_{d+1}} \left({}_{\beta_1}S^{(1)} * {}_{\beta_2}S^{(2)} * \cdots * {}_{\beta_{d+1}}S^{(d+1)} \right)$$
$$= {}_{\alpha_{d+1}}S^{(d+1)} * \cdots * {}_{\alpha_2}S^{(2)} * {}_{\alpha_1}S^{(1)} \tag{21}$$

where the sum over repeated indices β_k is understood and the $*$ product between operators ${}_{\alpha_k}S^{(k)}$ is defined as in eq.(19). Note that if $R^{\beta_1\cdots\beta_{d+1}}_{\alpha_1\cdots\alpha_{d+1}} = \delta^{\beta_1}_{\alpha_1} \cdots \delta^{\beta_{d+1}}_{\alpha_{d+1}}$, eq.(21) would reduce to eq.(20) for any choice of the indices α_k.

In addition, this tensor will depend on $\frac{1}{2}d(d+1)$ hyperspherical angles which are functions of those present in eq.(18), cf. [14]. As a consequence of the fact that the system of equations given in (21) is overdetermined, the tensor R has to satisfy, as a consistency condition, the $(d+1)$-simplex equation of [11]. Thus, the obstruction to the d-simplex equation as given in equation (21) has to satisfy the $(d+1)$-simplex equation. The most simple example of this mechanism is given by considering the case of an obstruction to the 1-simplex (commutativity) equation. Eq.(21) leads in that case to :

$$R^{\beta_1\beta_2}_{\alpha_1\alpha_2} \; {}_{\beta_1}A \cdot {}_{\beta_2}B = {}_{\alpha_2}B \cdot {}_{\alpha_1}A \,, \tag{22}$$

where the dot product between the matrices A and B stands for the standard matrix product. A consistency condition arises now when we impose that an equation of the type (22) holds for any pair of matrices out of three families of matrices ${}_{\alpha}A_{12}$, ${}_{\beta}B_{13}$, ${}_{\gamma}C_{23}$. In fact, a combination of the form

$${}_{\alpha}A_{12} \cdot {}_{\beta}B_{13} \cdot {}_{\gamma}C_{23}$$

can be inverted in two different way, leading to the Yang-Baxter equation, for R, which is the 2-simplex equation. In the same way, the object that realizes an obstruction to the Yang-Baxter equation has to verify, as a consistency condition, the tetrahedron (3-simplex) equation of Zamolodchikov [6]. It is instructive to see how this works in detail. So on this level we build in an obstruction to the 2-simplex equation by giving the corresponding objects again a family index which is coupled to a new matrix $R^{\beta_1\beta_2\beta_3}_{\alpha_1\alpha_2\alpha_3}$, namely

$$R^{\beta_1\beta_2\beta_3}_{\alpha_1\alpha_2\alpha_3}\ ({}_{\beta_1}A_{12} \cdot {}_{\beta_2}B_{13} \cdot {}_{\beta_3}C_{23}) = {}_{\alpha_3}C_{23} \cdot {}_{\alpha_2}B_{13} \cdot {}_{\alpha_1}A_{12}. \tag{23}$$

There are now two different ways to bring the following arrangement of matrices

$${}_{\alpha}A_{12} \cdot {}_{\beta}B_{13} \cdot {}_{\gamma}C_{23} \cdot {}_{\delta}D_{14} \cdot {}_{\epsilon}E_{24} \cdot {}_{\zeta}F_{34}$$

into reverse order by using successively eq. (23) and similar equations for the triples A, D, E, and B, D, F and C, E, F. (The rule of the game here is that any such triple can only be reversed by an intertwining matrix R if the product between the members of the triple are coupled in the same way as in the 2-simplex equation). In fact, this is nothing but the action of the generalized braidings, introduced in [12]. Applying now the successive steps

$$
\begin{array}{c}
{}_{\alpha}A_{12} \cdot {}_{\beta}B_{13} \cdot {}_{\gamma}C_{23} \cdot {}_{\delta}D_{14} \cdot {}_{\epsilon}E_{24} \cdot {}_{\zeta}F_{34} \\
\downarrow \\
{}_{\gamma'}C_{23} \cdot {}_{\beta'}B_{13} \cdot {}_{\alpha'}A_{12} \cdot {}_{\delta}D_{14} \cdot {}_{\epsilon}E_{24} \cdot {}_{\zeta}F_{34} \\
\downarrow \\
{}_{\gamma'}C_{23} \cdot {}_{\beta'}B_{13} \cdot {}_{\epsilon'}E_{24} \cdot {}_{\delta'}D_{14} \cdot {}_{\alpha''}A_{12} \cdot {}_{\zeta}F_{34} \\
\| \\
{}_{\gamma'}C_{23} \cdot {}_{\epsilon'}E_{24} \cdot {}_{\beta'}B_{13} \cdot {}_{\delta'}D_{14} \cdot {}_{\zeta}F_{34} \cdot {}_{\alpha''}A_{12} \\
\downarrow \\
{}_{\gamma'}C_{23} \cdot {}_{\epsilon'}E_{24} \cdot {}_{\zeta'}F_{34} \cdot {}_{\delta''}D_{14} \cdot {}_{\beta''}B_{13} \cdot {}_{\alpha''}A_{12} \\
\downarrow \\
{}_{\zeta''}F_{34} \cdot {}_{\epsilon''}E_{24} \cdot {}_{\gamma''}C_{23} \cdot {}_{\delta''}D_{14} \cdot {}_{\beta''}B_{13} \cdot {}_{\alpha''}A_{12}
\end{array}
$$

by acting with the combination

$$R^{\gamma'\epsilon'\zeta'}_{\gamma''\epsilon''\zeta''}\ R^{\beta'\delta'\zeta}_{\beta''\delta''\zeta'}\ R^{\alpha'\delta\,\epsilon}_{\alpha''\delta'\epsilon'}\ R^{\alpha\,\beta\,\gamma}_{\alpha'\beta'\gamma'}$$

and in a similar way

$$ {}_{\alpha}A_{12} \cdot {}_{\beta}B_{13} \cdot {}_{\gamma}C_{23} \cdot {}_{\delta}D_{14} \cdot {}_{\epsilon}E_{24} \cdot {}_{\zeta}F_{34} $$

$$ \downarrow $$

$$ {}_{\alpha}A_{12} \cdot {}_{\beta}B_{13} \cdot {}_{\delta}D_{14} \cdot {}_{\zeta'}F_{34} \cdot {}_{\epsilon'}E_{24} \cdot {}_{\gamma'}C_{23} $$

$$ \downarrow $$

$$ {}_{\alpha}A_{12} \cdot {}_{\zeta''}F_{34} \cdot {}_{\delta'}D_{14} \cdot {}_{\beta'}B_{13} \cdot {}_{\epsilon'}E_{24} \cdot {}_{\gamma'}C_{23} $$

$$ \| $$

$$ {}_{\zeta''}F_{34} \cdot {}_{\alpha}A_{12} \cdot {}_{\delta'}D_{14} \cdot {}_{\epsilon'}E_{24} \cdot {}_{\beta'}B_{13} \cdot {}_{\gamma'}C_{23} $$

$$ \downarrow $$

$$ {}_{\zeta''}F_{34} \cdot {}_{\epsilon''}E_{24} \cdot {}_{\delta''}D_{14} \cdot {}_{\alpha'}A_{12} \cdot {}_{\beta'}B_{13} \cdot {}_{\gamma'}C_{23} $$

$$ \downarrow $$

$$ {}_{\zeta''}F_{34} \cdot {}_{\epsilon''}E_{24} \cdot {}_{\delta''}D_{14} \cdot {}_{\gamma''}C_{23} \cdot {}_{\beta''}B_{13} \cdot {}_{\alpha''}A_{12} $$

by acting with the combination

$$ R^{\alpha'\,\beta'\,\gamma'}_{\alpha''\beta''\gamma''}\; R^{\alpha\;\delta'\,\epsilon'}_{\alpha'\,\delta''\epsilon''}\; R^{\beta\;\delta\;\zeta'}_{\beta'\,\delta'\,\zeta''}\; R^{\gamma\;\epsilon\;\zeta}_{\gamma'\,\epsilon'\,\zeta'}, $$

we can identify these two different combinations of matrices R, that make it possible to invert the order of the matrices $A, \cdots F$, and obtain the tetrahedron equation for R. In a similar way one obtains for $d > 3$ the higher-order simplex equations.

6 Conclusions

We have formulated a possible scenario for relating directly the tetrahedron equations (and along the same lines the higher d-simplex equations) to ternary algebras as introduced in [22]. The main idea is to study decompositions of the tetrahedral object R into elementary simplex objects G, suggesting that these generalize in fact the object F that defines the deformation quantization in quantum groups. Our conjecture is that the elementary pentagonal move will enable one to –as it were– turn inside out the hypercube that represents the configuration in the tetrahedron equations. We are working on ways to visualize this procedure by computer. If this works out as we think, it will serve as a combinatorial tool to develop further the algebraic picture that relates ternary algebras to simplex equations on the one hand. On the other hand it may lead to new classes of non-trivial solutions of simplex equations, namely in terms of entirely new objects which are the d-associative d-algebras. As it stands there are no definite answers yet, but we hope that in the near future more precise statements can be made. Furthermore, we mention the interesting developments relating simplex equations to n-categories, cf. e.g. [30], cf. also [21, 31]. Finally, we mention that there seem to be promising possibilities of obtaining solutions of the simplex equations via computer algorithms, [32]. All these developments indicate clearly that a new chapter in the theory of integrable systems, and related algebraic structures, is about to be opened.

References

[1] L.D. Faddeev, in *Développements Récents en Théorie des Champs et Mécanique Statistique*, eds. R. Stora and J.B. Zuber, (North-Holland, Amsterdam, 1983), p.561.

[2] R.J. Baxter, *Exactly Solved Models in Statistical Mechanics*, (Academic Press, London, 1982).

[3] V.G. Drinfel'd, *Quantum Groups*, Proc. ICM Berkeley, 1986, ed. A.M. Gleason, (AMS, Providence,1987), p. 789.

[4] L.D. Faddeev, N.Yu. Reshetikhin and L.A. Takhtadzhyan, Algebra i Analiz, **1** # 1 (1988) 178 (in Russian).

[5] L.H. Kauffman, *Knots and Physics*, (World Scientific, 1991).

[6] A.B. Zamolodchikov, Sov. Phys. JETP **52** (1980) 325.

[7] A.B. Zamolodchikov, Commun. Math. Phys. **79** (1981) 489.

[8] R.J. Baxter, Commun. Math. Phys. **88** (1983) 185.

[9] R.J. Baxter, Phys. Rev. Lett. **53** (1984) 1795; Physica **18D** (1986) 321.

[10] V.V. Bazhanov and R.J. Baxter, *New Solvable Models in Three Dimensions*, Preprint SMS-015-92/MRR-005-92; *Star-Triangle Relations for a Three-Dimensional Model*, Preprint SMS-079-92/ MRR-020-92.

[11] V.V. Bazhanov and Yu.G. Stroganov, Theor. Math. Phys. **52** (1982) 685.

[12] Yu.I. Manin and V.V. Shekhtman, Funct. Anal. Appl. **20** (1986) 148; in *Group Theoretical Methods in Physics, vol. I* (Yurmala, 1985), (VNU Sci. Press, Utrecht, 1986), p. 151.

[13] I. Frenkel and G. Moore, Commun. Math. Phys. **138** (1991) 411.

[14] J.M. Maillet and F.W. Nijhoff, Phys. Lett. **A134** (1989) 221.

[15] F.W. Nijhoff and J.M. Maillet, in Proc. of the IVth International Conference on Nonlinear and Turbulent Processes in Physics, Kiev, October 1989.

[16] J.S. Carter and M. Saito, *On Formulations and Solutions of Simplex Equations*, Preprint 1992.

[17] F.W. Nijhoff and J.M. Maillet, in *Nonlinear Evolutions*, Proc. of the VIth Int. Workshop on Nonlinear Evolution Equations and Dynamical Systems, ed. J. Léon, (World Sci. Publ. Co., Singapore, 1988),p.281.

[18] J.M. Maillet and F.W. Nijhoff, in Proc. of the Int. Workshop on Nonlinear Evolution Equations: Integrability and Spectral Methods, Como, Italy 1988, ed. A.P. Fordy, (Manchester University Press, 1989).

[19] J.M. Maillet and F.W. Nijhoff, Phys. Lett. **224B** (1989) 389.

[20] J.M. Maillet and F.W. Nijhoff, Phys. Lett. **229B** (1989) 71.

[21] C. Ehresmann, *Catégories et Structures*, (Dunod, Paris, 1965).

[22] R.J. Lawrence, *Algebras and Triangle Relations*, Preprint Harvard University, June 1991.

[23] A.N. Kirillow and N. Yu. Reshetikhin, *Representations of the Algebra* $U_q(sl(2))$, *q-orthogonal polynomials and invariants of links*, LOMI Preprint E-9-88.

[24] V.G. Turaev, O.Y. Viro, *State sum invariant of 3-manifolds and quantum 6j-symbols*, Preprint (1990).

[25] G. Moore and N. Seiberg, Commun. Math. Phys. **123** (1989) 177, Phys. Lett **212B** (1988) 451.

[26] V.G. Drinfel'd, Leningrad J. Math. **1** (1990) 1419.

[27] J.D. Stasheff, Trans. AMS **108** (1963) 275.

[28] L.A. Takhtajan, in *Introduction to Quantum Groups and Integrable Massive Models of Quantum Field Theory*, eds. M.L. Ge and B.H. Zhao, (World Scientific, 1990).

[29] V.G. Drinfel'd, Sov. Math. Dokl. **28** (1983) 667.

[30] M.M. Kapranov and V.A. Voevodsky, *Braided monoidal 2-categories, 2-vector spaces and Zamolodchikov's Tetrahedra equations*, Preprint 1992.

[31] R. Street, J. Pure and Appl. Algebra **49** (1987) 283.

[32] J. Hietarinta, *Some Constant Solutions to Zamolodchikov's Tetrahedron Equations*, Preprint Turku University, 1992.

The Chern-Simons Character of a Lattice Gauge Field

Anthony V. Phillips*
Mathematics Department,
State University of New York at Stony Brook,
Stony Brook NY 11794-3651

David A. Stone†
Mathematics Department,
Brooklyn College, City University of New York,
Brooklyn NY 11210

Abstract

Let u be a generic $SU(2)$-valued lattice gauge field on a triangulation Λ of a manifold X of dimension ≥ 3. An algorithm is given for constructing from u a principal $SU(2)$-bundle ξ over X, and in ξ a piecewise smooth connection ω with the following property. The Chern-Simons character of the gauge field (ξ, ω) can be computed as the sum of local contributions $\hat{\chi}(\sigma)$, one for each 3-simplex σ of Λ, where $\hat{\chi}(\sigma)$ is the sum of a one-dimensional integral and five other terms which are volumes of elementary regions in the 3-sphere. For any 3-cycle $\sum_i a_i\sigma_i$, the sum $\sum_i a_i\hat{\chi}(\sigma_i)$, reduced mod $\mathbf{Z}$, is gauge-invariant. This character is related in the appropriate way to real and integral 4-cycles representing the second Chern class of ξ. If u has sufficiently small plaquette products, the isomorphism class of ξ is shown to depend only on u, and not on details of the construction.

*Partially supported by NSF grant DMS-8907753
†Partially supported by a grant from PSC-CUNY and by NSF grant DMS-8805485

1 Introduction

The Chern-Simons forms on the total space of a principal bundle with connection link the Chern-Weil-theoretic and the obstruction-theoretic approaches to characteristic classes. (We shall take up this theme in Theorem C.) When pulled down to the base, they give rise to the Chern-Simons characters [3, 2], with many applications to differential geometry. Recently interest in the Chern-Simons theory has grown remarkably, on the part of physicists as well as mathematicians. The reason is Witten's work [20], where the integral of the Chern-Simons form of an $SU(2)$-valued gauge field on a 3-manifold serves as the action in the path integrals of a topological quantum field theory. This turns out to have extensive applications to knot theory and to the theory of 3-manifolds.

Thus two well-explored parts of mathematics are joined by a chain with several fuzzy links, since the path-integral formalism, while powerful and intuitively very appealing, is not yet mathematically complete, and does not admit direct computation in general. Now lattice gauge fields (see Section 3.4) were invented by Wilson [19] as an attack on this very problem. It is in this spirit that we propose a lattice-theoretic construction of the Chern-Simons character, and an algorithm for its computation.

Our method involves a universal construction: following Dupont [4] we define a canonical connection in the total space of Milnor's universal G–bundle $\xi_\Delta G = (\pi_\Delta : E_\Delta G \to B_\Delta G)$, and a corresponding canonical Chern-Simons form. Then we construct from the lattice gauge field a coherent set of maps $\Gamma_\sigma : C_\sigma \to E_\Delta G$, one for each simplex σ in the lattice Λ (C_σ is a cube of the same dimension as σ), and use these maps to obtain a classifying map from Λ to $B_\Delta G$. For these maps to pull back the canonical Chern-Simons form in a computable way, the Γ_σ's must be defined compatibly with the natural cellular structure of $E_\Delta G$. This structure was established by Eilenberg and MacLane [5] and is called the bar construction. Finally the contribution to the Chern-Simons character from each 3–simplex σ is computed by integrating the pulled-back Chern-Simons form over a pseudo-section ψ (constructed from the lattice gauge field) above σ. For recent, related work on this topic, see [6, 7, 9, 21]. This paper amplifies, with some modifications, the preliminary account given in [16].

In our previous work [13, 14, 15] we have explained various methods for constructing a bundle (and, when appropriate, a connection) from lattice data. The method used here is a technical (but important) refinement of the method of [15]. The difference between the Γ_σ's and the analogous H_σ's [15] is that H_σ uses the product structure of C_σ, whereas Γ_σ is defined using an appropriate decomposition of C_σ into bar-products of lower-dimensional cubes. The H_σ's form a pseudo-section (called the "canonical pseudo-section" in [15]) but neither they nor the Γ_σ's have the property, of lifting a cycle to a cycle, necessary for the definition of the Chern-Simons character; hence the need for the more elaborate construction of ψ, which in fact contains the Γ_σ's (see Figure 5).

The rest of this section contains basic definitions and states the main results of this paper.

1.1 *The Chern-Simons form as a differential character*

Here we summarize results of Chern and Simons [3] and of Cheeger and Simons [2]. Let (ξ, ω) be an $SU(2)$-valued gauge field on a manifold X of dimension ≥ 3. That is, $\xi = (\pi : E \to X)$ is a principal $SU(2)$-bundle, and ω is an $su(2)$-valued 1-form on E satisfying the usual conditions [8] (see also Prop. 4.2).

The real-valued 4-form $\tilde{p}$ on E given by

$$\tilde{p} = \frac{1}{8\pi^2} Tr(\Omega \wedge \Omega),$$

where Ω is the curvature of ω, is the lift of a unique 4-form p on X:

$$\pi^*(p) = \tilde{p}.$$

The form p (the "topological charge density" of the gauge field when $\dim X = 4$) represents the second Chern class of ξ.

The form $\tilde{p}$ is always exact, no matter what the bundle ξ may be. In fact, a simple calculation shows that it is the coboundary of the 3-form $\tilde{\chi}$ defined on E by [3]

$$\tilde{\chi} = \frac{1}{8\pi^2} Tr(\omega \wedge \Omega - \frac{1}{3}\omega \wedge \omega \wedge \omega).$$

This is the Chern-Simons form. Now it is impossible to find a 3-form α on X such that $\pi^*(\alpha) = \tilde{\chi}$; for on any $SU(2)$ fibre $\pi^*(\alpha)$ would vanish, while $\tilde{\chi}$ restricts to the volume form. However, $\tilde{\chi}$ can be represented as a "differential character" on X, as follows.

Let $\mathcal{Z}$ be the additive group of 3-cycles on X. For $Z \in \mathcal{Z}$, the restriction $\xi|_Z$ is trivial (for topological reasons) and so admits a section $\psi : Z \to \pi^{-1}(Z)$. Now set

$$\chi(Z) = \int_{\psi(Z)} \tilde{\chi}, \quad \textit{reduced mod } \mathbf{Z}.$$

It can be shown [2] that if $\psi' : Z \to \pi^{-1}(Z)$ is another section of $\xi|_Z$, then $\int_{\psi(Z)} \tilde{\chi} - \int_{\psi'(Z)} \tilde{\chi}$ is an integer; hence $\chi(Z)$ is independent of ψ. We will call this element of $\mathbf{R}/\mathbf{Z}$ the *Chern–Simons character of the gauge field* $(\xi, \tilde{\omega})$.

Remark 1.1 Suppose M is a connected 4-chain on X with $\partial M = Z$; if $Z \neq \emptyset$ then $\xi|_M$ is trivial, so let $\varphi : M \to \pi^{-1}(M)$ be a section; then

$$\int_M p = \int_{\varphi(M)} \tilde{p} = \int_{\varphi(Z)} \tilde{\chi} \equiv \chi(Z) \ \ mod\ \mathbf{Z}.$$

Thus $\chi : \mathcal{Z} \to \mathbf{R}/\mathbf{Z}$ is indeed a differential character as defined in [2].

1.2 The Chern-Simons character of a lattice gauge field

By a *(simplicial) lattice* on X we mean a simplicial triangulation Λ of X, as in [14]. In the sections following this one we shall need an ordering (which can be arbitrary) of the vertices of Λ, say as $<0>, <1>, <2>, \ldots$. We shall call a simplicial lattice together with such an ordering an *ordered (simplicial) lattice*. Every simplex σ of an ordered lattice has an orientation induced by the ordering of its vertices, and when we write a simplex as $\sigma = <i_0, \ldots, i_k>$, we list the vertices according to their order.

Henceforth we restrict attention to 3-cycles Z which are chains on Λ; that is, $Z = \sum a_i \sigma_i$, where $\sigma_i \in \Lambda, \dim \sigma_i = 3$, σ_i has the orientation just described, $a_i \in \mathbf{Z}$, and $\sum a_i \partial \sigma_i = 0$.

Let $\Lambda^{(3)} = \bigcup\{\sigma : \dim \sigma = 3\}$, i.e. the the 3-skeleton of Λ. Then $\xi|_{\Lambda^{(3)}}$ is trivial, again for dimensional reasons. Let $\psi : \Lambda^{(3)} \to \pi^{-1}(\Lambda^{(3)})$ be a section or a pseudosection (see Section 5).

For every 3-simplex σ, set

$$\hat{\chi}(\sigma) = \int_{\psi(\sigma)} \tilde{\chi}.$$

This is a real-valued 3-cochain on Λ. The values $\hat{\chi}(\sigma)$ depend on ψ (as well as ω); but for any 3-cycle $Z = \sum a_i \sigma_i$, the element $\chi(Z) \in \mathbf{R}/\mathbf{Z}$, defined as

$$\chi(Z) = \sum a_i \hat{\chi}(\sigma_i) \ \ reduced\ \ mod\ \mathbf{Z},$$

is independent of ψ.

Now let $\mathbf{u} = \{u_{ij} :< ij > \in \Lambda\}$ be a generic (see below) $SU(2)$-valued lattice gauge field (see Section 3.4) on the ordered simplicial lattice Λ.

Our procedure will be to construct from $\mathbf{u}$ a principal $SU(2)$-bundle ξ, a connection in ξ and the derived Chern-Simons form $\tilde{\chi}$, and a (pseudo)section of ξ over the 3–skeleton $\Lambda^{(3)}$ for which the calculation of the values $\hat{\chi}(\sigma)$ is as simple as possible. In order to state our result we need some preliminary notation and constructions.

Notation. We take $SU(2)$ as the unit sphere S^3 in the quaternionic space $\mathbf{H}$. For a quaternion $u = a\mathbf{1} + b\mathbf{i} + c\mathbf{j} + d\mathbf{k}$, $(a, b, c, d \in \mathbf{R})$, we set as usual $\Re(u) = a$, $\Im(u) = b\mathbf{i} + c\mathbf{j}f + d\mathbf{k}$. The matrix trace of u becomes $2\Re(u)$.

For a product of transporters along consecutive 1-simplexes we use the abbreviation

$$u_{i_0 i_1 \dots i_k} = u_{i_0 i_1} u_{i_1 i_2} \dots u_{i_{k-1} i_k}.$$

The following constructions make extensive use of shortest geodesics in S^3 between various points constructed from $\mathbf{u}$, and certain families of such geodesics. For almost all $\mathbf{u}$, every such shortest geodesic or family is unique. Henceforth we shall assume that $\mathbf{u}$ is ***generic*** in this sense.

For every 1-simplex $< ij >$ of Λ, let $\mathbf{g}_{\alpha ij}$ be the geodesic in S^3 from $\mathbf{1}$ to u_{ij}, parametrised by $0 \le s_i \le 1$; for every 2-simplex $< ijk >$, let $\mathbf{g}_{ijk}$ be the geodesic from u_{ik} to u_{ijk}, parametrised by $0 \le s_j \le 1$.

Now let σ be a 3-simplex of Λ. We construct five 3-dimensional, oriented regions in S^3 associated to $\mathbf{u}$ and σ: four tetrahedra and what we have called a "pyramid" [14]. To simplify the notation, write σ as $< 0123 >$.

Let Q be the doubly ruled quadrilateral in S^3 defined by $Q(s_0, s_2) = \mathbf{g}_{\alpha 01}(s_0)\mathbf{g}_{123}(s_2), 0 \le s_0, s_2 \le 1$. Our ***pyramid*** P is the cone from $\mathbf{1}$ on Q; that is, P is the union of the shortest geodesics from $\mathbf{1}$ to $Q(s_0, s_2)$ for all $s_0, s_2 \in [0, 1]$. P is oriented by the following sequence of its vertices: $\mathbf{1}$, $u_{13} = Q(0,0)$, $u_{013} = Q(1,0)$, $u_{0123} = Q(1,1)$.

This leaves the four tetrahedra. By a ***tetrahedron*** with vertices u,v,w and x we mean the geodesic convex hull of these four points, which we denote by $cx(u, v, w, x)$. The tetrahedron is oriented by the given ordering of its vertices. Now set

$$\begin{aligned} \Delta_1 &= cx(\mathbf{1}, u_{03}, u_{013}, u_{0123}), \\ \Delta_2 &= cx(\mathbf{1}, u_{03}, u_{023}, u_{0123}), \\ \Delta_3 &= cx(\mathbf{1}, u_{23}, u_{023}, u_{0123}), \\ \Delta_4 &= cx(\mathbf{1}, u_{23}, u_{123}, u_{0123}). \end{aligned}$$

Theorem A Let $\mathbf{u}$ be a generic $SU(2)$-valued lattice gauge field on an ordered simplicial lattice Λ of dimension ≥ 3. Let ξ be the principal $SU(2)$-

bundle assigned to $\mathbf{u}$ by our algorithm. Then there exist a piecewise smooth connection ω in ξ and a pseudo-section ψ of $\xi|_{\Lambda^{(3)}}$ such that the localisation of the Chern-Simons character, when applied to $\sigma =< 0123 >$, has the form

$$\hat{\chi}(\sigma) = \int_P dv + \sum_{i=1}^{4}(-1)^i \int_{\Delta_i} dv + \frac{1}{4\pi^2}\int_0^1 \Re[U_{01}\cdot Ad_{\mathbf{g}_{\alpha 01}(s_0)}W_{123}]ds_0.$$

In this expression dv is the normalised volume form on S^3, so $\int_{S^3} dv = 1$, and $U_{01} \in \mathfrak{su}_2$ is the matrix such that $\exp(U_{01}) = u_{01}$ (so that $\mathbf{g}_{\alpha 01}(s_0) = \exp(s_0 U_{01})$); similarly $U_{3123} \in \mathfrak{su}_2$ is the matrix such that $\exp(U_{3123}) = u_{3123}$ (so that $\mathbf{g}_{123}(s_2) = u_{13}\exp(s_2 U_{3123})$, and $W_{123} = Ad_{u_{13}} U_{3123}$.

The value of $\hat{\chi}(\sigma)$ will clearly change if $\mathbf{u}$ is replaced by a gauge-equivalent field. Nevertheless we have as much gauge-invariance as can be expected (compare Section 1.1):

Theorem B For any 3-cycle $Z = \sum_i a_i\sigma_i$, the sum $\chi(Z) = \sum_i a_i\hat{\chi}(\sigma_i)$, reduced mod $\mathbf{Z}$, is gauge-invariant.

Remarks on uniqueness. To what extent do these results depend on $\mathbf{u}$ alone, and not on details of the algorithm? If $\mathbf{u}$ is continuous, i.e. satisfies the following *continuity hypothesis*:

For every 2-simplex $< ijk > \in \Lambda$, $d(u_{ij}u_{jk}, u_{ik},) < \pi/n$,
where d is the unit-sphere metric on $SU(2)$ and $n = \dim\Lambda$,

then (Proposition 3.14) $\mathbf{u}$ determines a unique principal $SU(2)$-bundle $\xi = \xi(\mathbf{u})$. For a discussion of the relation between the lattice gauge field we start with, and the connection used to define its Chern=Simons character, see Section 4.4.

1.3 The Chern-Simons character and the calculation of topological charge

Here let $\dim\Lambda \geq 4$. Let $\mathbf{u}$ be a continuous, generic $SU(2)$-valued l.g.f. on Λ, and let ξ be the corresponding principal bundle. Let ω be the piecewise smooth connection in ξ constructed in Theorem A.

The Chern-Simons form in classical $SU(2)$ theory serves to connect the differential-geometric (Chern-Weil) and the obstruction-theoretic definitions of the second Chern class. The point of Theorem C is that the Chern-Simons character developed in this paper is related in almost exactly this way to the Chern form p pulled back from the universal example constructed from Dupont's connection in Milnor's classifying space and to the integral 4-cocycle N we constructed in [14] to compute the topological charge of an $SU(2)$-valued lattice gauge field. The "almost" is necessary because in [14] we augmented the lattice Λ by adding a new vertex ordered *above* the original ones, whereas in this paper it was more convenient to follow the opposite convention.

Let N be the 4-cocycle representative on Λ of the second Chern class of $\xi(\mathbf{u})$ constructed according to the algorithm of [14]. For any 4-simplex $\tau \in \Lambda$, by definition, $N(\tau)$ is the degree of a certain map from the boundary of a 4-cube into S^3.

On the other hand both $\int_\tau p$ and $\hat{\chi}(\partial\tau)$ give $\chi(\partial\tau)$ when reduced mod $\mathbf{Z}$, so their difference $\int_\tau p - \hat{\chi}(\partial\tau)$ is an integer.

Theorem C $\int_\tau p - \chi(\partial\tau)$ is equal to $\bar{N}(\bar{\tau})$, for every 4-simplex $\tau \in \Lambda$.

Here $\bar{N}(\bar{\tau})$ is the obstruction calculated as above, but when the order of the vertices is reversed. For details, see Section 6.4.

The rest of this paper: Section 3 summarizes material from Milnor [11, 12], Eilenberg-MacLane [5], and our own previous work [17]; Section 4 introduces a connection in Milnor's universal bundle and calculates the Chern-Simons character of this connection. The pseudosection associated to a lattice gauge field is constructed in Section 5. Section 6 contains the proofs of the main theorems; one computation is postponed to Section 7.

2 Milnor's universal bundle and the bar construction

2.1 Let $\xi_\Delta G = (\pi_\Delta : E_\Delta G \to B_\Delta G)$ be Milnor's model of a universal principal G–bundle [12]. Here $E_\Delta G = G_0 * G_1 * \cdots$, the join of countably many copies of G: a point of $E_\Delta G$ is $\sum_{i=0}^{\infty} t_i g_i$ where $g_i \in G_i$, $0 \le t_i \le 1$, all but finitely many t_i are 0, and $\sum t_i = 1$; it is to be understood that whenever $t_i = 0$, the term $t_i g_i$ is redundant. G acts on the left on $E_\Delta G$ by $L_g \sum t_i g_i = g \cdot (\sum t_i g_i) = \sum t_i g \cdot g_i$; and $B_\Delta G$ is the quotient space.

It is convenient to use the vectors $\mathbf{e}_0 = (1,0,0,\ldots)$, $\mathbf{e}_1 = (0,1,0,\ldots)$, etc. to index the various copies of G, by identifying $\mathbf{e}_i$ with the identity element in the i–th copy; thus we will write $\sum t_i g_i$ as $\sum t_i g_i \mathbf{e}_i$, where now all $g_i \in G$.

A feature that distinguishes $\xi_\Delta G$ from the more geometric Grassmannian models is the existence of *right*–automorphisms: if $\mathbf{h} = \{h_0, h_1, \ldots\}, h_i \in G$, set $R_\mathbf{h}(\sum t_i g_i) = \sum t_i g_i h_i$. Since left and right multiplication commute, $R_\mathbf{h}$ will take fibres to fibres, so there is a unique $r_\mathbf{h} : B_\Delta G \to B_\Delta G$ such that $r_\mathbf{h} \circ \pi_\Delta = \pi_\Delta \circ R_\mathbf{h}$.

There is a natural cell structure on $E_\Delta G$ which has been studied under the name of the *bar construction* [5].

Definition 2.2 Let $C_0, \ldots, C_k$ be cubes (of any dimension). The ***geometric bar product*** $C_0[C_1|\cdots|C_k]$ is defined by induction on k.

$C_0[\,]$ is just C_0; the brackets are included only to simplify certain formulas.

If $C_1[C_2|\cdots|C_k]$ has been defined, then

$$C_0[C_1|\cdots|C_k] = C_0 \times \mathit{cone}(C_1[C_2|\cdots|C_k]).$$

We extend this linearly to define the geometric bar product of cubical chains.

Remark 2.3 **1.** If $C_0, \ldots, C_k$ are all 0–dimensional, then $C_0[C_1|\cdots|C_k]$ is just the simplex $< C_0, \ldots, C_k >$. Furthermore in general we have the following "barycentric" coordinates on $C_0[C_1|\cdots|C_k]$: it sits inside the r-fold join

$$C_0 * (C_0 \times C_1) * (C_0 \times C_1 \times C_2) * \cdots * (C_0 \times \cdots \times C_k);$$

and if $\vec{s}_i = (s_{i1}, \ldots, s_{ip_i})$ is a positively oriented coordinate system on C_i (so $0 \le s_{ij} \le 1$ and $p_i = \dim C_i$), then $C_0[C_1|\cdots|C_k]$ is the subset

$$\{t_0\vec{s}_0 + t_1(\vec{s}_0, \vec{s}_1) + t_2(\vec{s}_0, \vec{s}_1, \vec{s}_2) + \cdots + t_k(\vec{s}_0, \vec{s}_1, \ldots, \vec{s}_k)\}$$

where $0 \le t_i \le 1$ and $\sum t_i = 1$, positively oriented (following Milgram [10]) by the coordinates $t_0, \ldots, t_k, s_{01}, \ldots, s_{0p_0}, \ldots, s_k, \ldots, s_{k,p_k}$.

2.

$$\dim C_0[C_1|\cdots|C_k] = k + \sum_{i=0}^{k} \dim C_i.$$

3.

$$\begin{aligned}\partial(C_0[C_1|\cdots|C_k]) &= \partial C_0[C_1|\cdots|C_k] \\ &+\sum_{i=1}^{k}(-1)^{k+\delta(i)}C_0[C_1|\cdots|\partial C_i|\cdots|C_k] \\ &+(C_0 \times C_1)[C_2|\cdots|C_k] \\ &+\sum_{i=1}^{k-1}(-1)^i C_0[C_1|\cdots|C_{i-1}|C_i \times C_{i+1}|C_{i+2}|\cdots|C_k] \\ &+(-1)^k \epsilon(C_k)C_0[C_1|\cdots|C_{k-1}],\end{aligned}$$

where $\delta(i) = \sum_{j<i} \dim C_j$, and $\epsilon(C_k) = 1$ if $\dim C_k = 0$, and $= 0$ otherwise (See Figure 1); note that our convention for orienting the boundary of X is that the outward normal to ∂X, followed by a positive basis for ∂X, gives a positive basis for X.

We extend these notions to certain singular cubes as follows.

Definition 2.4 Given $C_0, \ldots, C_k$ as before, together with indices $j(0) < j(1) < \cdots < j(k)$, and a collection of piecewise smooth maps $g_i : C_i \to G$, which we make into a collection of singular cubes $\gamma_i : C_i \to E_\Delta G$ by $\gamma_i(\vec{s_i}) = g_i(\vec{s_i}) \cdot \mathbf{e}_{j(i)}$, the *bar product*

$$\gamma_0[\gamma_1|\cdots|\gamma_k] : C_0[C_1|\cdots|C_k] \to E_\Delta G,$$

is given by

$$\gamma_0[\gamma_1|\cdots|\gamma_k](t_0\vec{s}_0 + t_1(\vec{s}_0, \vec{s}_1) + \cdots + t_k(\vec{s}_0, \ldots, \vec{s}_k))$$

$$= t_0(g_0(\vec{s}_0))\mathbf{e}_{j(0)} + t_1(g_0(\vec{s}_0) \cdot g_1(\vec{s}_1))\mathbf{e}_{j(1)} + \cdots + t_k(g_0(\vec{s}_0) \cdot \cdots \cdot g_k(\vec{s}_k))\mathbf{e}_{j(r)}.$$

Remark 2.5 1.

$$\dim \gamma_0[\gamma_1|\cdots|\gamma_k] = k + \sum_{j=0}^{k} \dim \gamma_j.$$

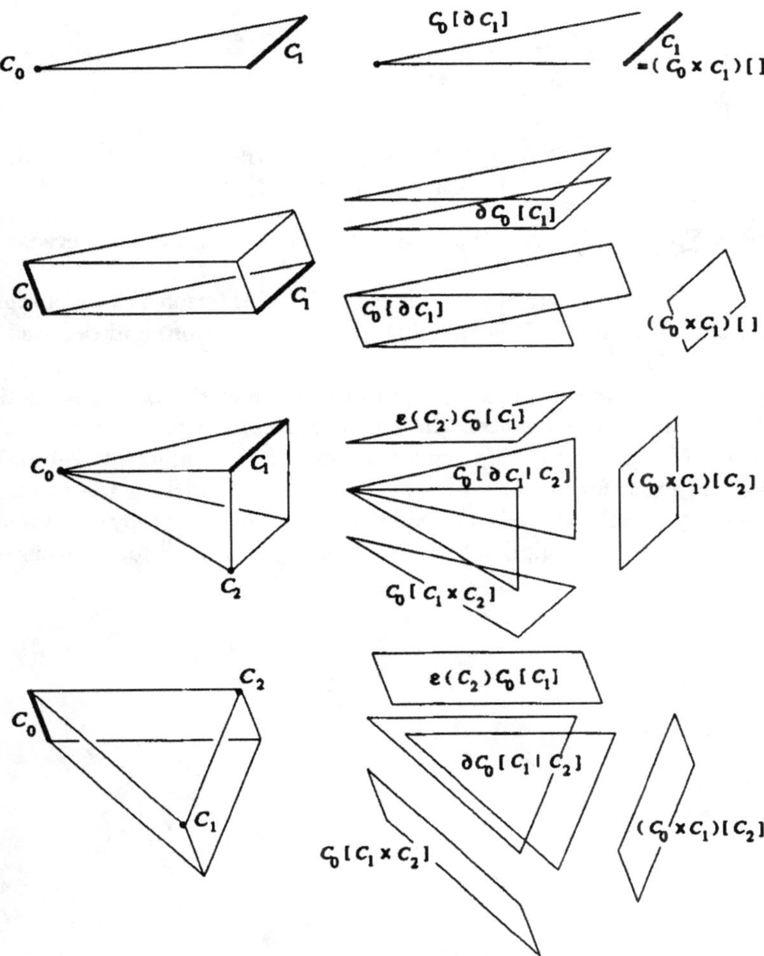

Figure 1: Some geometric bar-products and their boundaries. (a) $C_0[C_1]$, with $\dim C_0 = 0$, $\dim C_1 = 1$. (b) $C_0[C_1]$, with $\dim C_0 = 1$, $\dim C_1 = 1$. (c) $C_0[C_1|C_2]$, with $\dim C_0 = 0$, $\dim C_1 = 1$, $\dim C_2 = 0$. (d) $C_0[C_1|C_2]$, with $\dim C_0 = 1$, $\dim C_1 = 0$, $\dim C_2 = 0$.

2.

$$\begin{aligned}\partial(\gamma_0[\gamma_1|\cdots|\gamma_k]) &= (\partial\gamma_0)[\gamma_1|\cdots|\gamma_k] \\ &+\textstyle\sum_{i=1}^{k}(-1)^{k+\delta(i)}\gamma_0[\gamma_1|\cdots|\partial\gamma_i|\cdots|\gamma_k] \\ &+(g_0\cdot\gamma_1)[\gamma_2|\cdots|\gamma_k] \\ &+\textstyle\sum_{i=1}^{k-1}(-1)^{i}\gamma_0[\gamma_1|\cdots|\gamma_{i-1}|g_i\cdot\gamma_{i+1}|\gamma_{i+2}|\cdots|\gamma_k] \\ &+(-1)^k\epsilon(\gamma_k)\gamma_0[\gamma_1|\cdots|\gamma_{k-1}],\end{aligned}$$

where $\delta(i)=\sum_{j<i}\dim\gamma_j$, and $\epsilon(\gamma_k)=1$ if $\dim\gamma_k=0$, and $=0$ otherwise.

So the set of bar products of singular cubical chains forms a chain complex of singular chains on $E_\Delta G$. It is called the bar construction, and denoted by $\mathcal{E}_*$.

Note that this construction is not quite the same as the bar construction $\mathcal{E}_*$ of [17]. The difference between the two chain complexes corresponds to that between Milnor's and Milgram's versions of the universal G–bundle. The important point for us is that (either) $\mathcal{E}_*$ gives a model for the topology of $E_\Delta G$. More precisely, there is a G-equivariant chain homotopy equivalence from $\mathcal{E}_*$ to the singular (simplicial) complex of $E_\Delta G$. See [18] for a discussion of the two models.

3 The classifying map of a lattice gauge field

We start with a finite simplicial complex Λ and a total ordering of its vertices. For notational convenience we will renumber the vertices 0,1,2,... accordingly. (We will be using the consequent identification of Λ with a subset of Δ^∞.) When we write a simplex of Λ as $< i_0, \ldots, i_r >$, it will always be understood that $i_0 < i_1 < \ldots < i_r$ in this ordering.

3.1 *Simplexes, cubes, and bar-products*

For each r–simplex $\sigma = < i_0, \ldots, i_r > \in \Lambda$, let $C_\sigma \subset \mathbf{R}^\infty$ be the r–cube given by $0 \le s_{i_1} \le 1, \ldots, 0 \le s_{i_r} \le 1$, with the other coordinates zero, and similarly c_σ the $(r-1)$–cube given by $0 \le s_{i_1} \le 1, \ldots, 0 \le s_{i_{r-1}} \le 1$. If $r = 0$, set $C_\sigma = (0, 0, \ldots) = 0$. Furthermore (for future reference) for $1 \le j \le r$, let $\partial^j C_\sigma = \{s_{i_j} = 1\}$, and $\partial^+ C_\sigma = \bigcup_{j=1}^r \partial^j C_\sigma$; and similarly define $\partial^j c_\sigma$ and $\partial^+ c_\sigma$.

To simplify notation in the rest of this section, suppose $\sigma = < 0, \ldots, r >$.

Definition 3.1 (See [15]) The *standard projection* $\pi_\sigma : C_\sigma \to \sigma$ is defined by induction on r. When $r = 0$, C_σ and σ are both points; π_σ takes the first to the second. Now assume that for $\tau = < 0, \ldots, r-1 >$, $\pi_\tau : C_\tau \to \tau$ has been constructed. Then set

$$\pi_\sigma(s_1, \ldots, s_r) = (1 - s_r)\pi_\tau(s_1, \ldots, s_{r-1}) + s_r < r > .$$

3.2 The *bar-product decomposition* of the r–cube C_σ is a useful way of writing it as a union of bar-products of some of its subcubes. (The case $r = 3$ is shown in Figure 2; the case $r = 2$ appears in Figure 4). We continue with the convention that $\sigma = < 0, \ldots, r >$.

For each pair of indices (i, j), with $0 \le i < j \le r$, let $C_\sigma(i, j)$ represent the $(j - i - 1)$-face of C_σ given by

$$\left\{ \begin{array}{ll} s_k = 1 & k \le i \\ s_j = 1 & \\ s_k = 0 & k > j \end{array} \right\}$$

and let $C_\sigma(0)$ represent the point $(0, 0, \ldots) \in C_\sigma$ (this will be $C_\sigma(i_0)$ if i_0 is the first-ordered vertex of σ).

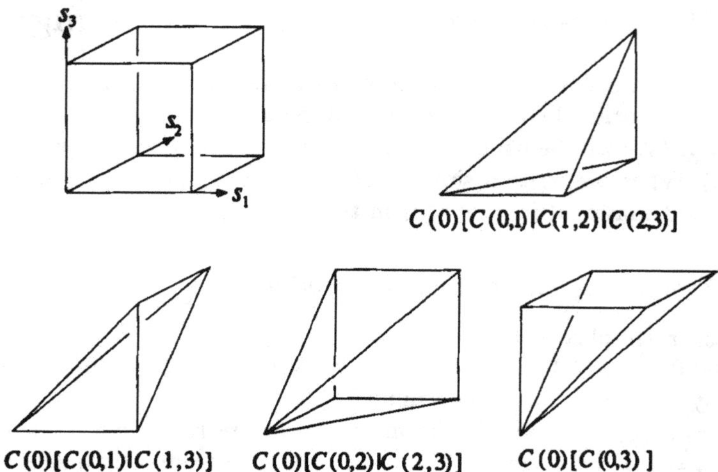

Figure 2: The bar-product decomposition of a 3–cube.

For each set of indices $J = \{0 = j_0 < j_1 < \cdots < j_p = r\}$ define $C_{\sigma,J}$ be the subpolyhedron of C_σ given by the inequalities

$$\begin{array}{ll} s_j \leq s_{j_{k+1}} & \text{for } j_k < j < j_{k+1} \quad k = 1, \ldots, p-1 \\ s_{j_1} \geq \ldots \geq s_{j_p}. & \end{array}$$

Proposition 3.3 (1) $C_{\sigma,J} = C_\sigma(0)[C_\sigma(0, j_1)|C_\sigma(j_1, j_2)| \cdots |C_\sigma(j_{p-1}, r)]$.
(2) As J varies, the $C_{\sigma,J}$ and their faces form a cellular decomposition of C_σ. More precisely, as oriented r-chains: $C_\sigma = \sum_J (-1)^{\theta_J} C_{\sigma,J}$, where $\theta_J = \sum_{i=1}^{p} (j_i - i)$.

Proof: (1) The construction begins with $C_\sigma(j_{p-2}, j_{p-1})[C_\sigma(j_{p-1}, r)]$. We embed this product in C_σ as follows. We construct the cone on $C_\sigma(j_{p-1}, r)$ from the vertex $o_p = \{s_j = 1 \ \textit{for}\ j \leq j_{p-1}, s_j = 0 \ \textit{for}\ j \geq j_p\}$. This cone and $C_\sigma(j_{p-2}, j_{p-1})$ have exactly the point o_p in common, so their cartesian product embeds in C_σ *via* the map $C_\sigma(j_{p-2}, j_{p-1}) \times \textit{cone}\{C_\sigma(j_{p-1}, r)\} \to C_\sigma$ given by $(a, b) \to a + b - o_p$, with addition defined componentwise as usual.

For the next step, i.e. $C_\sigma(j_{p-3}, j_{p-3})[C_\sigma(j_{p-2}, j_{p-1})|C_\sigma(j_{p-1}, r)]$ we cone from $o_{p-1} = \{s_j = 1 \ \textit{for}\ j \leq j_{p-2}, s_j = 0 \ \textit{for}\ j \geq j_{p-1}\}$. This cone only has o_{p-1} in common with $C_\sigma(j_{p-3}, j_{p-3})$, so we can embed their cartesian product in C_σ as before.

The beginning of this construction may be made clearer by the following table, where we abbreviate $C_\sigma(j_{p-1}, r)$ by C_p, etc. Coordinates marked with

$*$ vary between 0 and 1. The coning variables are labelled $T_1, \ldots, T_p$; for their relation to the "barycentric" coordinates of Remark 2.3 see Remark 3.4 below.

coordinates	$[1, j_{p-2}-1]$	j_{p-2}	$[j_{p-2}+1, j_{p-1}-1]$	j_{p-1}	$[j_{p-1}+1, r-1]$	r
C_p	1	1	1	1	$*$	1
o_p	1	1	1	1	0	0
$\text{cone}\{C_p\}$	1	1	1	1	T_p*	T_p
C_{p-1}	1	1	$*$	1	0	0
$C_{p-1} \times \text{cone}\{C_p\}$	1	1	$*$	1	T_p*	T_p
o_{p-1}	1	1	0	0	0	0
$\text{cone}\{C_{p-1}[C_p]\}$	1	1	$T_{p-1}*$	T_{p-1}	$T_{p-1}T_p*$	$T_{p-1}T_p$

Finally we arrive at $C_\sigma(0)[C_\sigma(0,j_1)|C_\sigma(j_1,j_2)|\cdots|C_\sigma(j_{p-1},r)]$, where we have already embedded $C_\sigma(0,j_1)[C_\sigma(j_1,j_2)|\cdots|C_\sigma(j_{p-1},r)]$ as

$$\{(s_1,\ldots,s_{j_1-1},1,T_2 s_{j_1+1},\ldots,T_2 s_{j_2-1},T_2,\ldots,\ldots,T_2T_3\cdots T_p)\}.$$

We cone from $(0,\ldots,0) = o_1$ but this time the cartesian product is with $C_\sigma(0) = o_1$ itself, so leaves the cone unchanged, and gives us the geometric bar-product $C_\sigma(0)[C_\sigma(0,j_1)|C_\sigma(j_1,j_2)|\cdots|C_\sigma(j_{p-1},r)]$ as

$$\{(T_1 s_1,\ldots,T_1 s_{j_1-1},T_1,T_1T_2 s_{j_1+1},\ldots,T_1T_2 s_{j_2-1},T_1T_2,\ldots,\ldots,T_1T_2T_3\cdots T_p)\}$$

which is clearly $C_{\sigma,J}$.

For the proof of (2) we refer to [17]; see Figure 2 for the cases $r = 2,3$ which will be the ones of most importance for this work.□

Remark 3.4 A point $x \in C_\sigma(0)[C_\sigma(0,j_1)|C_\sigma(j_1,j_2)|\cdots|C_\sigma(j_{p-1},r)]$ has two sets of coordinates: the "barycentric" coordinates of Remark 2.3

$$x = t_0\vec{s}_0 + t_1(\vec{s}_0,\vec{s}_1) + t_2(\vec{s}_0,\vec{s}_1,\vec{s}_2) + \cdots + t_p(\vec{s}_0,\vec{s}_1,\ldots,\vec{s}_p)$$

(with $\sum t_i = 1$) and the "interior" coordinates just calculated, which may be written

$$x = T_1\vec{s}_1, T_1, T_1T_2\vec{s}_2, T_1T_2, \ldots, T_1\cdots T_p\vec{s}_p, T_1\cdots T_p.$$

For convenience in calculating the bar-product of singular cubes in Section 3.3 we note the relation

$$\begin{aligned} t_0 &= 1 - T_1 \\ t_1 &= T_1(1-T_2) \\ t_2 &= T_1T_2(1-T_3) \\ \vdots & \quad \vdots \\ t_p &= T_1T_2\cdots T_p. \end{aligned}$$

3.2 Parallel transport functions

Definition 3.5 A G-valued *parallel transport function* (p.t.f.; G a Lie group) over a locally ordered simplicial complex Λ consists of a family $\mathbf{V}$ of piecewise smooth maps of cubes into G, one $V_\sigma: c_\sigma \to G$ for each r-simplex σ of Λ, $r \geq 1$, such that, for every $\sigma = < i_0, \ldots, i_r >$ the following compatibility conditions hold:

1. Cocycle condition

$$V_\sigma(s_{i_1}, ..., s_{i_{p-1}}, s_{i_p} = 1, s_{i_{p+1}}, ..., s_{i_{r-1}}) =$$
$$V_{<i_0,...i_p>}(s_{i_1}, ..., s_{i_{p-1}}) \cdot V_{<i_p,...i_r>}(s_{i_{p+1}}, ..., s_{i_{r-1}}),$$

2. Compatibility with faces

$$V_\sigma(s_{i_1}, ..., s_{i_{p-1}}, s_{i_p} = 0, s_{i_{p+1}}, ..., s_{i_{r-1}}) =$$
$$V_{<i_0,...i_{p-1},i_{p+1},...i_{r-1}>}(s_{i_1}, ..., s_{i_{p-1}}, s_{i_{p+1}}, ..., s_{i_{r-1}}).$$

(See [15] for the relation of p.t.f.'s to coordinate bundles, etc.)

From a G-valued p.t.f. $\mathbf{V}$ over Λ we shall construct a map

$$f = f(\mathbf{V}) : |\Lambda| \to B_\Delta G;$$

this map induces a principal G–bundle $\xi = \xi(\mathbf{V})$ over $|\Lambda|$. We shall give an algorithm that from a generic lattice gauge field $\mathbf{u}$ over Λ constructs a p.t.f. $\mathbf{V}$; if in addition $\mathbf{u}$ satisfies a certain "continuity condition", the resulting bundle $\xi(\mathbf{V})$ will turn out to depend only on $\mathbf{u}$.

3.3 The classifying map of a p.t.f.

Let $\mathbf{V}$ be a G-valued p.t.f. on Λ, and consider an r–simplex $\sigma \in \Lambda$ which we will take again as $< 0, \ldots, r >$. For each pair (i, j) with $0 \leq i < j \leq r$, set $\tau(i,j) =< i, i+1, \ldots, j-1, j >$. We can identify $C_\sigma(i,j)$ with c_τ since both are parametrized by $(s_{i+1}, \ldots, s_{j-1})$. In particular, since $\tau(0,r) = \sigma$, we identify the domain c_σ of V_σ with the face $C_\sigma(0,r) = \{s_r = 1\}$ of C_σ.

Define $\gamma_\sigma(i,j) : C_\sigma(i,j) \to G_j$ by

$$\gamma_\sigma(i,j)(s_{i+1}, \ldots, s_{j-1}) = V_{\tau(i,j)}(s_{i+1}, \ldots, s_{j-1}) \cdot \mathbf{e}_j,$$

where G_j is the jth factor of $E_\Delta G = G_0 * G_1 * \cdots$. Also, for the first-ordered vertex 0, define $\gamma_\sigma(0)$ to take the point $C_\sigma(0) = (0, 0, ...)$ to the identity element $\mathbf{e}_0$ in G_0.

Now (compare Proposition 3.3) for each $J = (0 = j_0 < j_1 < \cdots < j_p = r)$ set $\gamma_{\sigma,J} : C_{\sigma,J} \to E_\Delta G$ to be the bar product

$$\gamma_{\sigma,J} = \gamma_\sigma(0)[\gamma_\sigma(0,j_1)|\cdots|\gamma_\sigma(j_{p-1},r)].$$

Proposition 3.6 As J varies, the maps $\gamma_{\sigma,J}$ agree on the intersections of their domains; so they combine to define a map $\Gamma_\sigma : C_\sigma \to E_\Delta G$. □

For examples derived from lattice gauge fields, see Figures 4 and 7.

Proposition 3.7 (1) There exists a map (necessarily unique) $f_\sigma : \sigma \to B_\Delta G$ such that the following diagram commutes:

$$\begin{array}{ccccc} & & \Gamma_\sigma & & \\ & C_\sigma & \rightarrow & E_\Delta G & \\ \pi_\sigma & \downarrow & & \downarrow & \pi_\Delta . \\ & \sigma & \rightarrow & B_\Delta G & \\ & & f_\sigma & & \end{array}$$

(2) If τ is a face of σ, then $f_\sigma|_\tau = f_\tau$. Thus the $\{f_\sigma : \sigma \in \Lambda\}$ combine to give a map $f = f(\mathbf{V}) : |\Lambda| \to B_\Delta G$. Moreover f is an embedding.

Proof: Direct check, or see [17]. The fact that f is an embedding follows from the fact that

$$f_\sigma(\sum t_i < i >) = \pi_\Delta(\sum t_i g_i \mathbf{e}_i)$$

for suitable g_i. □

Proposition 3.8 Continuing with the same notation, let $\xi = (\pi : E \to |\Lambda|)$ be the pull-back $f^*(\xi_\Delta G)$, and let $F = F(\mathbf{V}): E \to E_\Delta G$ be the bundle map covering f. Then ξ is uniquely determined by $\mathbf{V}$ up to isomorphism. Moreover, since F is an embedding, we may identify E with $F(E) = E_\Delta G|_{f(|\Lambda|)}$. □

We write $\xi = \xi(\mathbf{V})$.

Definition 3.9 Let $\mathbf{V}$ and $\mathbf{V}'$ be G-valued p.t.f.'s on Λ. A ***gauge-equivalence*** of $\mathbf{V}$ with $\mathbf{V}'$ is a family $\mathbf{g} = \{g_i\}_{<i>\in\Lambda}$ of elements of G such that for every $\sigma = < i_0, \dots, i_r >$ of dimension ≥ 1 and every $\vec{s} \in c_\sigma$,

$$V'_\sigma(\vec{s}\,) = (g_{i_0})^{-1} \cdot V_\sigma(\vec{s}\,) \cdot g_{i_r}.$$

We write $\mathbf{V}' = Ad_{\mathbf{g}}(\mathbf{V})$.

Proposition 3.10 Suppose $\mathbf{V}' = Ad_{\mathbf{g}}(\mathbf{V})$. Let $f, f' : X \to B_\Delta G$ be the classifying maps determined by $\mathbf{V}, \mathbf{V}'$ respectively, as in Prop. 3.7 and let $\xi(\mathbf{V}) = (\pi : E \to X)$, $\xi(\mathbf{V}') = (\pi' : E' \to X)$ be the corresponding bundles, with $X = |\Lambda|$ as usual. Set $\mathbf{h} = \{h_i\}, i = 0, 1, 2, \ldots$ to be the family of G–elements obtained by extending $\mathbf{g}$ by the identity for indices not corresponding to vertices of Λ, so $h_i = g_i$ if $< i > \in \Lambda$, and $h_i = e$ otherwise. Then the right-automorphism $R_{\mathbf{h}}$, restricted to $E = E_\Delta G|_{f(X)}$ gives a bundle isomorphism $\rho_{\mathbf{g}} : E \to E' = E_\Delta G|_{f'(X)}$. In particular, $\xi(\mathbf{V}) \approx \xi(\mathbf{V}')$.

Proof. For each $\sigma \in \Lambda$, let the $\gamma_{\sigma,J}$ and $\Gamma_\sigma : C_\sigma \to E_\Delta G$ be as in Proposition 3.6 for $\mathbf{V}$, and similarly let the $\gamma'_{\sigma,J}$ and Γ'_σ correspond to $\mathbf{V}'$. Write σ as $< 0, \ldots, r >$. By hypothesis, for every i and j with $0 \leq i < j \leq r$,

$$V'_{\tau(i,j)} = (g_i)^{-1} V_{\tau(i,j)} g_j,$$

which means that

$$\gamma'_\sigma(i,j) = (g_i)^{-1} \cdot \gamma_\sigma(i,j) \cdot g_j$$

where the dots refer to the global left and right actions of G on $E_\Delta G$. The corresponding relation for bar-products of the corresponding $\gamma_\sigma(i,j)$ is then given by

$$\gamma'_{\sigma,J} = (g_0)^{-1} \cdot R_{\mathbf{h}} \gamma_{\sigma,J},$$

where $J = (0 = j_0 < j_1 < \ldots < j_p = r)$. So

$$\pi_\Delta \circ \gamma'_{\sigma,J} = \pi_\Delta \circ R_{\mathbf{h}} \circ \gamma_{\sigma,J} = r_{\mathbf{h}} \circ \pi_\Delta \circ \gamma_{\sigma,J},$$

so $\pi_\Delta \circ \Gamma'_\sigma = r_{\mathbf{h}} \circ \pi_\Delta \circ \Gamma_\sigma$ and thus $f'_\sigma = r_{\mathbf{h}} \circ f_\sigma$, where $f_\sigma, f'_\sigma : \sigma \to B_\Delta G$ are as in Proposition 3.7 (1) for $\mathbf{V}$ and $\mathbf{V}'$ respectively. It follows that $f' = r_{\mathbf{h}} \circ f$, and that the automorphism $R_{\mathbf{h}} : E_\Delta G \to E_\Delta G$ carries $F(E)$ isomorphically onto $F'(E')$. Setting $\rho_{\mathbf{g}} = (F')^{-1} \circ R_{\mathbf{h}} \circ F$ completes the argument. □

3.4 Lattice gauge fields and p.t.f.'s

Definition 3.11 A G-valued ***lattice gauge field*** (l.g.f.) on Λ is a collection $\mathbf{u} = \{u_{ij}\}$ of group elements, one for each 1–simplex $< ij >$ of Λ, subject to the condition $u_{ji} = u_{ij}^{-1}$. If $\mathbf{V}$ is a p.t.f. on Λ satisfying $u_{ij} = V_{<ij>}(0)$ for each 1–simplex $< ij >$ ($V_{<ij>}(0)$ is the image of $V_{<ij>}$ in this case), we will say $\mathbf{V}$ ***is consistent with*** $\mathbf{u}$.

Proposition 3.12 Let $\mathbf{u}$ be an $SU(2)$–valued l.g.f. on Λ. Let K be a subcomplex of Λ, and let a p.t.f. $\mathbf{V}$ on K be given which is consistent with $\mathbf{u}|_K$. Assume every simplex of Λ which is not in K has dimension ≤ 4. Then there exists a p.t.f. $\mathbf{V}'$ on Λ which is consistent with $\mathbf{u}$ and is such that $V'_\sigma = V_\sigma$ for every $\sigma \in K$.

Proof. We use induction on $r = \dim \sigma$ to construct $\{V'_\sigma\}$ for $\sigma \notin K$, $\dim \sigma \geq 1$. When $r = 1$, then V'_σ is determined by the consistency requirement. For the general step, suppose $\sigma =< 0, \ldots, r >$ and that V'_τ has been constructed for every proper face τ if σ, so as to satisfy the axioms for a p.t.f. and the requirements of the proposition. By those axioms, V'_σ is determined on ∂c_σ:

$$V'_\sigma = \begin{cases} V'_{<0\ldots\hat{i}\ldots r>} & on\ \{s_i = 0\}, i = 1, \ldots, r-1 \\ V'_{<0\ldots i>} V'_{<i\ldots r>} & on\ \{s_i = 1\}, i = 1, \ldots, r-1. \end{cases}$$

(We omit the verification that these formulae are consistent.) Now $r \leq 4$ implies $\dim(\partial c_\sigma) \leq 2$; and since $SU(2)$ is 2–connected, there is no obstruction to extending V'_σ to a map $V'_\sigma : c_\sigma \to SU(2)$. This completes the inductive step. □

We shall make use of this result in Section 5. Its weakness is that the isomorphism class of the bundle $\xi(\mathbf{V}')$ may depend on the choice of $\mathbf{V}'$, not just on $\mathbf{u}$ and $\mathbf{V}$. To obtain a bundle independent of choices, one must restrict the class of l.g.f.'s considered.

Definition 3.13 $\mathbf{u}$ is ***continuous*** if, for every 2-simplex $< ijk > \in \Lambda$,

$$d(u_{ij}u_{jk}, u_{ik}) < \pi/n$$

where $n = \dim \Lambda$, and the distance $d(\cdot,\cdot)$ is measured in the unit–sphere metric on $SU(2)$. (This condition is somewhat weaker than the ones used to define "continuous" in [14] and in [15]).

For the computations to follow, we shall need an explicit choice of $\mathbf{V}$, for a given continuous $\mathbf{u}$. This p.t.f. will be denoted $\mathbf{V}(\mathbf{u})$, and called the p.t.f. obtained from $\mathbf{u}$ by ***iterated coning***.

Proposition 3.14 Let $\mathbf{u}$ be a continuous l.g.f. on Λ. Set

$$\delta(\mathbf{u}) = \max_{<ijk>\in\Lambda} \{d(u_{ij}u_{jk}, u_{ik})\}.$$

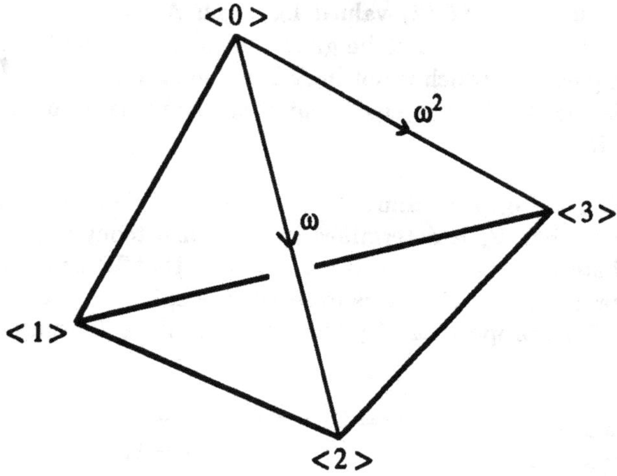

Figure 3: A $U(1)$-valued l.g.f. on the surface of a tetrahedron. ω and ω^2 are complex cube roots of 1. Unmarked bonds carry the transporter 1.

(1) The iterated coning procedure yields a p.t.f. V on Λ such that:
(i) V is consistent with u
(ii) Whenever $\sigma =< i \ldots j >$ has dimension $r \geq 2$, then

$$d(V_\sigma(\vec{s}), u_{ij}) \leq \delta(\mathbf{u})(r-1),$$

for every $\vec{s} \in c_\sigma^{r-1}$.
(2) Moreover if V′ is another p.t.f. on Λ for which (i) and (ii) hold, then the bundles $\xi(\mathbf{V})$ and $\xi(\mathbf{V}')$ are isomorphic.

We write $\xi(\mathbf{u})$ for any $\xi(\mathbf{V})$ as in the proposition.

Figures 3, 4, 5 show how this procedure works in the more easily illustrated case $G = U(1)$. In these figures Λ is the surface of a tetrahedron $< 0123 >$, and **u** is made up of the third roots of unity $1, \omega, \omega^2$. Figure 6 gives a topological proof that the induced bundle is non-trivial: collapsing the face $< 123 >$ to a single vertex $< \infty >$ allows the use of $S^3 \to S^2$ as the universal bundle; the resulting map of Λ to S^2 is patently essential. It is instructive to repeat this exercise with the trivial l.g.f. $u_{ij} \equiv 1$. In that case the image of f is a single meridian of S^2.

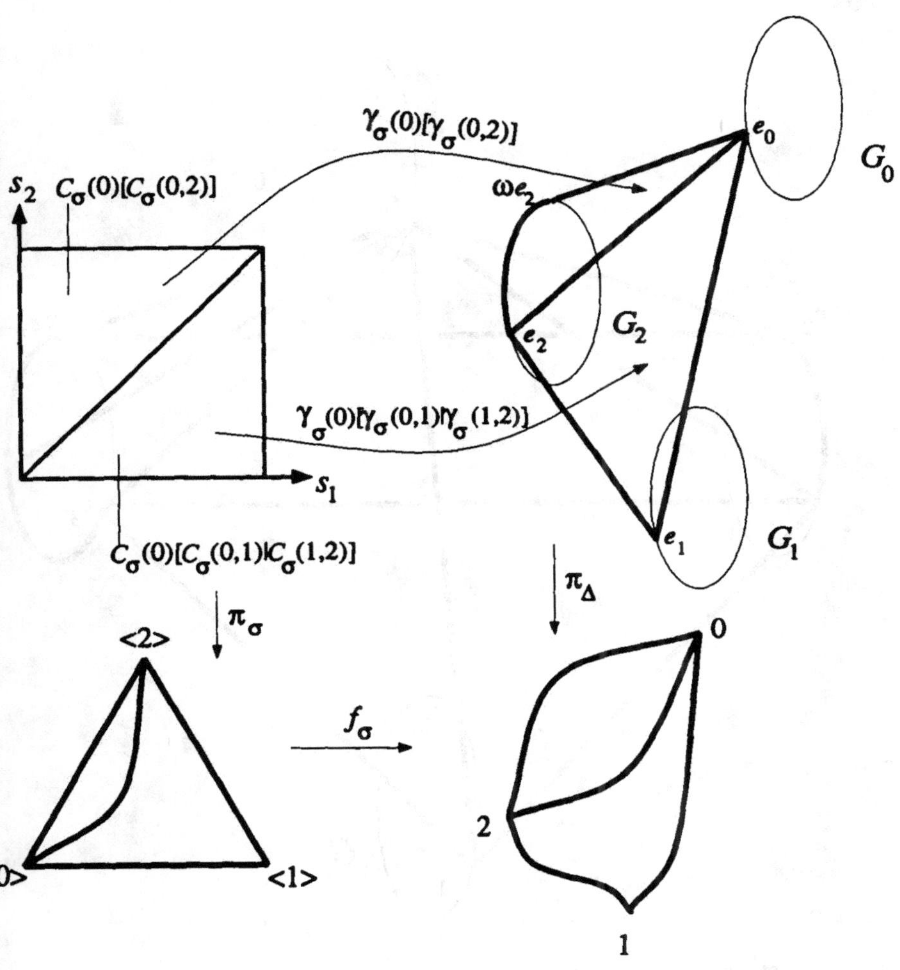

Figure 4: The maps Γ_σ and f_σ for $\sigma = < 012 >$.

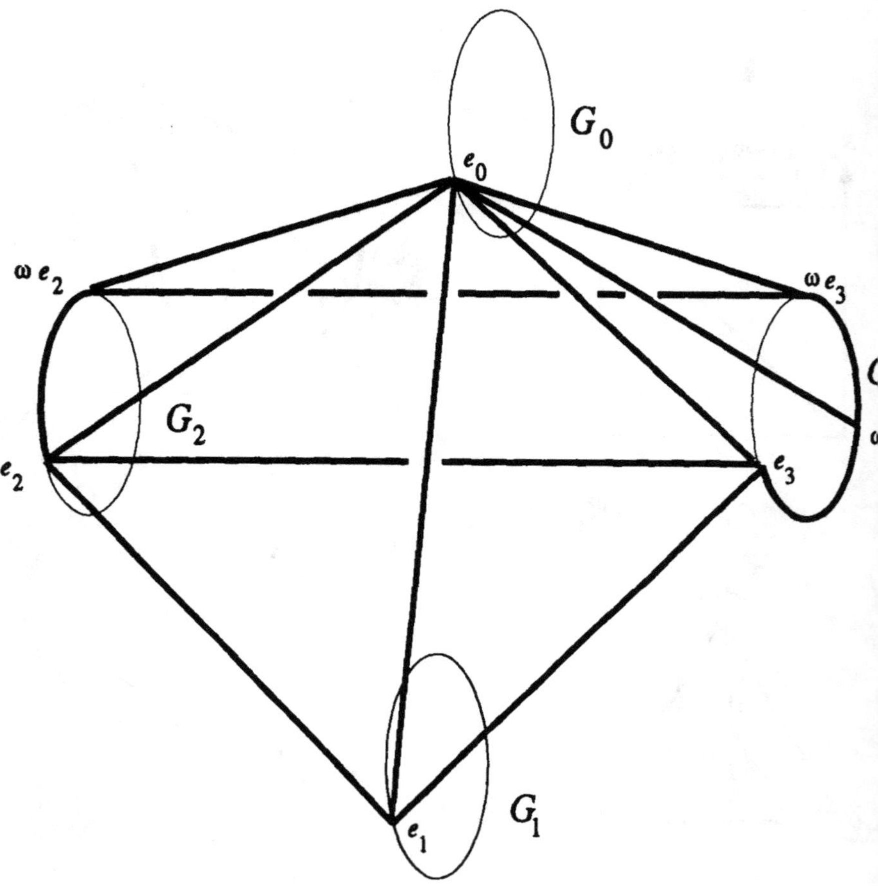

Figure 5: The four Γ_σ's; the horizontal lines in the rear of the figure are distinct, but have the same projection in $B_\Delta G$.

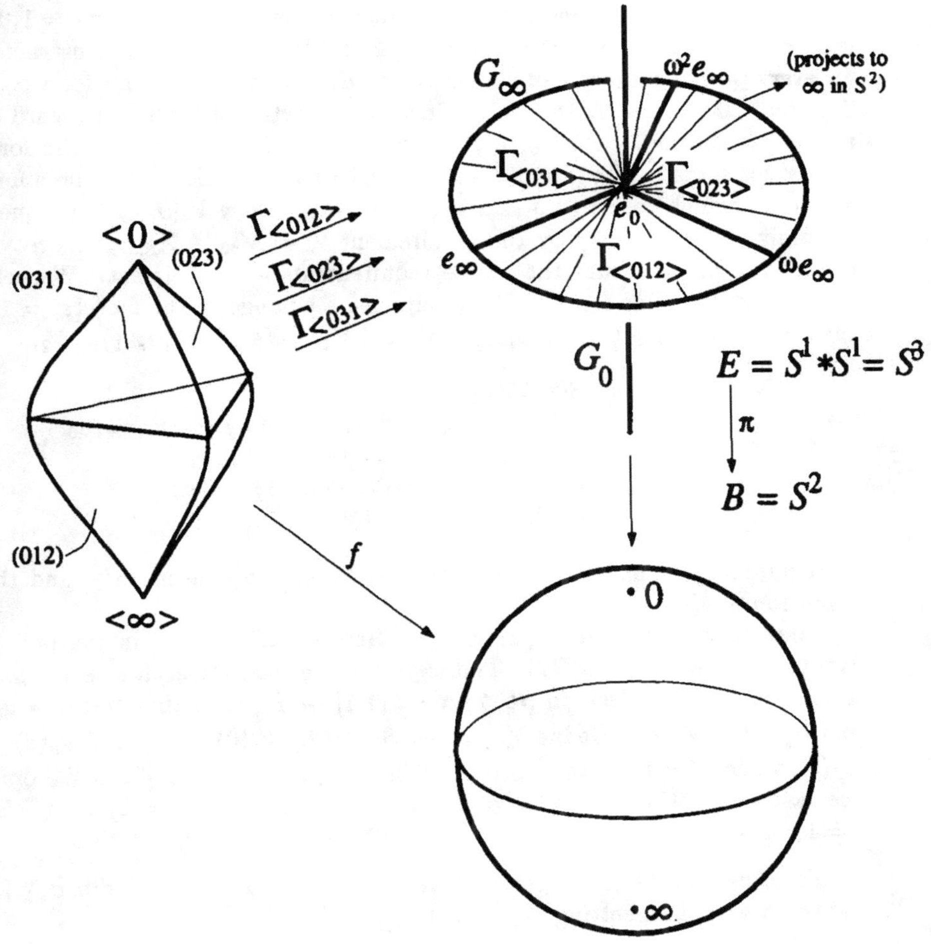

Figure 6: Collapse $< 123 >$ to a single vertex $< \infty >$. The Γ's now map to $G_0 * G_\infty = S^3 \subset E_\Delta G$. The induced $f\colon |\Lambda| \to S^2 = S^3/U(1)$ covers the sphere exactly once.

Proof of (1). V_σ is constructed by induction on $r = \dim\sigma$. For $r = 1$, V_σ is specified by (i). Now assume that $r \geq 2$ and that V_τ has been constructed for every proper subface τ of σ so as to satisfy (ii) and the axioms for a p.t.f. Say $\sigma = < 0 \ldots r >$. With $\partial^+ c_\sigma$ defined as in Section 3.1, we can regard c_σ as the cone from $\vec{0}$ on $\partial^+ c_\sigma$, i.e. every $\vec{s} \in c_\sigma$ can be written in the form $\vec{s} = t\vec{s'}$ for some $t \in [0,1]$ and $\vec{s} \in \partial^+ c_\sigma$; and this expression is unique unless $\vec{s} = \vec{0}$ (in which case $t = 0$ and $\vec{s'}$ is arbitrary). Now $V_\sigma|\partial^+ c_\sigma$ is uniquely determined on $\{s_i = 1\}$ by the requirement $V_\sigma = V_{<0\ldots i>} V_{<i\ldots r>}$; moreover the p.t.f. axioms ensure that these requirements are consistent. We next check that (ii) holds for $\vec{s} \in \partial^+ c_\sigma$. Suppose $\vec{s}$ belongs to the face $\{s_i = 1\}$; then, writing $V_{<0\ldots i>}(s_1, \ldots, s_{i-1}) = V_1$ and $V_{<i\ldots r>}(s_{i+1}, \ldots, s_{r-1}) = V_2$,

$$\begin{aligned} d(V_\sigma(\vec{s}), u_{0r}) &= d(V_1 V_2, u_{0r}) \\ &\leq d(V_1 V_2, u_{0i} V_2) + d(u_{0i} V_2, u_{0i} u_{ir}) + d(u_{0i} u_{ir}, u_{0r}) \\ &= d(V_1, u_{0i}) + d(V_2, u_{ir}) + d(u_{0i} u_{ir}, u_{0r}) \\ &\leq \delta(\mathbf{u})(i-1) + \delta(\mathbf{u})(r-i-1) + \delta(\mathbf{u}) \\ &= \delta(\mathbf{u})(r-1), \end{aligned}$$

as required. (We have used the fact that V_2 and u_{0i} are unitary, and the definition of $\delta(\mathbf{u})$.)

Since $\mathbf{u}$ is continuous, $\delta(\mathbf{u}) < \pi/r$. Hence $V_\sigma(\partial^+ c_\sigma)$ lies in the ball of radius π about u_{0r} in $SU(2)$. Therefore there exists, for each $\vec{s'} \in \partial^+ c_\sigma$, a unique shortest geodesic in $SU(2)$, $\alpha_{\vec{s'}} : [0,1] \to S^3$, such that $\alpha_{\vec{s'}}(0) = u_{0r}$ and $\alpha_{\vec{s'}}(1) = V_\sigma(\vec{s'})$. Define $V_\sigma : c_\sigma \to SU(2)$ by $V_\sigma(0) = u_{0r}$ and $V_\sigma(\vec{s}) = \alpha_{\vec{s'}}(t)$, where $\vec{s} = t\vec{s'}$ with $\vec{s'} \in \partial^+ c_\sigma$. Then (ii) holds on all of c_σ. We omit the last verification: that $V_\sigma = V_{<0\ldots\hat{i}\ldots r>}$ on the face $\{s_i = 0\}$ of c_σ, for $i = 1, \ldots, r$.

Proof of (2). We shall construct $\mathbf{W} = \{W_\sigma : c_\sigma \times I \to SU(2), \dim\sigma \geq 2\}$ in such a way that, setting $W_\sigma^t = W_\sigma|_{c_\sigma \times \{t\}}$,

- $W_\sigma^0 = V_\sigma$; $W_\sigma^1 = V'_\sigma$
- $\mathbf{W}^t = \{W_\sigma^t, \dim\sigma \geq 2\}$ is a p.t.f. on Λ which satisfies (i) and (ii) of (1).

Then for each t we get a principal $SU(2)$-bundle $\xi^t = \xi(\mathbf{W}^t)$ over $|\Lambda|$. It follows from the construction of the classifying map $f^t = f(\mathbf{W}^t) : |\Lambda| \to B_\Delta SU(2)$ that f^t varies continuously in t. Hence the ξ^t are all isomorphic bundles; in particular $\xi = \xi'$.

We construct W_σ by induction on $r = \dim\sigma$.

For $r = 1$, W_σ^t is determined by (i).

Now assume that $r \geq 2$ and that W_τ has been constructed for every proper face of σ so as to satisfy (ii) and the axioms of a p.t.f. Say $\sigma = < 0, \ldots, r >$. The induction hypothesis determines W_σ on $\partial(c_\sigma \times I)$ in the following way:

(a) on $c_\sigma \times 0$ and $c_\sigma \times 1$ by V_σ and V'_σ, respectively;

(b) on $\{s_i = 0\}$, for $i = 1, \ldots, r-1$ by $W_{\tau(i)}$, where $\tau(i) = < 0, \ldots, \hat{i}, \ldots, r >$;

(c) on $\{s_i = 1\}$, for $i = 1, \ldots, r-1$ by $W_{<0,\ldots,i>} \cdot W_{<i,\ldots,r>}$.

It follows that $W_\sigma(\partial(c_\sigma \times I))$ is contained in the ball $B(u_{0r}, \delta(\mathbf{u})(r-1))$ about u_{0r} of radius $\delta(\mathbf{u})(r-1)$ in $SU(2)$: in case (a) this is the hypothesis of (2); in (b) it follows from the induction hypothesis and (ii); and in case (c) by the argument used in the proof of (1).

Since $\delta(\mathbf{u}) < \pi/n$ by the continuity of $\mathbf{u}$, it follows that $\delta(\mathbf{u})(r-1) < \pi$; so $B(u_{0r}, \delta(\mathbf{u})(r-1))$ is contractible in $SU(2)$, and the required extension W_σ thus exists.□

Remark 3.15 This algorithm can be performed, even if $\mathbf{u}$ is not continuous, on simplexes of dimension ≤ 4, provided $\mathbf{u}$ is "generic." (No uniqueness guarantees, however.)

Definition 3.16 Let $\mathbf{u}$ and $\mathbf{u}'$ be l.g.f.'s on Λ. They are called ***gauge equivalent*** if there exists $\mathbf{g} = \{g_i \in G\}_{<i> \in \Lambda}$ such that for every $< ij > \in \Lambda$, $u'_{ij} = g_i^{-1} u_{ij} g_j$. We write $\mathbf{u}' = Ad_{\mathbf{g}}(\mathbf{u})$.

Proposition 3.17 Let $\mathbf{u}' = Ad_{\mathbf{g}}(\mathbf{u})$. Then:
(1) If $\mathbf{u}$ is continuous, then so is $\mathbf{u}'$.
(2) The p.t.f.'s $\mathbf{V}(\mathbf{u})$ and $\mathbf{V}(\mathbf{u}')$ constructed by iterated coning are related by

$$\mathbf{V}(\mathbf{u}') = Ad_{\mathbf{g}}(\mathbf{V}(\mathbf{u})).$$

In particular, $\xi(\mathbf{u}') \approx \xi(\mathbf{u})$. □

4 Dupont's canonical connection in Milnor's universal bundle

4.1 Dupont's connection

If G is a Lie group, the total space $E_\Delta G$ of Milnor's universal bundle is a stratified space. The strata can be labelled by sets of indices:

$$\Sigma_{i_0,\dots,i_m} = \{\sum t_i g_i \mathbf{e}_i | \begin{matrix} t_i = 0 \ if \ i \neq i_0,\dots,i_m \\ t_i \neq 0 \ if \ i = i_0,\dots,i_m \end{matrix} \}.$$

These strata are invariant under the global left and right–actions of G.

Definition 4.1 The *canonical connection* ω_Δ on $E_\Delta G$ is defined as follows:

Let $x \in TE_\Delta G$ be a vector tangent to one of the strata, say $\Sigma_{i_0,\dots,i_m}$; i.e. let

$$x(s) = \sum_{j=0}^{m} t_{i_j}(s) g_{i_j}(s) \mathbf{e}_{i_j}$$

be a smooth curve in $\Sigma_{i_0,\dots,i_m}$, defined on some open interval containing 0, such that $x'(0) = x$.

Then for each j, $g_{i_j}(s) \cdot g_{i_j}^{-1}(0)$ is a curve through e in G. So

$$\omega_{i_j}(x) = \frac{d}{ds}[g_{i_j}(s) \cdot g_{i_j}^{-1}(0)]|_{s=0}$$

is an element of $TG|_e$, i.e. of the Lie Algebra $\mathfrak{g}$. Now set

$$\omega_\Delta(x) = \sum t_{i_j}(0) \omega_{i_j}(x).$$

Thus we may write $\omega_\Delta = \sum t_{i_j} dg_{i_j} g_{i_j}^{-1} = \sum t_{i_j} R_{g_{i_j}*}^{-1} dg_{i_j}$, using $R_g : G \to G$ for right-multiplication by g as usual. We leave the reader to check that the constructions of $\omega_{i_j}(x)$ and $\omega_\Delta(x)$ are independent of the choice of curve $x(s)$ satisfying (1) and (2). This connection is due to Dupont [4] who gives a general construction of connections in semi-simplicial bundles and specializes it to the Milgram model of the universal bundle. We will briefly develop its properties as they are manifested in $E_\Delta G$.

The next lemma shows that ω_Δ does indeed have the characteristic properties of a connection.

Proposition 4.2 ω_Δ satisfies the following conditions.

(1) Let $k \in \mathfrak{g}$, and let $p \to k_p$ be the corresponding fundamental vectorfield (see below) on $E_\Delta G$. Then, for any $p \in E_\Delta G$,

$$\omega_\Delta(k_p) = k.$$

(2) Let $g \in G$, and let $L_g : E_\Delta G \to E_\Delta G$ represent left-multiplication by g. Consider a vector $x \in T_p E_\Delta G$. Then

$$\omega_\Delta(L_{g*}x) = Ad_g(\omega_\Delta(x)).$$

Proof: (1) Take k and $p \to k_p$ as in the statement. Thus if $k(s)$ is a curve in G with $k(0) = e$ and $k'(0) = k$ (for example, $k(s) = \exp(sk)$), then at $p = \sum t_i g_i \mathbf{e}_i \in E_\Delta G$ the vector k_p is given by

$$k_p = \frac{d(k(s) \cdot p)}{ds}|_{s=0} = \frac{d \sum t_i k(s) g_i \mathbf{e}_i}{ds}|_{s=0},$$

the dot referring to the global left-action of G on $E_\Delta G$. It follows that

$$\omega_j(k_p) = \frac{d}{ds}[k(s)g_i \cdot g_i^{-1}] = k$$

$$\omega_\Delta(k_p) = \sum_j t_j k = k,$$

as required.

(2) To study $L_g^* \omega_\Delta$ we examine its value on a test vector $v \in TE_\Delta G|_p$, say $v = c'(0)$ where $c(s) = \sum t_i(s) g_i(s) \mathbf{e}_i$ is a curve in $E_\Delta G$ with $c(0) = p$. Then $L_{g*}(v)$ is the tangent vector to the curve $L_g \circ c(s) = \sum t_i(s) g g_i(s) \mathbf{e}_i$. Now

$$\omega_\Delta(v) = \sum t_i(0) R_{g_i(0)*}^{-1} \frac{dg_i(s)}{ds}|_{s=0}$$

where $\frac{dg_i(s)}{ds}|_0 \in TG|_{g_i(0)}$ and the linear combination takes place in $TG|_e = \mathfrak{g}$. So

$$\omega_\Delta(L_{g*}(v)) = \sum t_i(0) R_{gg_i(0)*}^{-1} \frac{dgg_i(s)}{ds}|_{s=0} = \sum t_i(0) R_{g*}^{-1} R_{g_i(0)*}^{-1} L_{g*} \frac{dg_i(s)}{ds}|_{s=0}.$$

Since $R_{g_i(0)}^{-1}$ and L_g commute, this is

$$R_{g*}^{-1} L_{g*} \omega_\Delta(v) = Ad_g \omega_\Delta(v),$$

as required. □

For $\mathbf{h} = \{h_0, h_1, \ldots\}, h_i \in G$, let $R_\mathbf{h} : E_\Delta G \to E_\Delta G$ be the right automorphism given in Paragraph 2.1.

Proposition 4.3 The automorphism $R_{\mathfrak{h}}$ preserves ω_Δ.

Proof: The identity $R_{\mathfrak{h}}^*\omega_\Delta = \omega_\Delta$ is established in the same way as (2) of Proposition 4.2.□

4.2 *Its relation to the classical standard connection when $G = SU(2)$*

Dupont shows how the curvature and Chern-Simons forms may be defined for connections in general semi-simplicial bundles. In the interest of explicitness, we will show how, when $G = SU(2) =$ the group of unit quaternions, and $\mathfrak{g} =$ the vector space of pure imaginary quaternions, these forms may also be defined directly, in terms of the standard, classical representatives.

The standard, classical model of the universal $SU(2)$–bundle is $(E \to B) = \lim_k(E_k \to B_k)$ where $E_k = S^{4k+3}$, the unit sphere in quaternionic $(k+1)$–space, B_k is quaternionic projective k–space, and the quotient is by the action of $SU(2)$ as the group of unit quaternions.

There is a standard connection in this bundle. We will define it for a fixed k but it clearly is compatible with the inclusions $E_k \subset E_{k+1} \subset \cdots$ and so extends to the limit. Consider $S^{4k+3} \subset R^{4(k+1)} = H^{k+1}$, so a point in S^{4k+3} can be labelled by its quaternionic coordinates $p_0, \ldots, p_k$, where $p_i = x_i + y_i\mathbf{i} + z_i\mathbf{j} + w_i\mathbf{k}$ with $\sum_0^k(x_i^2 + y_i^2 + z_i^2 + w_i^2) = 1$. Similarly a tangent vector to S^{4k+3} at $(p_0, \ldots, p_k)$ is given by a quaternionic k–tple $(q_0, \ldots, q_k)$ where $q_i = \kappa_i + \lambda_i\mathbf{i} + \mu_i\mathbf{j} + \nu_i\mathbf{k}$ with

$$(*) \quad \sum_0^k \kappa_i x_i + \lambda_i y_i + \mu_i z_i + \nu_i w_i = 0.$$

With this notation, the standard connection form ω assigns to the vector $(q_0, \ldots, q_k)$ at $(p_0, \ldots, p_k)$ the pure imaginary quaternion

$$\omega_{(p_0,\ldots,p_k)}(q_0, \ldots, q_k) = \Im \sum q_i\bar{p}_i.$$

Proposition 4.4 The homeomorphism $\Phi : E_\Delta G \to E$ defined by

$$\Phi(t_0g_0 + t_1g_1 + \cdots) = (\sqrt{t_0}g_0, \sqrt{t_1}g_1, \ldots)$$

is smooth on strata and, restricted to any stratum, pulls back the standard connection ω to the canonical connection ω_Δ .

Proof: Smoothness on strata is clear. Suppose for simplicity we are in the top–dimensional stratum of the $(k+1)$–fold join, at the point $t_0g_0 + \cdots + t_kg_k$. The tangent space to the stratum at that point is spanned by $\partial/\partial t_0, \partial/\partial t_1, \ldots, \partial/\partial t_k$ (note that $\sum \partial/\partial t_i = 0$) and by the tangent spaces to $G_0, G_1, \ldots, G_k$, with total dimension $k + 3(k+1) = 4k+3$.

Consider first a vector in the span of $\partial/\partial t_0, \partial/\partial t_1, \ldots, \partial/\partial t_k$, say $\mathbf{v} = \sum_0^k a_i\partial/\partial t_i$, with $\sum a_i = 0$. This vector is automatically in the kernel of ω_Δ, since ω_Δ does not involve any of the dt_i's. To calculate $\Phi_*(\mathbf{v})$, consider the curve $c(s) = (t_0 + a_0s)g_0 + \cdots (t_k + a_ks)g_k$ which has $c'(0) = \mathbf{v}$. Substituting we find $\Phi \circ c(s) = (\sqrt{t_0 + a_0s}g_0, \ldots, \sqrt{t_k + a_ks}g_k) \in S^{4k+3}$, and $\Phi_*(\mathbf{v}) = (\Phi \circ c)'(0) = \frac{a_0}{2\sqrt{t_0}}g_0 + \cdots + \frac{a_k}{2\sqrt{t_k}}g_k$. Now this vector is at $(\sqrt{t_0}g_0, \sqrt{t_1}g_1, \ldots)$, a point with conjugate coordinates $(\sqrt{t_0}g_0^{-1}, \ldots, \sqrt{t_k}g_k^{-1})$. Applying ω gives $(1/2)(a_0 + \cdots + a_k) = 0$.

Now suppose the test vector is in the complement of the span of the $\partial/\partial t_i$'s. This would be the tangent vector to a curve $c(s) = t_0g_0(s) + \cdots + t_kg_k(s)$, and $\omega_\Delta(c'(0))$ would be

$$(**) \quad t_0g_0'(0)g_0(0)^{-1} + \cdots + t_kg_k'(0)g_k(0)^{-1}.$$

Proceeding as above, we calculate $\Phi \circ c(s) = (\sqrt{t_0}g_0(s), \ldots, \sqrt{t_k}g_k(s)) \in S^{4k+3}$ and $\Phi_*(c'(0)) = (\Phi \circ c)'(0) = \sqrt{t_0}g_0'(0) + \cdots + \sqrt{t_k}g_k'(0)$. Now the calculation of $\omega(\Phi_*(c'(0)))$ gives $\sum_i \sqrt{t_i}g_i'(0)\sqrt{t_i}g_i^{-1}(0)$, which is exactly(**).□

The standard connection has curvature form $\Omega = d\omega + \frac{1}{2}[\omega, \omega]$, with Chern form $\tilde{p}$ and Chern-Simons $\tilde{\chi}$ form defined as in Section 1.1. (In quaternionic notation, we have $\tilde{\chi} = \frac{1}{4\pi^2}Re(\omega \wedge \Omega - \frac{1}{3}\omega \wedge \omega \wedge \omega)$.) These last two are related by $\tilde{p} = d\tilde{\chi}$.

On any stratum of $E_\Delta G$ we can use Proposition 4.4 to define $\Omega_\Delta, \tilde{p}_\Delta$ and $\tilde{\chi}_\Delta$ as Φ^* of the corresponding standard forms.

Theorem 4.5 $\tilde{\chi}_\Delta$ is invariant under the automorphisms R_h of $\xi_\Delta G$.

Proof: this follows directly from the invariance of ω_Δ.□

4.3 The connection and Chern-Simons forms of a p.t.f.

Let $\mathbf{V}$ be a G-valued p.t.f. on an ordered simplicial complex Λ with underlying space $|\Lambda| = X$. We assume that each V_σ is piecewise smooth. Following Prop. 3.7 and Prop. 3.8 we construct $\xi = \xi(\mathbf{V})$, the principal bundle corresponding to $\mathbf{V}$; say $\xi = (\pi : E \to X)$, with classifying map $f = f(\mathbf{V}) : X \to B_\Delta G$; let $F : E \to E_\Delta G$ be the corresponding bundle map.

Set

$$\omega = \omega_{\mathbf{V}} = F^*\omega_\Delta,$$
$$\tilde{\chi} = \tilde{\chi}_{\mathbf{V}} = F^*\tilde{\chi}_\Delta.$$

Proposition 4.6 $\omega_{\mathbf{V}}$ is a piecewise smooth, $\mathfrak{g}$-valued 1-form on E which has the two characteristic properties of a connection (see Proposition 4.2). □

Definition 4.7 $\tilde{\chi}_{\mathbf{V}}$ is the *Chern-Simons form* of $\mathbf{V}$.

When $\mathbf{V}$ is constructed from a lattice gauge field $\mathbf{u}$ as in Proposition 3.14, we will write these forms as $\omega_{\mathbf{u}}$ and $\tilde{\chi}_{\mathbf{u}}$.

4.4 *The relation between* $\omega_{\mathbf{u}}$ *and* $\mathbf{u}$.

The construction of $\omega_{\mathbf{u}}$ from $\mathbf{u}$ proceeds in a roundabout way: $\mathbf{u}$ is used to construct a bundle $\xi_{\mathbf{u}}$ (*which we will think of here as a subbundle of* $\xi_\Delta G$, see Proposition 4.4), and $\omega_{\mathbf{u}}$ is taken to be the restriction to $\xi_{\mathbf{u}}$ of the canonical connection ω_Δ. Nevertheless the two are very closely related.

Lemma 4.8 Any join-line in $E_\Delta G = G_0 * G_1 * \cdots$, i.e. any line of the type $t \mapsto (1-t)x_i + tx_j$, where $x_i \in G_i$ and $x_j \in G_j$, is horizontal with respect to ω_Δ.

Proof: Immediate since $\omega_\Delta = \sum t_i dg_i \cdot g_i^{-1}$ does not involve the differentials dt_i of the join variables.□

Proposition 4.9 Parallel transport by $\omega_{\mathbf{u}}$ around the edges of a plaquette $< 012 >$ of Λ takes an element x in the fiber over vertex $< 2 >$ back to xu_{2012}.

Proof: We identify $< 012 > = \sigma$ with its image in $B_\Delta G$ under the classifying map constructed from $\mathbf{u}$ (compare Figures 4 and 5). That image is covered by $\Gamma_\sigma : C_\sigma \to E_\Delta G$, and in particular the boundary of σ is covered by three of the four edges of Γ_σ: the path $< 2 > \to < 0 > \to < 1 > \to < 2 >$ in $\partial\sigma$ is covered by the path $\Gamma_\sigma(s_1 = 0, s_2), 1 \geq s_2 \geq 0$ followed by $\Gamma_\sigma(s_1, s_2 = 0), 0 \leq s_1 \leq 1$ followed by $\Gamma_\sigma(s_1 = 1, s_2), 0 \leq s_2 \leq 1$. These paths are, respectively, $(1-s_2)\mathbf{e}_0 + s_2 u_{02}\mathbf{e}_2$ leading from $u_{02}\mathbf{e}_2$ (when $s_2 = 1$) to $\mathbf{e}_0$, $(1-s_1)\mathbf{e}_0 + s_1 u_{01}\mathbf{e}_1$ leading from there to $u_{01}\mathbf{e}_1$, and $(1-s_2)u_{01}\mathbf{e}_1 + s_2 u_{012}\mathbf{e}_2$, ending at $u_{012}\mathbf{e}_2$. By Lemma 4.8 this entire path is horizontal with respect to ω_Δ, and so $u_{012}\mathbf{e}_2$ is the result of parallel transport of $u_{02}\mathbf{e}_2$ around $\partial\sigma$. The corresponding holonomy element is $u_{02}^{-1}u_{012}$, which is precisely the "plaquette" (2–simplex) product u_{2012}. By left-invariance of parallel transport any $x \in G_2$ will end up at $x \cdot u_{2012}$. □

This means that **u** essentially determines all the curvature of $\omega_{\mathbf{u}}$: any nullhomotopic loop in Λ, when deformed into the 1–skeleton, can be written (by a standard argument) as a product of paths of the form $c \cdot \partial\sigma \cdot c^{-1}$; so the holonomy around that loop is a product of plaquette products and images of plaquette products under conjugation in $SU(2)$.

5 Construction of the pseudosection ψ

On an ordered simplicial complex Λ, with underlying space $X = |\Lambda|$, let **V** be a parallel transport function with values in $G = SU(2)$, and $f : X \to B_\Delta G$ the classifying map constructed from **V** in Proposition 3.7. It will be convenient to use the standard explicit form of the bundle $f^*\xi_\Delta G$: it has total space $E = \{(e,x) \in E_\Delta G \times X : \pi_\Delta e = f(x)\}$; the projection $\pi : E \to X$ and the bundle map $F : E \to E_\Delta G$ covering f are given by $\pi(e,x) = x$ and $F(e,x) = e$.

The Chern-Simons form of **V** on E is $\tilde{\chi} = F^*\tilde{\chi}_\Delta$. To calculate the value of the Chern-Simons character on a 3-cycle $Z \subset \Lambda$ we will construct from **V** in E over Z a pseudo-section on which $\tilde{\chi}$ can be integrated.

This construction has several steps, and can be summarized in this diagram:

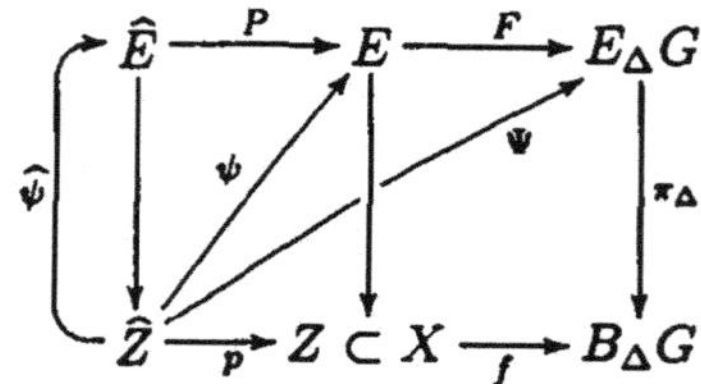

5.1 A partial cone on Λ.

Suppose for simplicity that Λ is a complex of dimension 3 or 4. Let Λ^α represent the union of Λ with the cone on the 3-skeleton $\Lambda^{(3)}$ from a new vertex α. For $\sigma = \langle 0,\ldots,r \rangle \in \Lambda^{(3)}$, set $\sigma^\alpha = \langle \alpha, 0,\ldots,r \rangle$, so $\Lambda^\alpha = \Lambda \cup \bigcup_{\sigma\in\Lambda^{(3)}} \sigma^\alpha$. Order the vertices of Λ^α by ranking α below all the vertices of Λ, and let $\mathbf{u}^\alpha$ be the lattice gauge field on Λ^α defined by

$$\mathbf{u}^\alpha_{ij} = V_{\langle ij\rangle} \in G \quad for \quad \langle i,j \rangle \in \Lambda$$

$$\mathbf{u}^\alpha_{\alpha i} = \; identity \quad for \quad \langle i \rangle \in \Lambda.$$

Following Proposition 3.12, **V** can be extended to a p.t.f. $\mathbf{V}^\alpha$ defined on Λ^α and consistent with $\mathbf{u}^\alpha$. For any simplex $\lambda \in \Lambda^\alpha$ let $\Gamma_\lambda : C_\lambda \to E_\Delta G$ be defined from $\mathbf{V}^\alpha$ as in Proposition 3.6.

5.2 A cycle $\widehat{Z}$ covering Z.

Set

$$\widehat{Z} = \bigcup_{\sigma\in Z} \partial^+ C_{\sigma^\alpha},$$

using the notation of Section 3.1.

Proposition 5.1

$$\partial\partial^+ C_{\sigma^\alpha} = \sum(-1)^i \partial^+ C_{\sigma_i^\alpha},$$

where σ_i is the i-th face of σ.□

It follows that $\widehat{Z}$ is again a cycle. Moreover, $\widehat{Z}$ admits a projection map

$$p : \widehat{Z} \to Z$$

defined by

$$p|_{\partial^+ C_{\sigma^\alpha}} = \pi_{\sigma^\alpha}|_{\partial^+ C_{\sigma^\alpha}},$$

where $\pi_{\sigma^\alpha} : C_{\sigma^\alpha} \to \sigma^\alpha$ is the standard projection of Section 3.1. This map is clearly compatible with the the face relation among simplexes of Z, and therefore extends to all of $\widehat{Z}$ as a well-defined, continuous map. To see that p actually maps $\partial^+ C_{\sigma^\alpha}$ to $\sigma \subset \sigma^\alpha$ note that for $I = \{\alpha, 0, \ldots, r\}$ the cell

$$C_{\sigma^\alpha, I} = C_{\sigma^\alpha}(\alpha)[C_{\sigma^\alpha}(\alpha, 0)|C_{\sigma^\alpha}(0,1)|\cdots|C_{\sigma^\alpha}(r-1, r)] \subset C_{\sigma^\alpha}$$

contributes $C_{\sigma^\alpha}(\alpha, 0)[C_{\sigma^\alpha}(0,1)|\cdots|C_{\sigma^\alpha}(r-1, r)]$ to $\partial^+ C_{\sigma^\alpha}$; this face is an r-simplex which is mapped by p onto σ. The other faces of $\partial^+ C_{\sigma^\alpha}$ are collapsed onto various faces of σ.

5.3 A section over $\widehat{Z}$ ...

Use $f \circ p$ to pull back $\xi_\Delta G$ to a bundle $\widehat{\xi}$ over $\widehat{Z}$. The total space is $\widehat{E} = \{(e, y) \in E_\Delta G \times \widehat{Z} : \pi_\Delta e = f(p(y))\}$. Define a map $\Psi = \Psi(\mathbf{V}) : \widehat{Z} \to E_\Delta G$ by

$$\Psi|\partial^+ C_{\sigma^\alpha} = \Gamma_{\sigma^\alpha}|\partial^+ C_{\sigma^\alpha}.$$

Then it is straightforward to check that $\pi_\Delta \circ \Psi = f \circ p$ and that therefore $y \mapsto (\Psi(y), y) \in \widehat{E}$ defines a section $\widehat{\psi}$ in $\widehat{\xi}$.

5.4 ... becomes a pseudosection over Z.

Let $P : \widehat{E} \to E$ be given by $P((e, y)) = (e, p(y))$. The map $\psi = \psi(\mathbf{V})$ given by

$$\psi = P \circ \widehat{\psi} : \widehat{Z} \to E$$

is what we will call *the pseudosection associated to* $\mathbf{V}$. Since $\pi(\psi(y)) = p(y)$, it follows that ψ of the face $C_{\sigma^\alpha}(\alpha, 0)[C_{\sigma^\alpha}(0,1)|\cdots|C_{\sigma^\alpha}(r-1, r)] \subset \partial^+ C_{\sigma^\alpha}$ lies "over" σ. This justifies the name. See Figure 7, which illustrates the

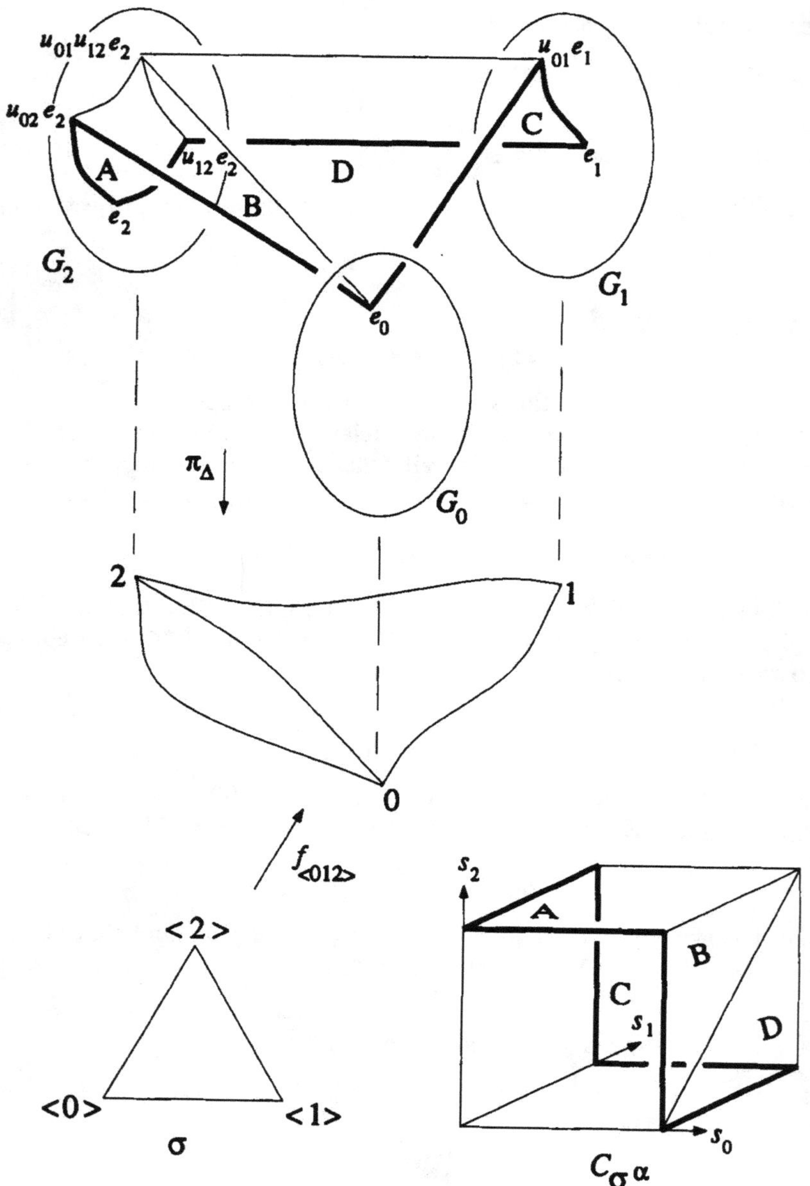

Figure 7: The pseudosection ψ constructed from a l.g.f. **u**, shown for a 2-simplex $< 012 >$. Note that parts B and D make up the map $\Gamma_{<012>}$.

construction over a 2–simplex $\sigma = <012>$. In this figure, the cells marked **A, B, C, D** make up $\partial^+ C_{\sigma^\alpha}$. The top of the figure represents their image, under Ψ, in $E_\Delta G$.

Definition 5.2 For a 3–cycle $Z \subset \Lambda$, the value of the Chern-Simons character on Z is defined to be

$$\chi(Z) = \sum_{\sigma \in Z} \hat{\chi}_{\mathbf{u}}(\sigma) = \sum_{\sigma \in Z} \int_{\psi(\partial^+ C_{\sigma^\alpha})} F^* \tilde{\chi}_\Delta = \sum_{\sigma \in Z} \int_{\Gamma_{\sigma^\alpha}(\partial^+ C_{\sigma^\alpha})} \tilde{\chi}_\Delta,$$

reduced mod $\mathbf{Z}$.

Remark 5.3 In the case of a simplex of dimension 3, say $\sigma = <0,1,2,3>$, then $\sigma^\alpha = <\alpha,0,1,2,3>$, and

$$\begin{aligned}
\Gamma_{\sigma^\alpha}(\partial^+ C_{\sigma^\alpha}) \quad = \quad & \gamma_{\sigma^\alpha}(\alpha,0)[\gamma_\sigma(0,1) \mid \gamma_\sigma(1,2) \mid \gamma_\sigma(2,3)] \\
& + \gamma_{\sigma^\alpha}(\alpha,0)[\gamma_\sigma(0,2) \mid \gamma_\sigma(2,3)] \\
& - \gamma_{\sigma^\alpha}(\alpha,0)[\gamma_\sigma(0,1) \mid \gamma_\sigma(1,3)] \\
& + \gamma_{\sigma^\alpha}(\alpha,0)[\gamma_\sigma(0,3)] \\
& - \gamma_{\sigma^\alpha}(\alpha,1)[\gamma_\sigma(1,2) \mid \gamma_\sigma(2,3)] \\
& - \gamma_{\sigma^\alpha}(\alpha,1)[\gamma_\sigma(1,3)] \\
& + \gamma_{\sigma^\alpha}(\alpha,2)[\gamma_\sigma(2,3)] \\
& - \gamma_{\sigma^\alpha}(\alpha,3)[\ ],
\end{aligned}$$

continuing with the notation of Proposition 3.3 and Remark 2.3.

5.5 *Preparation for the proof of Theorem C.*

In the proof of Theorem C we shall need an extension of the construction of $\psi_{\mathbf{V}}$. Let $\tau = <01234>$ be a 4-simplex of Λ, and $\tau^\alpha = <\alpha 01234>$. Continuing with the notation of Section 3.1, set $W(\tau^\alpha)$ to be the chain $\partial^+ C_{\tau^\alpha} - \partial^4 C_{\tau^\alpha}$, so that, as a subset of C_{τ^α}, $W(\tau^\alpha) = \{\vec{s} \in C^5_{\tau^\alpha} : s_j = 1 \text{ for some } j \neq 4\}$. Note that as before the standard projection $\pi_{\tau^\alpha} : C_{\tau^\alpha} \to \tau^\alpha$ restricts to a surjective map $p: W(\tau^\alpha) \to \tau$.

The obstruction to defining Γ_{τ^α} stems from the impossibility in general of extending $V_{\tau^\alpha} : c_{\tau^\alpha} \to SU(2)$ over the interior of the cell, since ∂c_{τ^α} is a 3-sphere. Examining the definition of Γ_{τ^α} as a union of singular bar-products (see Section 3.3) shows that V_{τ^α} only occurs in $\gamma_{\tau^\alpha}(\alpha)[\gamma_{\tau^\alpha}(\alpha,4)]$ via the singular cell $\gamma_{\tau^\alpha}(\alpha,4): C_{\tau^\alpha}(\alpha,4) \to SU(2)$. It follows that Γ_{τ^α} may be defined on all of C_{τ^α} except the bar-cell $C_{\tau^\alpha}(\alpha)[C_{\tau^\alpha}(\alpha,4)]$, and in particular, since $C_{\tau^\alpha}(\alpha,4) = \{s_4 = 1\}$, on all the bar-cells of $W(\tau^\alpha)$. Define a map

$$\Psi^\square_{\tau^\alpha}: W(\tau^\alpha) \to E_\Delta G$$

by

$$\Psi^\square_{\tau^\alpha} = \Gamma_{\tau^\alpha}|W(\tau^\alpha).$$

As before, we pull back $\xi_\Delta G$ by $f_\tau \circ p$ to a bundle $\hat{E}_\tau$ over $W(\tau^\alpha)$, and we interpret $\Psi^\square_{\tau^\alpha}$ as a section in that bundle, which we project to a pseudosection $\psi^\square(\tau)$ in $E|_\tau$, where E is the bundle defined by $\mathbf{V}$ over Λ.

Lemma 5.4 (1) $\pi_*\psi^\square(\tau) = \tau$ as a 4-chain.
(2) $\partial\psi^\square(\tau) = \psi(\partial\tau) + \psi^\square_{\tau^\alpha} \mid \partial^4 C_{\tau^\alpha}$.
(3) Moreover the last term is a map of a 3-sphere into the single fibre $E|_{<4>}$.

Proof: Statement (1) was handled above. For statement (2), observe that

$$\begin{aligned} \partial W(\tau^\alpha) &= \partial\partial^+ C_{\tau^\alpha} - \partial\partial^4 C_{\tau^\alpha} \\ &= \Sigma(-1)^i \partial^+ C_{\tau_i^\alpha} - \partial\partial^4 C_{\tau^\alpha}, \end{aligned}$$

(using Proposition 5.1) where $\partial\tau = \Sigma(-1)^i\tau_i$. Statement (3) follows from the definition of Γ_{τ^α}. □

6 Proofs of Main Theorems

We now turn to the computation of $\hat{\chi}_{\mathbf{V}}$ for a p.t.f. $\mathbf{V}$ obtained from a generic l.g.f. $\mathbf{u}$ by iterated coning. We write $\hat{\chi}_{\mathbf{u}}$ for $\hat{\chi}_{\mathbf{V}_{\mathbf{u}}}$ in this case.

6.1 *Integration over a bar-product*

For $i = 0, \ldots, r$ let C_i be a cube of dimension $p(i)$, parametrised by $\vec{s^i} = (s^i_1, \ldots, s^i_{p(i)})$, and let $\mathbf{C} = C_0[C_1 \mid \ldots \mid C_r]$ be their bar-product. Set $p = \dim \mathbf{C} = r + \sum_{i=0}^{r} p(i)$. We identify the p-cube C^p with $C^r \times (\prod_{i=0}^{r} C_i)$, parametrised by $(s_1, \ldots, s_r, \vec{s^0}, \ldots, \vec{s^r})$.

Definition 6.1 The *standard bar-projection* $\pi^C : C^p \to \mathbf{C}$ is given by

$$\pi^C(s_1, \ldots, s_r, \vec{s^0}, \ldots, \vec{s^r}) =$$
$$t_0\vec{s^0} + t_1(\vec{s^0}, \vec{s^1}) + \ldots + t_r(\vec{s^0}, \vec{s^1}, \ldots, \vec{s^r}),$$

where $(t_0, \ldots, t_r)$ is the image of $(s_1, \ldots, s_r)$ under the standard projection $\pi_{<0,\ldots,r>} : C^r \to < 0, \ldots, r >$ of Section 3.1.

Now let $\gamma : \mathbf{C} \to E_\Delta G$ be the bar-product of singular cubes $\gamma_i : C_i \to G_{j(i)}$, for $i = 0, \ldots, r$; so γ is a generator of $\mathcal{E}_{**}$. Here as in Definition 2.4 $\gamma_i(\vec{s^i}) = g_i(\vec{s^1})\mathbf{e}_{j(i)}$, with g_i mapping C_i to G. Set $\gamma^C = \gamma \circ \pi^C : C^p \to E_\Delta$, and define

$$\int_\gamma \lambda = \int_{C^p} (\gamma^C)^* \lambda$$

for any p-form λ on E_Δ.

In the next section we shall prove the following result.

Proposition 6.2 Let $\gamma : \mathbf{C} \to E_\Delta$ be as above, with dimension $p = 3$. Then $\int_\gamma \tilde{\chi}_\Delta = 0$ except in these two cases
(1) If $\mathbf{C} = C_0[\,]$ with $\dim C_0 = 3$, then $\int_\gamma \tilde{\chi}_\Delta = \int_{\gamma_0} dv$, where dv is the bi-invariant volume form on the unit sphere in quaternionic space, normalised so that $\int_{S^3} dv = 1$.
(2) If $\mathbf{C} = C_0[C_1]$ with $\dim C_0 = \dim C_1 = 1$, then γ_0 and γ_1 are curves, and

$$\int_\gamma \tilde{\chi}_\Delta = -\frac{1}{4\pi^2} \int_0^1 \int_0^1 \Re(g_0'(s_0^1) g_0(s_0^1)^{-1} \cdot Ad_{g_0(s_0^1)}(g_1'(s_1^1) g_1(s_1^1)^{-1}) ds_0^1 ds_1^1.$$

6.2 Proof of Theorem A

We begin by restating the theorem.

Theorem 6.3 Let $\mathbf{u}$ be a generic $SU(2)$–valued l.g.f. on Λ, with Chern-Simons character $\hat{\chi}_{\mathbf{u}}$ given as above. Then for every 3-simplex $\sigma =< 0123 >\in \Lambda$,

$$\hat{\chi}_{\mathbf{u}}(\sigma) = \int_P dv + \sum_{i=1}^{4}(-1)^i \int_{\Delta_i} dv + \frac{1}{4\pi^2}\int_0^1 \Re[U_{01} \cdot Ad_{\mathbf{g}_{\alpha 01}(s_0)} W_{123}] ds_0$$

where

$$\begin{aligned} P &= cone\ (1, \mathbf{g}_{\alpha 01} \cdot \mathbf{g}_{123}), \\ \Delta_1 &= cx(1, u_{03}, u_{013}, u_{0123}), \\ \Delta_2 &= cx(1, u_{03}, u_{023}, u_{0123}), \\ \Delta_3 &= cx(1, u_{23}, u_{023}, u_{0123}), \\ \Delta_4 &= cx(1, u_{23}, u_{123}, u_{0123}), \end{aligned}$$

where $\mathbf{g}_{\alpha 01}(s_0) = \exp(s_0 U_{01})$ is the shortest geodesic from $\mathbf{1}$ to u_{01}, where $\mathbf{g}_{123}(s_2) = u_{13}\exp(s_2 U_{3123})$ is the shortest geodesic from u_{13} to $u_{12}u_{23}$, and $W_{123} = Ad_{u_{13}} U_{3123}$.

Proof. By Proposition 6.2, Remark 5.3 and the definition of $\hat{\chi}$,

$$\hat{\chi}(\sigma) = -\int_{\gamma_{\sigma^\alpha}(\alpha,3)} dv - \int_{\gamma_{\sigma^\alpha}(\alpha,1)[\gamma_\sigma(1,3)]} \chi_\Delta, \tag{1}$$

and the second term on the right-hand side of this equation equals

$$+\frac{1}{4\pi^2}\int_0^1\int_0^1 \Re(\mathbf{g}'_{\alpha 01}(s_0)\mathbf{g}_{\alpha 01}(s_0)^{-1} \cdot Ad_{\mathbf{g}_{\alpha 01}(s_0)}(\mathbf{g}'_{123}(s_2)\mathbf{g}_{123}(s_2)^{-1}) ds_0 ds_2.$$

The expressions $\mathbf{g}_{\alpha 01}(s_0) = \exp(s_0 U_{01})$ and $\mathbf{g}_{123}(s_2) = u_{13}\exp(s_2 U_{3123})$ introduced above reduce $\mathbf{g}'_{\alpha 01}(s_0)\mathbf{g}_{\alpha 01}(s_0)^{-1}$ to U_{01} and $\mathbf{g}'_{123}(s_2)\mathbf{g}_{123}(s_2)^{-1}$ to $Ad_{u_{13}} U_{3123}$. The variable s_2, which no longer appears in the integrand, integrates out to 1; the integral now matches the last term in the statement of the theorem.

Turning now to the first integral on the right hand side of Equation 1, we see that the proof of the theorem will be complete once we show that, as oriented 3-chains,

$$V^\alpha_{\sigma^\alpha}(\alpha,3) = -P - \sum_{i=1}^{4}(-1)^i\Delta_i. \tag{2}$$

Now $V^\alpha_{\sigma^\alpha}(\alpha,3)$ is the image of the fundamental chain on $c^3_{\sigma^\alpha}$ under the map $V^\alpha_{\sigma^\alpha} : c^3_{\sigma^\alpha} \to G$. And $V^\alpha_{\sigma^\alpha}$ was constructed so that line segments from $\vec{0}$ to points of $\partial^+ c^3_{\sigma^\alpha}$ are mapped into geodesics from **1** to points of $V^\alpha_{\sigma^\alpha}(\partial^+ c^3_{\sigma^\alpha})$. Thus it suffices to show that, as oriented 2-chains,

$$\begin{aligned} V^\alpha_{\sigma^\alpha}(\partial^+ c^3_{\sigma^\alpha}) &= -V^\alpha_{\alpha 01} \cdot V_{123} \\ &\quad + cx(u_{03}, u_{013}, u_{0123}) \\ &\quad - cx(u_{03}, u_{023}, u_{0123}) \\ &\quad + cx(u_{23}, u_{023}, u_{0123}) \\ &\quad - cx(u_{23}, u_{123}, u_{0123}). \end{aligned}$$

On $\partial^+ c^3_{\sigma^\alpha}$, $V^\alpha_{\sigma^\alpha}$ is defined *as a function* by:

$$V^\alpha_{\sigma^\alpha}(s_0, s_1, s_2) = \begin{cases} V_\sigma(s_1, s_2), & \text{if } s_0 = 1 \\ V^\alpha_{\alpha 01}(s_0) \cdot V_{123}(s_2), & \text{if } s_1 = 1 \\ V^\alpha_{\alpha 012}(s_0, s_1) \cdot u_{23}, & \text{if } s_2 = 1 \end{cases}$$

(since $V^\alpha_{\alpha 0}$ and V_{23} map 0-cubes into **1** and u_{23} respectively). Allowing for orientations, this gives that, *as oriented 2-chains,*

$$V^\alpha_{\sigma^\alpha}(\partial^+ c^3_{\sigma^\alpha}) = V_\sigma - V^\alpha_{\alpha 01} \cdot V_{123} + V^\alpha_{\alpha 012} \cdot u_{23}.$$

But $V_\sigma(c^2_\sigma)$ is the 2-chain $cx(u_{03}, u_{013}, u_{0123}) - cx(u_{03}, u_{023}, u_{0123})$; and $V^\alpha_{\alpha 012}(c^2_{<012>^\alpha})$ is the 2-chain $cx(\mathbf{1}, u_{02}, u_{012}) - cx(\mathbf{1}, u_{12}, u_{012})$. Now Equation 2 follows, and with it, the theorem. □

6.3 *Proof of Theorem B*

Suppose on a lattice Λ we are given two $SU(2)$–valued lattice gauge fields $\mathbf{u}$ and $\mathbf{u}'$ which are related by a change of gauge: there exists a family (see Definition 3.16) $\mathbf{g} = \{g_0, g_1, \ldots\}$ of elements of $G = SU(2)$ such that $\mathbf{u}' = Ad_{\mathbf{g}}\mathbf{u}$.

Suppose $\mathbf{u}$ and $\mathbf{u}'$ are generic, so that they may be extended by iterated coning (see Remark 3.15) to parallel transport functions $\mathbf{V}$ and $\mathbf{V}'$; let $f, f' \colon |\Lambda| \to B_\Delta G$ be the corresponding classifying maps, constructed as in Section 3.3; they pull back $\xi_\Delta G$ to bundles ξ, ξ' for which we will use the explicit form described in Section 5.

Given a 3–cycle $Z \subset \Lambda$, let $\chi_{\mathbf{u}}$ and $\chi_{\mathbf{u}'}$ be the values on Z of the Chern-Simons characters defined from $\mathbf{u}$ and $\mathbf{u}'$ as in Section 5. With this notation, Theorem B is equivalent to the following statement.

Theorem 6.4 $\chi_{\mathbf{u}}(Z) \equiv \chi_{\mathbf{u}'}(Z) \bmod \mathbf{Z}$.

Proof: We adapt to this context an argument of Cheeger and Simons [2], continuing with the notation from Section 5: $\xi = (\pi: E \to X)$, $F: E \to E_\Delta G$, $\widehat{\xi} = (f \circ p)^* \xi_\Delta G = (\widehat{\pi}: \widehat{E} \to \widehat{Z})$, $P: \widehat{E} \to E$, the section $\widehat{\psi}: \widehat{Z} \to \widehat{E}$, the pseudosection $\psi = P \circ \widehat{\psi}: \widehat{Z} \to E$, as well as $E', F', \ldots, \psi'$ the corresponding elements constructed from $\mathbf{u}'$. Note that the bundles ξ, ξ' inherit connections pulled back from the canonical connection in $\xi_\Delta G$.

Let $\mathbf{h} = \{h_0, h_1, \ldots\}$ extend $\mathbf{g}$ to the rest of the join-factors of $E_\Delta G$. Right multiplication by $\mathbf{h}$ defines an automorphism $R_{\mathbf{h}}: E_\Delta G \to E_\Delta G$ (see Paragraph 2.1) and the corresponding $r_{\mathbf{h}}: B_\Delta G \to B_\Delta G$. From Proposition 3.10 we know that $f' = r_{\mathbf{h}} \circ f$, leading to the following commutative diagram.

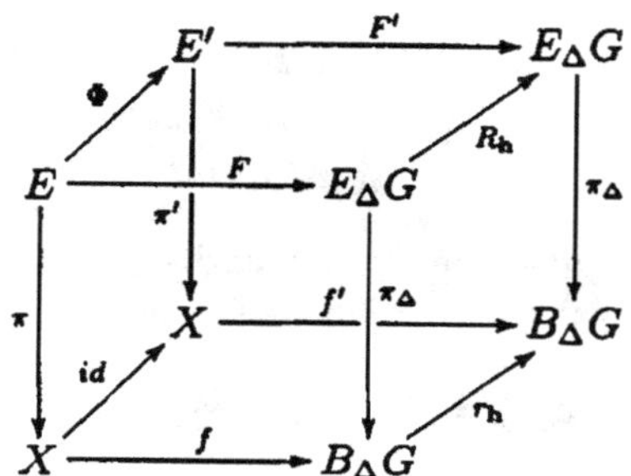

It follows that the right-automorphism $R_{\mathbf{h}}$ induces an isomorphism $\Phi: \xi \to \xi'$. On the total space E it is defined by $\Phi((e, x)) = (R_{\mathbf{h}} e, x)$. To check that $\Phi(E) = E'$ note that if $\pi_\Delta e = f(x)$ then $\pi_\Delta R_{\mathbf{h}} e = r_{\mathbf{h}} \pi_\Delta e = r_{\mathbf{h}} f(x) = f'(x)$. Since $R_{\mathbf{h}}$ preserves the canonical connection (Proposition 4.3) it follows that Φ is an isomorphism of bundles with connections.

Similarly, moving up to $\widehat{Z}$, this Φ, or $R_{\mathbf{h}}$, induce an isomorphism $\widehat{\Phi}: \widehat{E} \to \widehat{E}'$ of bundles with connections, and extending the diagram above, and the diagram at the beginning of Section 5, as shown below.

Furthermore $f \circ p$ and $f' \circ p$ are homotopic, since $r_{\mathbf{h}}$ is homotopic to the identity. If the homotopy between them is $H: \widehat{Z} \times [0,1] \to B_\Delta G$, then $H^* \xi_\Delta G$ is a bundle with connection which restricts to the bundles with connection $\widehat{\xi}$ over $\widehat{Z} \times \{0\}$ and $\widehat{\xi}'$ over $\widehat{Z} \times \{1\}$. Say $H^* \xi_\Delta G = (\tilde{\pi}: \tilde{E} \to \widehat{Z} \times [0,1])$. The fiber is $SU(2) = S^3$.

The cell decomposition of $\widehat{Z} \times \{0\}$ and $\widehat{Z} \times \{1\}$ extends to a cell decomposition of $\widehat{Z} \times [0,1]$, a 4-manifold with boundary, with 4-cells $c_1, \ldots, c_\ell$. The sections $\widehat{\psi}$ and $\widehat{\psi}'$ defined over the two ends extend over the 3-skeleton of this decomposition, and may be radially extended to a section $\widehat{\Psi}$ defined on the complement of one point b_i in the interior of each c_i, $i = 1, \ldots, \ell$.

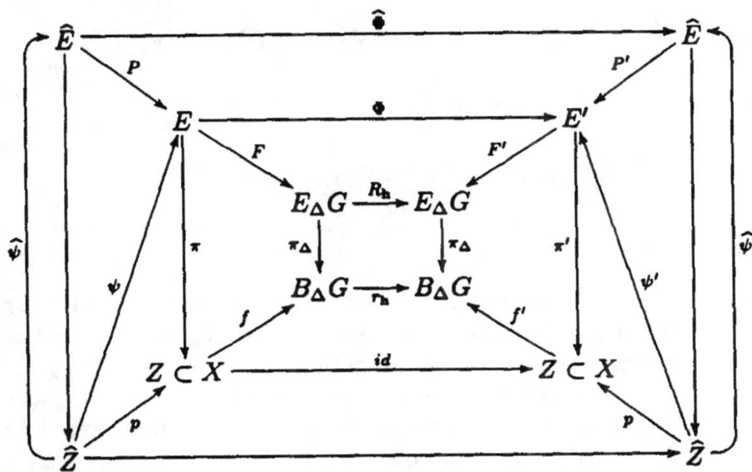

Puting in the radial limits makes $\widehat{\Psi}$ into a piecewise smooth map from $\widehat{Z} \times [0,1] - b_1 - \cdots - b_\ell \cup S_1^3 \cup \cdots \cup S_\ell^3$ into $\tilde{E}$ with the property that each S_i^3 is mapped into the fiber over the corresponding b_i.

Now let us use $\widehat{\Phi}$ to glue $\tilde{E}|_{\widehat{Z}\times\{0\}} = \widehat{E}$ to $\tilde{E}|_{\widehat{Z}\times\{1\}} = \widehat{E}'$. Since $\widehat{\Phi}$ is an isomorphism of bundles with connection, this gives a bundle-with-connection $\overline{\xi} = (\overline{\pi}: \overline{E} \to \widehat{Z} \times S^1)$, with base a 4-manifold. The "section" $\widehat{\Psi}$ gets wrapped around so that over $\widehat{Z} \times \{\textit{the basepoint}\}$ we see the sections $\widehat{\psi}'$ and $\widehat{\Phi} \circ \widehat{\psi}$

Let $\overline{p}$ and $\overline{\chi}$ be the Chern and Chern-Simons forms in $\overline{E}$, and consider $\int_{\widehat{\Psi}} \overline{p}$. By Stokes' Theorem this integral is equal to

$$\int_{\partial\widehat{\Psi}} \overline{\chi} = \int_{\widehat{\psi}'} \overline{\chi} - \int_{\widehat{\Phi}\circ\widehat{\psi}} \overline{\chi} + \sum_i \int_{\widehat{\Psi}(S_i^3)} \overline{\chi}.$$

The following three observations finish the argument.

- Each $\widehat{\Psi}(S_i^3)$ is a map of S^3 into a fiber, where $\overline{\chi}$ restricts to the volume form; so $\int_{\widehat{\Psi}(S_i^3)} \overline{\chi}$ is the winding number of $\widehat{\Psi}(S_i^3)$, an integer.

- $\int_{\widehat{\Psi}} \overline{p} = \int_{\widehat{\Psi}} \overline{\pi}^* p = \int_{\widehat{Z}\times S^1} p$
where p is the "downstairs" Chern form, is a Chern number of the bundle $\overline{\xi}$ and therefore an integer.

- $\int_{\widehat{\Phi}\circ\widehat{\psi}} \overline{\chi} = \int_{\widehat{\Phi}\circ\widehat{\psi}} (F' \circ P')^* \tilde{\chi}_\Delta = \int_{\widehat{\psi}} \widehat{\Phi}^*(P')^*(F')^* \tilde{\chi}_\Delta = \int_{\widehat{\psi}} P^* F^* R_h^* \tilde{\chi}_\Delta$

and, using Theorem 4.5,

$$\int_{\widehat{\psi}} P^* F^* R_{\mathbf{h}}^* \tilde{\chi}_\Delta = \int_{\widehat{\psi}} P^* F^* \tilde{\chi}_\Delta = \int_{\psi} F^* \tilde{\chi}_\Delta = \chi_{\mathbf{u}}(Z),$$

whereas

$$\int_{\widehat{\psi'}} \overline{\chi} = \int_{\widehat{\psi'}} (F' \circ P')^* \tilde{\chi}_\Delta = \int_{\psi'} (F')^* \tilde{\chi}_\Delta = \chi_{\mathbf{u}'}(Z).$$

6.4 Proof of Theorem C

Refering to Section 1.3 for motivation, let $\bar{\Lambda}$ and $\bar{\Lambda}^\alpha$ denote Λ and Λ^α (respectively) with the order of their vertices reversed. We shall denote vertex $< i >$ in the reverse ordering by $< -i >$, $< 0 >$ by $< -0 >$ and $< \alpha >$ by $< -\alpha >$. For $\sigma =< i_0, \ldots, i_r >$, set $\bar{\sigma} =< -i_r, \ldots, -i_0 >$, $\bar{\sigma}_\alpha =< -i_r, \ldots, -i_0, -\alpha >$.

Given a l.g.f. $\mathbf{u}$ on Λ, let $\bar{\mathbf{u}}$ denote the same l.g.f., but regarded as being on $\bar{\Lambda}$. Then $\bar{\mathbf{u}}$ is continuous if and only if $\mathbf{u}$ is; and in this case, $\delta(\bar{\mathbf{u}}) = \delta(\mathbf{u})$. By Proposition 3.14, Part 2, we have $\xi(\mathbf{u}) = \xi(\bar{\mathbf{u}})$.

Let $\bar{N}$ be the 4-cocycle representative on $\bar{\Lambda}$ of the second Chern class of $\xi(\bar{\mathbf{u}}) \approx \xi(\mathbf{u})$ constructed according to the algorithm of [14]. For any 4–simplex $\tau =< 0, \ldots, 4 > \in \Lambda$, by definition, $\bar{N}(\bar{\tau})$ is the degree of a certain map, which we shall here denote by $\bar{V}_{\bar{\tau}_\alpha} : \partial c^4_{\bar{\tau}_\alpha} \to S^3$, where $c^4_{\bar{\tau}_\alpha} = \{(s_{-3}, s_{-2}, s_{-1}, s_{-0}) : 0 \le s_{-i} \le 1\}$.

Theorem C may be restated as follows:

Theorem 6.5 For every 4-simplex τ of Λ, the value of the Chern-Simons character $\chi = \chi_{\mathbf{u}}$ on the 3–cycle $\partial\tau$ is given by

$$\chi(\partial\tau) = \int_\tau p - \bar{N}(\bar{\tau}).$$

Proof. Let $\mathbf{u}^\alpha$, Λ^α and $\mathbf{V}^\alpha$ be as at the beginning of this section. Let

$$\Psi^\square_{\tau^\alpha} : W(\tau^\alpha) \to E_\Delta G;$$

and let $\psi^\square_{\mathbf{V}}(\tau)$ be as in Lemma 5.4. Then

$$\begin{aligned} \chi(\partial\tau) &= \int_{\psi(\partial\tau)} F^* \tilde{\chi}_\Delta \\ &= \int_{\partial\psi^\square(\tau)} F^* \tilde{\chi}_\Delta - \int_{\Psi^\square_{\tau^\alpha}|\partial\partial^4 C_{\tau^\alpha}} F^* \tilde{\chi}_\Delta. \end{aligned}$$

The first integral, by Stokes' Theorem, naturality and Lemma 5.4 (1), is equal to $\int_\tau p$. The second, since $\tilde{\chi}_\Delta$ restricts to the normalized volume form on fibers, is equal to the degree of $\Psi^{\square}_{\tau^\alpha}|\partial\partial^4 C_{\tau^\alpha}$, which is the degree of $V_{\tau^\alpha}|\partial^4 c_{\tau^\alpha}$ (see Section 5.5). To compare this integer with the degree of $\bar{V}_{\bar{\tau}_\alpha}: \partial c_{\bar{\tau}_\alpha} \to S^3$, we use the orientation-preserving homeomorphism $h: c_{\tau^\alpha} \to c_{\bar{\tau}_\alpha}$ given by $h(s_0, s_1, s_2, s_3) = (s_3, s_2, s_1, s_0)$. Comparing the construction of V_{τ^α} and $\bar{V}_{\bar{\tau}_\alpha}$ (the latter in [15]) shows that $\bar{V}_{\bar{\tau}_\alpha} \circ h = V_{\tau^\alpha}$, so the degrees are the same. $\square$

7 Which cells contribute to the integral of $\tilde{\chi}$? Proof of Proposition 6.2.

Proof of Proposition 6.2. Continuing with the notation from Section 6.1, let

$$\partial_i = (\gamma^C)_* \frac{\partial}{\partial s_i} \quad \text{and} \quad \partial_j^i = (\gamma^C)_* \frac{\partial}{\partial s_j^i},$$

$i = 1, \ldots, r$, $j = 1, \ldots, p(i)$, denote the images in $TE_\Delta G$ of the coordinate vector fields on C^3. The proposition will follow from Lemmas 7.1, 7.2 and 7.3.

The first lemma is a compilation of useful formulae.

Lemma 7.1 (A) $\tilde{\chi}_\Delta = \frac{1}{4\pi^2}\Re(\omega_\Delta \wedge \Omega_\Delta - \frac{1}{3}\omega_\Delta \wedge \omega_\Delta \wedge \omega_\Delta)$.
(B) $\tilde{\chi}_\Delta = \frac{1}{4\pi^2}\Re(\omega_\Delta \wedge d\omega_\Delta + \frac{2}{3}\omega_\Delta \wedge \omega_\Delta \wedge \omega_\Delta)$.
(C) $\int_\gamma \tilde{\chi}_\Delta = \int_{C^3} \tilde{\chi}_\Delta(\ldots \partial_i \ldots \partial_j^i \ldots) |_{(\ldots s_i \ldots s_j^i \ldots)} \ldots ds_i \ldots ds_j^i \ldots$
(where $(\ldots s_i \ldots s_j^i \ldots)$ are the parameters of C^3).
(D) ∂_i is horizontal with respect to each ω_j, and hence also w.r.t. ω_Δ.
(E) ∂_j^0 is vertical, i.e. in the kernel of $\pi_{\Delta *}$, for $j = 1, \ldots, p(0)$.
(F) $\omega_\Delta \wedge \omega_\Delta \wedge \omega_\Delta(\ldots \partial_i \ldots \partial_j^i \ldots) \equiv 0$ except when $C = C_0[]$ ($\dim C_0 = 3$).
(G) $\omega_\Delta \wedge \Omega_\Delta(\ldots \partial_i \ldots \partial_j^i \ldots) \equiv 0$ if $\dim C_0 \geq 2$.
(H) $\omega_\Delta \wedge d\omega_\Delta = \sum_{k,l=0}^\infty t_k t_l \omega_k \wedge d\omega_l - \sum_{k,l=0}^\infty t_k dt_l \wedge \omega_k \wedge \omega_l$.
(J) $\omega_k \wedge d\omega_l(\ldots \partial_i \ldots \partial_j^i \ldots) \equiv 0$ if $r \geq 1$ (r the number of bar-factors in γ).
(K) $dt_l \wedge \omega_k \wedge \omega_l(\ldots \partial_i \ldots \partial_j^i \ldots) \equiv 0$ if $r \geq 2$.

Proof: (A), (B), and (C) are definitions given earlier. (D) is clear since ω_j ignores the t_i and consequently the s_i also. (E) follows from π_Δ being constant on left-equivalence classes, and γ_0 appearing on the left. (F) follows from (D); (G) follows from (E) and the vanishing of the curvature form when one of its vector arguments is vertical; (H) is elementary.

Proof of (J), in case $r = 1$ (the other cases are similar). The images under $(\gamma^C)_*$ of the coordinate vector fields on C^3 must be of types ∂_i, $\partial_{j'}^{i'}$, $\partial_{j''}^{i''}$. Now $\omega_k \wedge d\omega_l(\partial_i, \partial_{j'}^{i'}, \partial_{j''}^{i''}) = \sum_\alpha\{(-1)^\alpha \omega_k(\alpha\partial_i) d\omega_l(\alpha\partial_{j''}^{i''}, \alpha\partial_{j'}^{i'}) :$ α is a permutation of $\{\partial_i, \partial_{j''}^{i''}, \partial_{j'}^{i'}\}\}$. The first factor is 0 in those terms in which $\alpha\partial_{j'}^{i'} = \partial_i$, by (D). In the remaining terms, the second factor is, up to sign, of the form $d\omega_l(\partial_{j'}^{i'}, \partial_i) = \partial_i(\omega_l(\partial_{j'}^{i'}))$ (again by (D)). But $\omega_l(\partial_{j'}^{i'})$, as a function of s_i, $s_{j'}^{i'}$, and $s_{j''}^{i''}$, depends on only the last two variables; so when this function is acted upon by the vector field ∂_i, the result is 0.

Proof of (K), in case $r = 2$ (the case $r = 3$ is similar). The images under $(\gamma^C)_*$ of the coordinate vector fields on C^3 must be of types $\partial_{i'}$, $\partial_{i''}$

and ∂^i_j. Now $dt_l \wedge \omega_k \wedge \omega_l(\partial_{i'}, \partial_{i''}, \partial^i_j) = \sum_\alpha\{(-1)^\alpha dt_l(\alpha\partial_{i'})\omega_k(\alpha\partial_{i''})\omega_l(\alpha\partial^i_j) :$ α is a permutation of $\{\partial_{i'}, \partial_{i''}, \partial^i_j\}\}$. Since in each term one of $\alpha\partial_{i'}$ or $\alpha\partial_{i''}$ must be either $\partial_{i'}$ or $\partial_{i''}$, each term vanishes, by (D). □

The combinatorial types of 3-dimensional bar-product cells $\mathbf{C}$ are the following:

a. $\mathbf{C} = C_0[\,]$, with $\dim C_0 = 3$.

b. $\mathbf{C} = C_0[C_1]$, with

i. $\dim C_0 = 2, \dim C_1 = 0$

ii. $\dim C_0 = 1 = \dim C_1$

iii. $\dim C_0 = 0, \dim C_2 = 2$.

c. $\mathbf{C} = C_0[C_1 \mid C_2]$, with

i. $\dim C_0 = 1, \dim C_1 = \dim C_2 = 0$

ii. $\dim C_1 = 1, \dim C_0 = \dim C_2 = 0$

iii. $\dim C_2 = 1, \dim C_0 = \dim C_1 = 0$.

d. $\mathbf{C} = C_0[C_1 \mid C_2 \mid C_3]$, with $\dim C_i = 0, i = 0, \ldots, 3$.

Lemma 7.2 $\int_\gamma \tilde{\chi}_\Delta = 0$ in cases $(b)(i), (c)((i), (ii)$ and $(iii))$, and (d).

Proof. This follows from (A), (F) and (G) in case $(b)(i)$; and from (B), (F), (H), (J) and (K) in the other cases. □

Analysis of the remaining cases.

Case *(a)*: $\mathbf{C} = C_0[\,]$, $\dim C_0 = 3$.

By (A) and (G), $\int_\gamma^C \tilde{\chi}_\Delta = -\frac{1}{12\pi^2}\int_{C^3} \Re(\omega_\Delta \wedge \omega_\Delta \wedge \omega_\Delta(\partial^0_1, \partial^0_2, \partial^0_3))ds^0_1 ds^0_2 ds^0_3$. Say $\gamma_0 : C_0 \to G_{j(0)}$; on $G_{j(0)}$, ω_Δ reduces to the canonical left-invariant connection ω^G. Proposition 6.2(1) now follows from the next lemma.

Lemma 7.3 Let ω^G denote the canonical left-invariant connection on the group of unit quaternions, S^3, and let dv be the bi-invariant volume form on S^3 for which $\int_{S^3} dv = 1$. Then $\Re(\omega^G \wedge \omega^G \wedge \omega^G) = -12\pi^2 dv$.

Proof. Let $\varphi : S^3 \to \mathbf{R}$ be the function such that $\Re(\omega^G \wedge \omega^G \wedge \omega^G) \mid_x = \varphi(x) dv \mid_x$. By left-invariance of ω^G and dv, φ is constant. To evaluate φ, we set $x = 1$ and evaluate both $\Re(\omega^G \wedge \omega^G \wedge \omega^G) \mid_1$ and $dv \mid_1$ on the orthonormal basis $(\mathbf{i}, \mathbf{j}, \mathbf{k})$ of $T_1 S^3$. Now $dv \mid_1 (\mathbf{i}, \mathbf{j}, \mathbf{k}) = (1/\text{vol } S^3) = (1/2\pi^2)$. And

$$\begin{aligned}(\omega^G \wedge \omega^G \wedge \omega^G) \mid_1 (\mathbf{i}, \mathbf{j}, \mathbf{k}) &= \Sigma_\alpha \{(-1)^\alpha \omega^G(\alpha \mathbf{i}) \omega^G(\alpha \mathbf{j}) \omega^G(\alpha \mathbf{k}) : \\ &\qquad \alpha \text{ is a permutation of } (\mathbf{i}, \mathbf{j}, \mathbf{k})\} \\ &= \mathbf{ijk} - \mathbf{jik} + \ldots \\ &= -6.\end{aligned}$$

Hence $\Re(\omega^G \wedge \omega^G \wedge \omega^G) \mid_1 = -6$; and the lemma follows. □

Case *(b)(ii)*: $\mathbf{C} = C_0[C_1]$, with $\dim C_0 = 1 = \dim C_1$.

The parameters on C^3 are (s_1, s_1^0, s_1^1). Say $\gamma_i = g_i \mathbf{e}_{j(i)} : C_i \to G_{j(i)}$ for $i = 0, 1$. Then

$$\gamma^C(s_1, s_1^0, s_1^1) = (1 - s_1) g_0(s_1^0) \mathbf{e}_{j(0)} + s_1 g_0(s_1^0) g_1(s_1^1) \mathbf{e}_{j(1)}. \tag{3}$$

It follows from (B), (F), (H), and (J) that in this case

$$\int_\gamma \tilde{\chi}_\Delta = -\frac{1}{4\pi^2} \sum_{k,l=0}^{\infty} \int \Re(t_k dt_l \wedge \omega_k \wedge \omega_l)(\partial_1, \partial_1^0, \partial_1^1) ds_1 ds_1^0 ds_1^1.$$

In this sum, the only non-zero terms are those in which each of k and l equals either $j(0)$ or $j(1)$. Moreover, the real part of any commutator of quaternions is zero; this reduces the integrand to

$$\Re(t_{j(0)} dt_{j(1)} \wedge \omega_{j(0)} \wedge \omega_{j(1)} + t_{j(1)} dt_{j(0)} \wedge \omega_{j(1)} \wedge \omega_{j(0)})(\partial_1, \partial_1^0, \partial_1^1).$$

Expanding the definitions of $\partial_1, \partial_1^0, \partial_1^1$ in terms of the decomposition (valid off a set of measure zero) of $T(G_{j(0)} * G_{j(1)})$ as $\mathbf{R} \oplus TG_{j(0)} \oplus TG_{j(1)}$ we have

$\partial_1 = (1, 0, 0)$;
$\partial_1^0 = (0, (1 - s_1) g_0'(s_1^0), s_1 g_0'(s_1^0) \cdot g_1(s_1^1))$;
$\partial_1^1 = (0, 0, g_0(s_1^0) \cdot g_1'(s_1^1))$.

In particular $\omega_{j(0)}(\partial_1) = \omega_{j(1)}(\partial_1) = \omega_{j(0)}(\partial_1^1) = 0$. Also, since $t_{j(0)} = (1 - s_1)$ and $t_{j(1)} = s_1$, it follows that $dt_{j(0)}(\partial_1) = -1$ and $dt_{j(1)}(\partial_1) = 1$, whereas clearly $dt_{j(i)}(\partial_1^k) = 0$ for either choice of i or k. The integrand thus reduces to

$$\Re[(1 - s_1)\omega_{j(0)} \wedge \omega_{j(1)}(\partial_1^0, \partial_1^1) - s_1 \omega_{j(1)} \wedge \omega_{j(0)}(\partial_1^0, \partial_1^1)] = \Re[\omega_{j(0)}(\partial_1^0) \omega_{j(1)}(\partial_1^1)].$$

Applying $\omega_{j(0)} = dg_{j(0)} \cdot g_{j(0)}^{-1}$ to ∂_1^0 gives, following Definition 4.1

$$\omega_{j(0)}(\partial_1^0) = g_0'(s_1^0) \cdot g_0(s_1^0)^{-1},$$

and similarly

$$\omega_{j(1)}(\partial_1^1) = Ad_{g_0(s_1^0)}(g_1'(s_1^1) \cdot g_1(s_1^1)^{-1}),$$

and finally, since s_1 integrates out to 1,

$$\int_\gamma \tilde{\chi}_\Delta = -\frac{1}{4\pi^2} \int_0^1 \int_0^1 \Re(g_0'(s_0^1)g_0(s_0^1)^{-1} \cdot Ad_{g_0(s_0^1)}(g_1'(s_1^1)g_1(s_1^1)^{-1}) ds_0^1 ds_1^1,$$

which is Proposition 6.2(2).

Case *(b)(iii)*: $\mathbf{C} = C_0[C_1]$ with $\dim C_0 = 0$, $\dim C_1 = 2$.

In this case the parameters on C^3 are (s_1, s_1^1, s_2^1), with

$$\gamma^C(s_1, s_1^1, s_2^1) = (1 - s_1)\gamma_0 \vec{e}_{j(0)} + s_1 \gamma_0 \gamma_1(s_1^1, s_2^1)\vec{e}_{j(1)}.$$

The proof that in this case $\int_{\gamma^C} \chi = 0$ is similar to the argument in the previous Case *(b)(ii)*, replacing ∂_1^0 and ∂_1^1 by ∂_1^1 and ∂_2^1 respectively. The only difference appears in analysing the terms on the right-hand side of the following equation.

$$\begin{aligned} \sum_{k,l=j(0),j(1)} (t_k dt_l \wedge \omega_k \wedge \omega_l)(\partial_i, \partial_1^1, \partial_2^1) = \\ (1 - s_1)\{\omega_{j(0)}(\partial_1^1)\omega_{j(1)}(\partial_2^1) - \omega_{j(0)}(\partial_2^1)\omega_{j(1)}(\partial_1^1)\} \\ -s_1\{\omega_{j(1)}(\partial_1^1)\omega_{j(0)}(\partial_2^1) - \omega_{j(1)}(\partial_2^1)\omega_{j(0)}(\partial_1^1)\}. \end{aligned}$$

Here $\omega_{j(0)}(\partial_1^1)$ and $\omega_{j(0)}(\partial_2^1)$ both vanish identically for the reason given in the previous argument. Thus the sum is zero. This completes the proof of Proposition 6.2. □

References

[1] J. Böhm, E. Hertel, Polyedergeometrie in n-dimensionaler Räumen konstanter Krümmung, *Lehrbücher und Monographien aus dem Gebiete der exakten Wissenschaften* Vol. 70, Birkhäuser, Basel 1981

[2] J. Cheeger, J. Simons, Differential characters and geometric invariants, in Geometry and Topology, *Lecture Notes in Math.* Vol. 1167, Springer, NY 1985

[3] S.-S. Chern, J. Simons, Characteristic forms and geometric invariants, *Ann. of Math.* **99** (1974) 48-69

[4] J. Dupont, Curvature and Characteristic Classes, *Lecture Notes in Math.* Vol. 640, Springer, NY 1978

[5] S. Eilenberg, S. MacLane, On the groups $H(\pi, n)$, I, *Ann. of Math.* **51** (1953) 55-106

[6] M. Göckeler, A. Kronfeld, G. Schierholz, U.-J. Wiese, Continuum gauge fields from lattice gauge fields, HLRZ preprint HLRZ-92-34

[7] F. Karsch, M. Laursen, B. Plache, T. Neuhaus, U.-J. Wiese, Chern-Simons term in the 4-dimensional SU(2) Higgs model, HLRZ preprint HLRZ-92-55

[8] S. Kobayashi, K. Nomizu, Foundations of Differential Geometry I, Interscience, New York 1963

[9] M. Laursen, Chern-Simons term and topological charge on the lattice, HLRZ preprint HLRZ-92-61

[10] R. Milgram, The bar construction and abelian H-spaces, *Ill. J. Math.* **11** (1967) 242-250

[11] J. Milnor, Construction of universal bundles I, *Ann. of Math.*(2) **63** (1956) 272-284

[12] J. Milnor, Construction of universal bundles II, *Ann. of Math.*(2) **63** (1956) 430-436

[13] A. Phillips, Characteristic numbers of U_1-valued lattice gauge fields, *Ann. Phys.* **161** (1985) 399-422

[14] A. Phillips, D. Stone, Lattice gauge fields, principal bundles, and the calculation of topological charge, *Commun. Math. Phys.* **103** (1986) 599-636

[15] A. Phillips, D. Stone, The computation of characteristic classes of lattice gauge fields, *Commun. Math. Phys.* **131** (1990) 255-282

[16] A. Phillips, D. Stone, Chern-Simons on a lattice, *Nucl. Phys.* **B** (Proc. Suppl.) **20** (1991) 28-31

[17] A. Phillips, D. Stone, A topological Chern-Weil theory, to appear in *Memoirs Amer. Math. Soc.*

[18] G. Segal, Classifying spaces and spectral sequences, *Publ. Math. I.H.E.S.* **34** (1968) 105-112

[19] K. Wilson, Confinement of Quarks, *Phys. Rev.* **D 10** (1974) 2445-2459

[20] E. Witten, Quantum field theory and the Jones polynomial, *Commun. Math. Phys.* **121** (1989) 351-399

[21] P. Woit, Chern-Simons numbers and universal bundles on the lattice, Stony Brook preprint ITP-SB-87-56

Elementary conjectures in classical knot theory

JÓZEF H. PRZYTYCKI

ABSTRACT. We analyze some recently formulated elementary conjectures in classical knot theory. Most of them deals with twist moves (t_k and $\bar{t}_k$ moves) on link diagrams. We show also a relation between the tricoloring of Fox and the Jones polynomials of links (at sixth root of unity).

"One of the oldest notes by Gauss to be found among his papers is a sheet of paper with the date 1794. It bears the heading "A collection of knots" and contains thirteen neatly sketched views of knots with English names written beside them... With it are two additional pieces of paper with sketches of knots. One is dated 1819; the other is much later, ..." [Dun].

1 Introduction

This paper is based on three talks given in the summer of 1989 at the University of British Columbia as a part of the summer course: "Geometry of graphs and knots". It has partly survey and partly novel character and is, to great extent, on elementary, combinatorial level (the course has been directed toward undergraduate students).

The classical knot theory studies a position of the circle (knot) or of several circles (link) in $S^3(= R^3 \cup \infty)$. We say that two links L_1 and L_2 are ambient isotopic, or shortly *isotopic*, if there is a global isotopy of S^3 (i.e. a continuous deformation of the space) which transforms L_1 onto L_2. We work here exclusively with tame links i.e. links which are isotopic to smooth or piece-wise linear links. We often consider oriented links that is links which components have chosen orientations. If L is an oriented link then $-L$ denotes the link obtained from L by changing orientation of any component of L.

To allow elementary combinatorial treatment of links we project links into a plane[1]. We can always find a regular projection, i.e. a projection in which the only multiple points are double points as in Fig.1.1(a) and we do not allow positions as in Fig.1.1(b), (c) and (d). Regular projection (called also *universe* in [Ka-1]) is a 4-valent graph (i.e. each vertex of the graph has valency four).

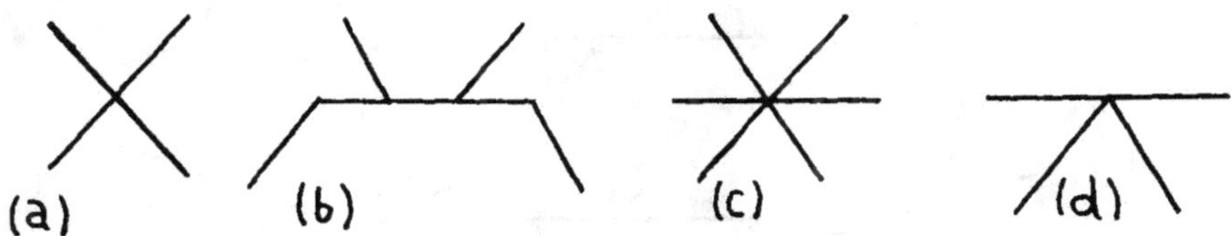

Fig. 1.1

[1]Our treatment is in the spirit of [Re] and [Ka-1]

If we specify which string of Fig.1.1(a) is over- and which under-crossing, we get a link diagram. From a link diagram one can recover the link up to isotopy.

We say that two diagrams are equivalent if they describe isotopic links. The following theorem of Reidemeister [Re,1927] and Alexander and Briggs [A-B,1927] (see also [Re] and [B-Z]), allows us to analyze links by working entirely with their diagrams.

Theorem 1.1 (Reidemeister theorem)

Two link diagrams are equivalent if and only if they are connected by a finite sequence of Reidemeister moves $R_i^{\pm 1}, i = 1, 2, 3$ (see Fig.1.2) and isotopy (deformation) of the plane of projection.

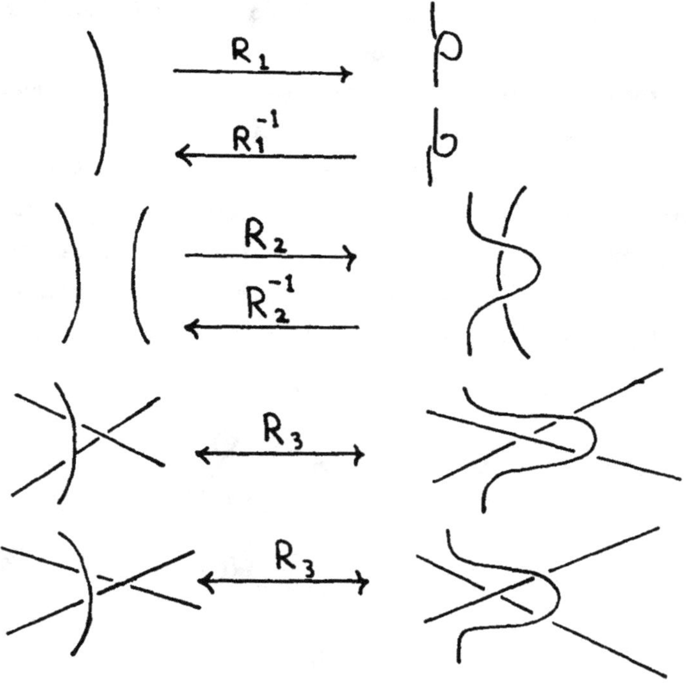

Fig. 1.2

Conjectures which we describe are related to t_k and $\bar{t}_k$ moves on link diagrams.

Definition 1.2 ([P-1])

(a) *The local change in an oriented link diagram which replaces* $\rightrightarrows$ *by* k *positive half-twists* ⋯ *is called a* t_k *move.*

(b) *For* k *even, the local change replacing* $\rightleftarrows$ *by* ⋯ *is called a* $\bar{t}_k$ *move.*

(c) *For an unoriented diagram, define a* k *move, replacing* $\asymp$ *by* k *positive half-twists,* ⋯ *, for any* k.

(d) *We say that two oriented links* L_1 *and* L_2 *are* t_i *(respectively* $\bar{t}_j$ *or* $t_i, \bar{t}_j$*) equivalent if there is a sequence of* $t_i^{\pm 1}$ *(respectively* $\bar{t}_j^{\pm 1}$ *or* $t_i^{\pm 1}, \bar{t}_j^{\pm 1}$*) moves and isotopies which converts* L_1 *to* L_2.

(e) *We say that two unoriented links* L_1 *and* L_2 *are* k *equivalent if there is a sequence of* k *moves and isotopies which converts* L_1 *to* L_2.

The following elementary conjectures were formulated after 1984; compare [P-1] and [Mo].

Conjecture 1.3 (Montesinos-Nakanishi) *Any link is 3-equivalent to a trivial link.*

Conjecture 1.4 ([P-1],[Mo]) *Any oriented link is* $t_3, \bar{t}_4$ *equivalent to a trivial link.*

Conjecture 1.5 ([P-2]) *Any oriented link is* $t_3, \bar{t}_8$ *equivalent to a trivial link.*

Conjecture 1.6 (Kawauchi-Nakanishi) *If two oriented links are homotopic (i.e. one can be converted to the other by a sequence of changes of over- and under-crossings, in self-crossings and isotopies) then they are* $t_4, \bar{t}_4$ *equivalent. In particular any oriented knot is* $t_4, \bar{t}_4$ *equivalent to the unknot.*

It has been observed in [P-1;p.650] that Conjectures 1.3 and 1.6 cannot be extended to $k > 4$. However, we will discuss later the possible conjecture for $t_3, \bar{t}_3$ equivalence.

We have proved in [P-2], Conjecture 1.4 for a class of knots. We present here the slight generalization of that result.

By a plane arcbody (compare [Lo]) we understand a pair (R, D) where R is a compact, connected subsurface of $S^2 = R^2 \cup \infty$ and D is a part of a link diagram being properly embedded in R (so that $D \cap \partial R = \partial D$ consists of $2n$ points ($n \geq 0$) called inputs and outputs of D); see Fig.1.3.

If R is a disc then (R, D) (sometimes D alone) is called a tangle (see Fig.1.3(a), (b)).

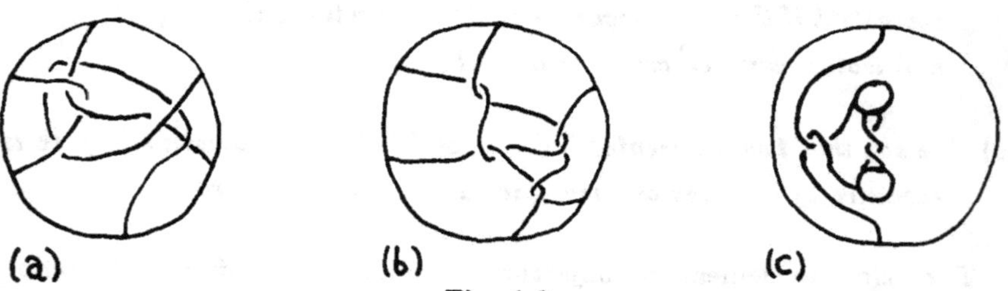

Fig. 1.3

Definition 1.7 ([A-P-R])

(a) *We say that a plane arcbody D is a matched arcbody if one can pair up the crossings in D so that each pair is connected as in Fig.1.4(a) (see Fig.1.3(b),(c) for examples of matched arcbodies).*

(b) *We say that an oriented plane arcbody D is an oriented matched arcbody (called coherently oriented in [P-2]) if it is obtained from an unoriented matched arcbody by the following construction: chessboard color the arcbody and then orient it according to the convention shown in Fig.1.4(b).*

Fig. 1.4

Notice that exchanging black and white colors changes the (global) orientation of the considered arcbody.

Theorem 1.8

(a) *If L_1 and L_2 are oriented links which are $t_3, \bar{t}_4$ equivalent to unlinks then their disjoint and connected sums ($L_1 \sqcup L_2$ and $L_1 \# L_2$) are $t_3, \bar{t}_4$ equivalent to unlinks.*

(b) *Let D_L be a diagram of an oriented link L which is neither disjoint nor connected sum. Assume that D_L is irreducible in the sense that any circle which cuts D in 2 points has interior or exterior composed of a simple arc (Fig.1.5). Assume additionally that there is a collection of circles, $\{C_i\}$, each C_i cutting D_L in exactly four points such that for each component E of $(R^2 \cup \infty) - \bigcup C_i$, the pair $(E, E \cap D_L)$ is one of the arcbodies (a)-(h) of Fig.1.6. Then L is $t_3, \bar{t}_4$ equivalent to a trivial link.*

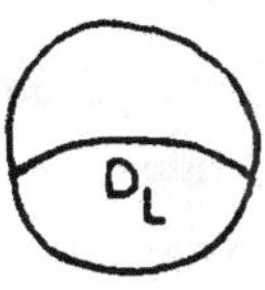

Fig. 1.5

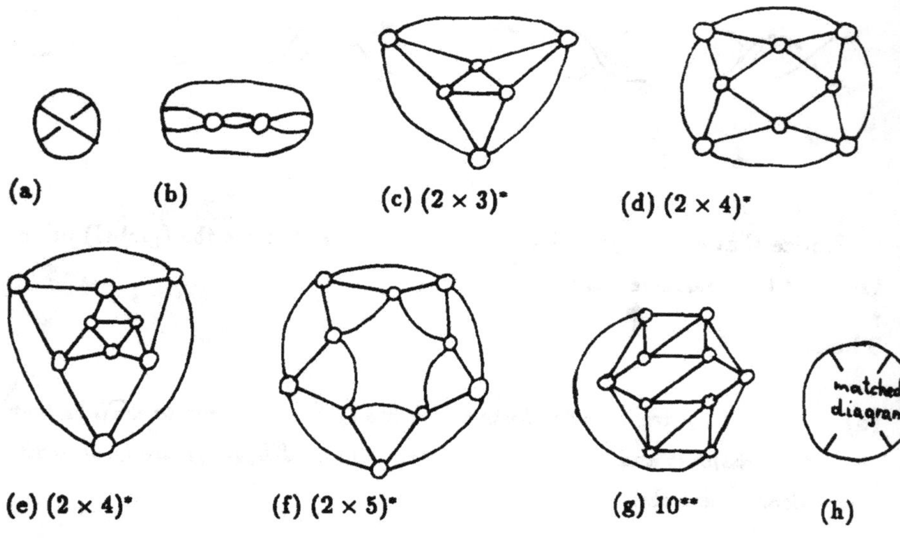

(a) (b) (c) $(2 \times 3)^*$ (d) $(2 \times 4)^*$

(e) $(2 \times 4)^*$ (f) $(2 \times 5)^*$ (g) 10^{**} (h)

Fig. 1.6

Corollary 1.9 *Let L be an oriented link which either*

(a) *is algebraic in the Conway sense (i.e. arborescent), or*

(b) *has a diagram with no more then 10 crossings* [2], *or*

(c) *has an oriented matched diagram, possibly outside a tangle with 9 (or less) crossings.*

Then L is $t_3, \bar{t}_4$ equivalent to a trivial link.

Proof: Proof of Corollary 1.9 from Theorem 1.8. By [B-S; Ch.2] a link is algebraic if it can be decomposed into plane arcbodies in Fig.1.6 (a) and (b). By [Co], any link diagram with no more than 10 crossings can be decomposed into plane arcbodies in Fig.1.6 (a)-(g). If a link has a diagram which is matched

---

[2] Our method works also for Conway graphs (arcbodies) with eleven vertices 11^*_A and 11^*_B which were described by Caudron [Ca] (see Fig.1.7). Therefore Corollary 1.9(b) can be extended to 11 crossings.

with possible exception of a tangle with at most 9 crossings than it is a matched diagram or can be decomposed into tangles (a)-(h) of Fig.1.6. In all described cases we can use Theorem 1.8. □

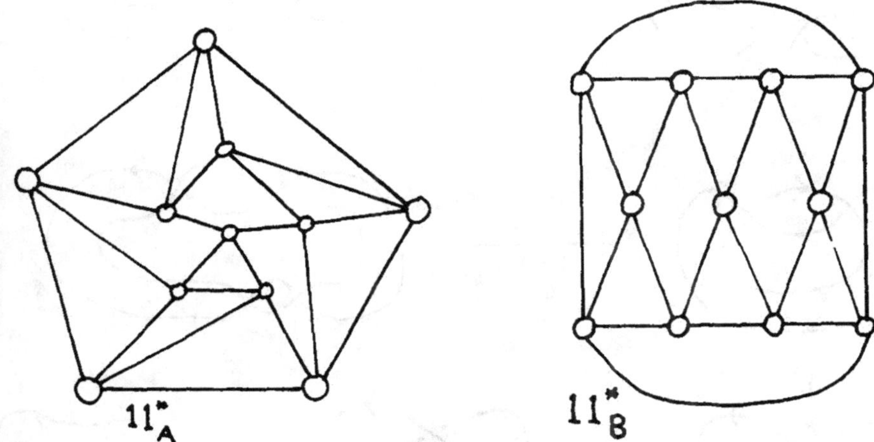

Fig. 1.7 (Conway graphs of eleven vertices)

It is unlikely that every oriented link has a matched diagram, however no counterexample is known. Nevertheless, the following weaker fact is much more likely.

Conjecture 1.10 *Any oriented link is t_3 equivalent to a link with matched diagram.*

Of course, by Theorem 1.8, Conjecture 1.4 follows from Conjecture 1.10.

Example 1.11 (Conjecture 1.5 for closed 3-braids) *The figure eight knot is $t_3, \bar{t}_4$ equivalent to the trivial knot, as illustrated in Fig.1.8. This, combined with the well known fact that any closed 3-braid is t_3 equivalent to an unlink or the figure eight knot, proves Conjecture 1.5 for closed 3-braids.*

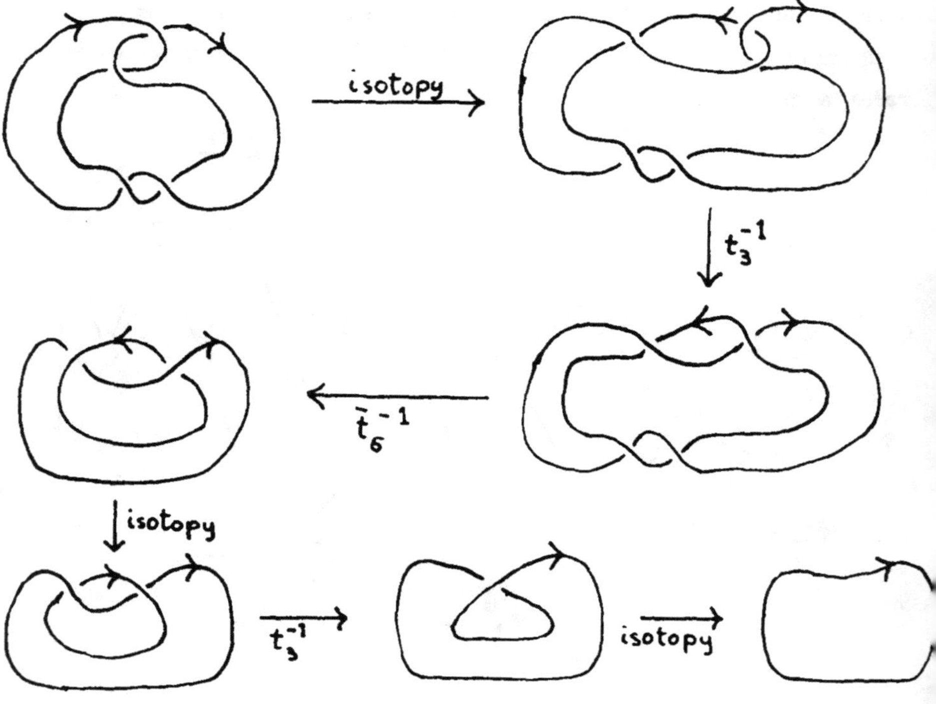

Fig. 1.8

2 Proof of Theorem 1.8

Consider the function Ψ from matched arcbodies to unoriented plane arcbodies defined as in Fig.2.1. We identify, for simplicity, oriented plane arcbody D with the arcbody with reversed global orientation, $-D$, so that unoriented matched arcbodies are in bijection with oriented ones.

Fig. 2.1

It was observed in [P-2] that for any unoriented plane arcbody D, elements of $\Psi^{-1}(D)$ are t_3 equivalent; see Fig.2.2.

Fig. 2.2

It has been proven in [P-2] that if unoriented plane arcbodies D_1 and D_2 are isotopic (modulo boundary), then elements of $\Psi^{-1}(D_1)$ and $\Psi^{-1}(D_2)$ are t_3 equivalent (modulo boundary); compare Lemma 2.1. Because a change of over-crossing and under-crossing in D corresponds to a $\bar{t}_4^{\pm 1}$ move in any element of $\Psi^{-1}(D)$, therefore a tangle of Fig.1.6(h) can be reduced by $t_3^{\pm 1}, \bar{t}_4^{\pm 1}$ moves to a tangle with no crossings or the tangles of Fig.2.3, each of which is a combination of tangles (a) and (b) of Fig.1.6.

Fig. 2.3

If we fill two circles of Fig.1.6(b) by tangles of type of Fig.1.6(a) or Fig.2.3 then, up to $t_3, \bar{t}_4$ moves and isotopy modulo boundary, we get either a tangle with no crossings or a tangle of type of Fig.1.6(a) or Fig.2.3. It remains to show that if we fill all circles but one, in arcbodies (c)-(g) of Fig.1.6 by tangles of Fig.1.6(a) (called singular tangles) or tangles of Fig.2.3 (called anticlasp tangles or A-tangles) then we can $t_3, \bar{t}_4$ reduce these tangles to singular or A-tangles. We proceed in several steps.

The following lemma is a slight generalization of the fact that if two unoriented plane arcbodies, D_1, D_2, differ by a third Reidemeister move then $\Psi^{-1}(D_1)$ and $\Psi^{-1}(D_2)$ are t_3 equivalent (compare [P-2]).

Lemma 2.1 *Consider an oriented tangle D with boundary S and three tangles inside: two of them being A-tangles and one, with boundary S_L filled by a tangle L such that the arcbody bounded by S and S_L is a matched arcbody (i.e. orientation is coherent); see Fig.2.4(a). Then D is $t_3, \bar{t}_4$ equivalent to the tangle of Fig.2.4(b).*

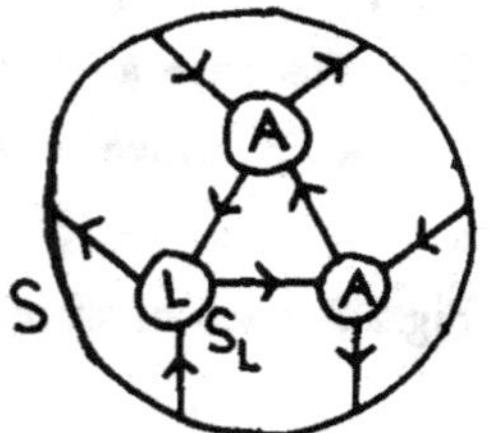

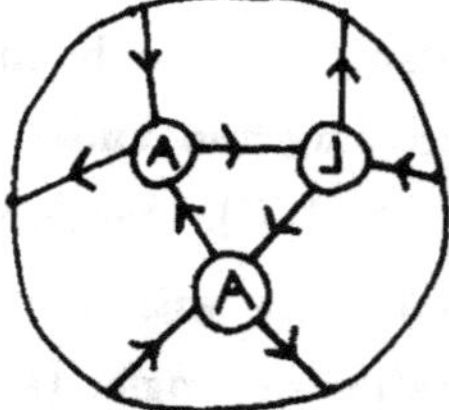

Fig. 2.4

Proof: We use appropriate anticlasps for A-tangles using $t_3, \bar{t}_4$ moves, and then the equivalence is illustrated in Fig.2.5. □

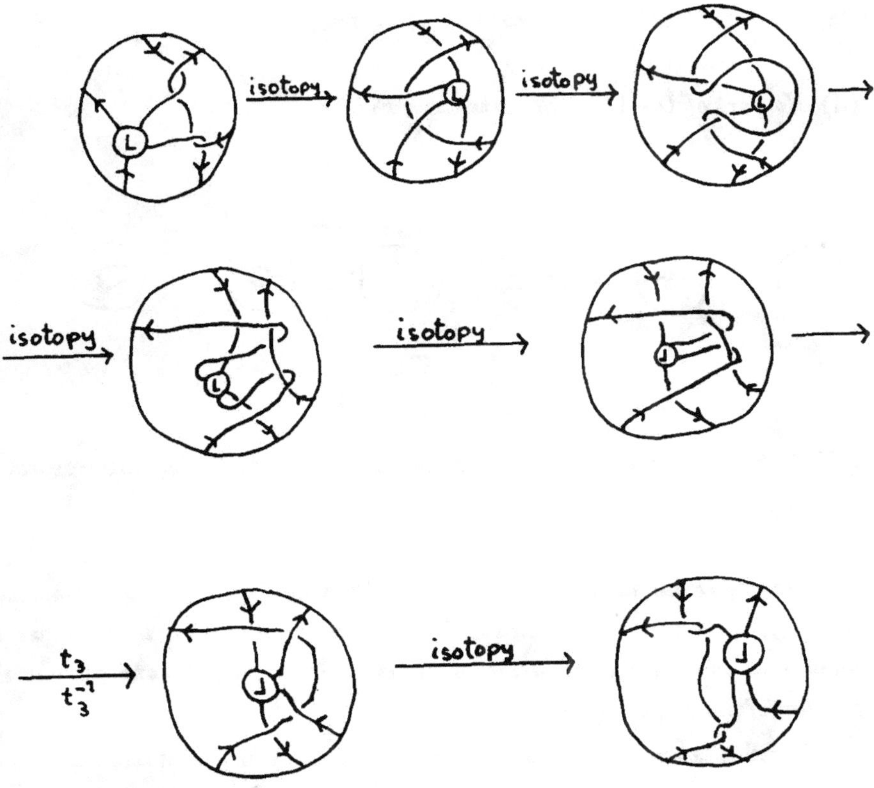

Fig. 2.5

We will show now how to simplify the given oriented tangle using t_3 and $\bar{t}_4$ moves, and isotopy. To do this we need a notion of a complexity of a tangle.

Definition 2.2 *Consider a collection C of disjoint circles in a tangle D (including $S = \partial D$) transversal to the tangle diagram, such that:*

(i) *For any $S_i \in C$, $S_i \cap D$ consists of at most 4 points.*

(ii) *No part of $D - \bigcup_{S_i \in C} S_i$ looks as in Fig.2.6.*

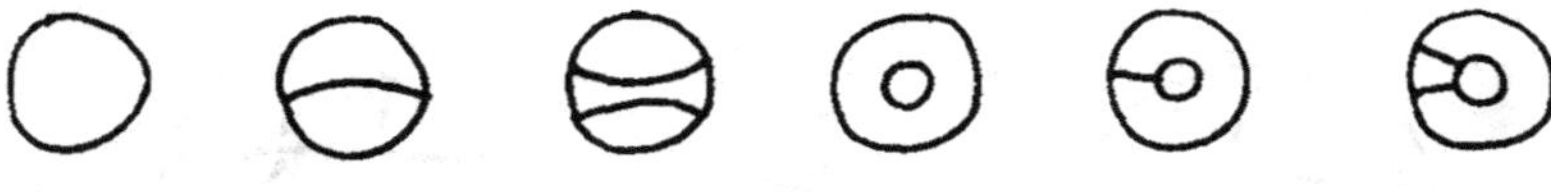

Fig. 2.6

(iii) *C has a maximal number of circles which satisfy (i)-(ii) with respect to D.*

For a tangle D and the collection of circles C satisfying (i)-(iii) the complexity $(s, n(s))$, ordered lexicographically, is defined as follows: s is the maximal number of boundary components of pieces of $D - \bigcup_{S_i \in C} S_i$, and $n(s)$ is the number of pieces (arcbodies) in $D - \bigcup_{S_i \in C} S_i$.

We call a tangle D *reducible* (respectively $t_3, \bar{t}_4$ *reducible*) if one can reduce the complexity of D by link isotopy (respectively $t_3, \bar{t}_4$ equivalence).

To finish the proof of Theorem 1.8(b) we need the following crucial lemma.

Lemma 2.3 *If a tangle D contains configuration of 3 triangles as in Fig.2.7 with "vertex" tangles filled by singular or A-tangles, then D is $t_3, \bar{t}_4$ reducible.*

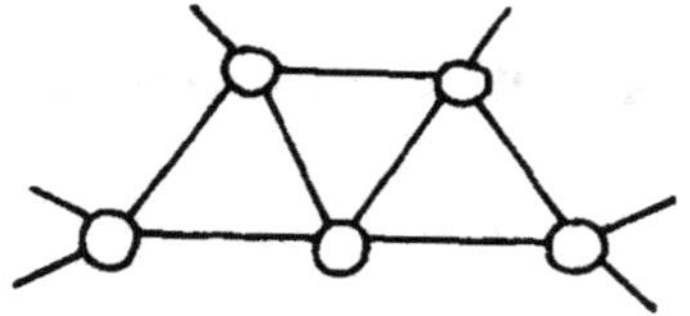

Fig. 2.7

Proof: Consider the following family of reducible configurations; see Fig.2.8-2.10:

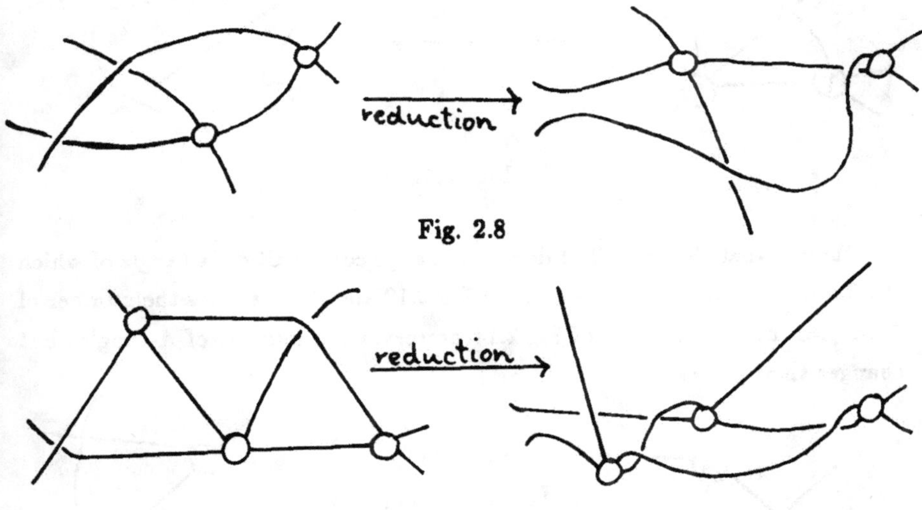

Fig. 2.8

Fig. 2.9

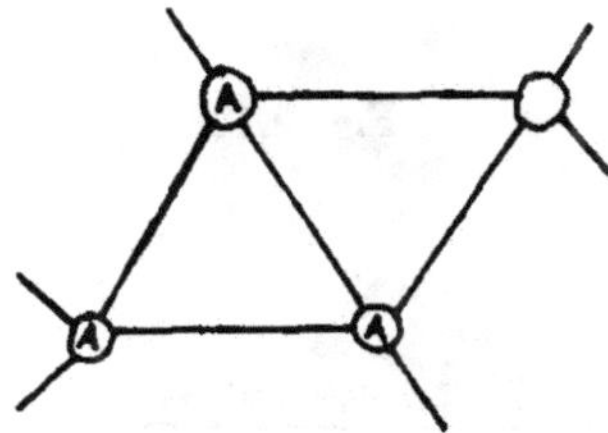

Fig. 2.10

Configuration in Fig.2.10 is reducible by Lemma 2.1. We will show now that configuration in Fig.2.7, up to $t_3, \bar{t}_4$ equivalence, always contains one of the configurations in Fig.2.8-2.10. To achieve this we use the following (isotopy) move:

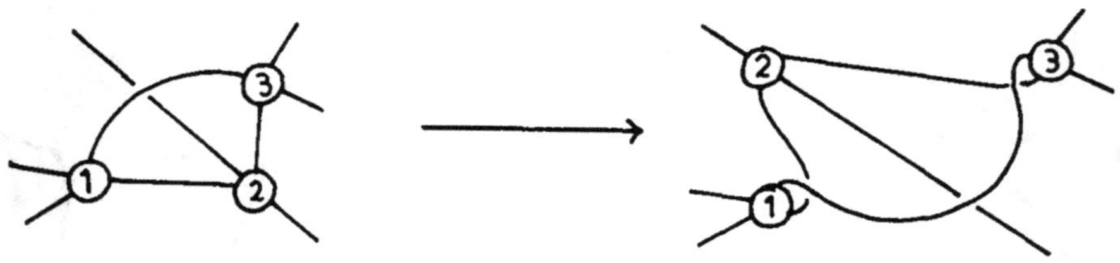

Fig. 2.11

Observe that the move 2.11 does not change complexity of a tangle of which the figure is a part, and in the case of Fig.2.12 and 2.13 reduces the number of A-tangles, and in the case of Fig.2.14 preserves the number of A-tangles but changes their places.

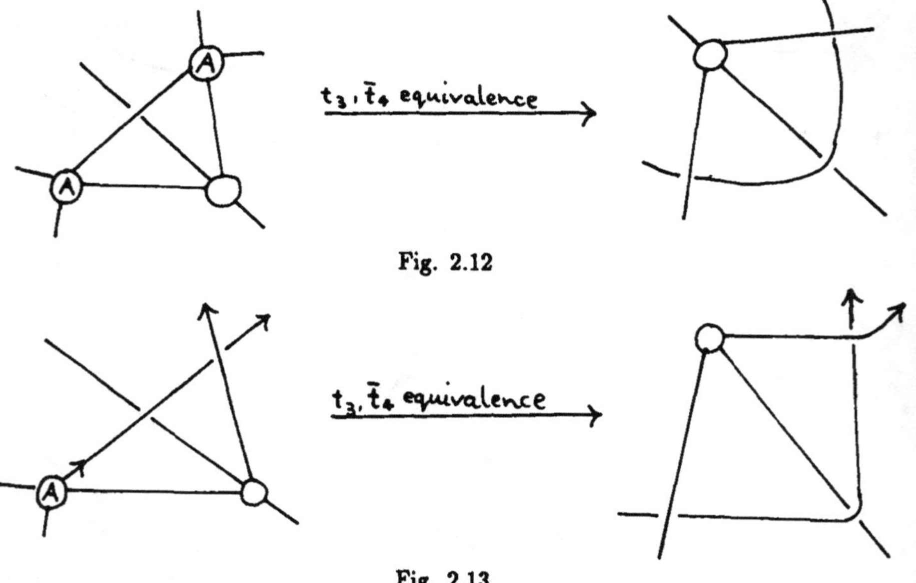

Fig. 2.12

Fig. 2.13

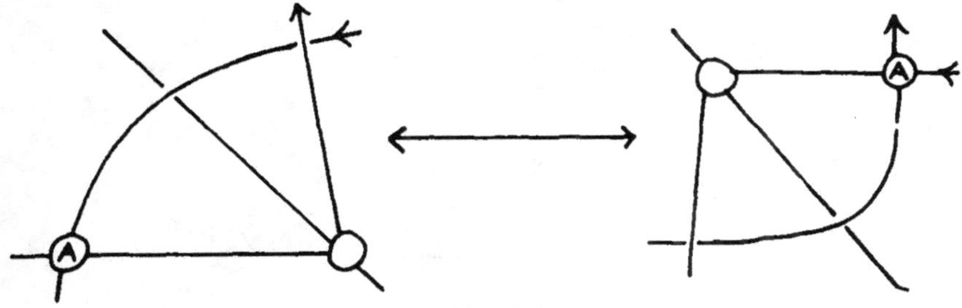

Fig. 2.14

Assume that Fig.2.7 has no configuration of Fig.2.8-2.10. In particular any triangle of Fig.2.7 has at least one singular crossing. We can use the move of Fig.2.11 to achieve one of described reducible configurations. The algorithm is as follows:

(1) Use the moves of Fig.2.12 or 2.13 to reduce the number of A-tangles.

(2) If (1) is impossible and Fig.2.8-2.10 is not achieved, then Fig.2.7 looks as Fig.2.15 (up to symmetry and mirror image).

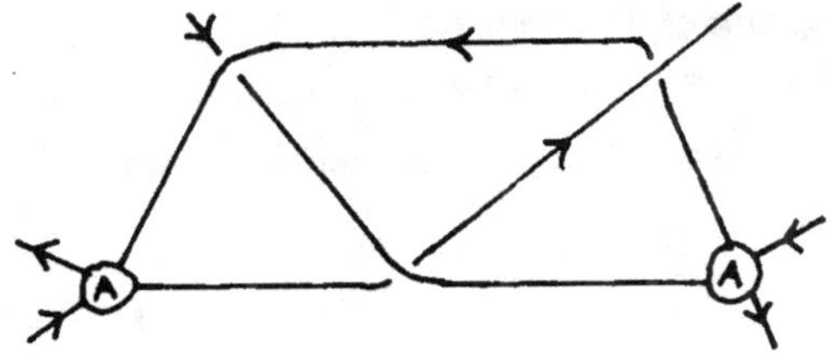

Fig. 2.15

Then we use the move of Fig 2.14, to obtain Fig.2.16 which can be $t_3, \bar{t}_4$ changed by the move of Fig.2.13 to a reducible configuration.

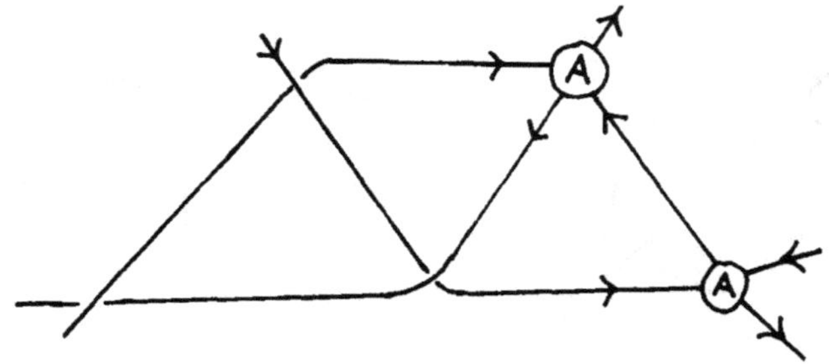

Fig. 2.16

This completes the proof of Lemma 2.3. Because Figures 1.6(c)-(g) contain the configuration 2.7, the proof of Theorem 1.8 is completed as well. □

Remark 2.4 *The proof of Theorem 1.8(b) works partially in the case of Conjecture 1.3. We have to exclude matched diagrams from the consideration, but the analysis of Conway graphs (arcbodies) is simpler than in the case of $t_3, \bar{t}_4$ equivalence. Namely we can obtain:*

Let D_L be a reduced diagram (on $R^2 \cup \infty$) of an unoriented link L which satisfies the assumption of Theorem 1.8(b) with $(E, E \cap D_L)$ of the form (a)-(g) of Fig.1.6 . Then L is 3 equivalent to a trivial link. In particular Conjecture 1.3 holds for any algebraic link and for any link with a diagram of no more than 10 crossings.

One can try to extend Theorem 1.8 using more configurations (Conway graphs) than on Fig.1.6. It still works for 11-vertex graphs but generally Lemma 2.3 is not sufficient and different reductions should be applied. The simplest example is given by the Conway graph $(3 \times 4)^*$; see Fig.2.17.

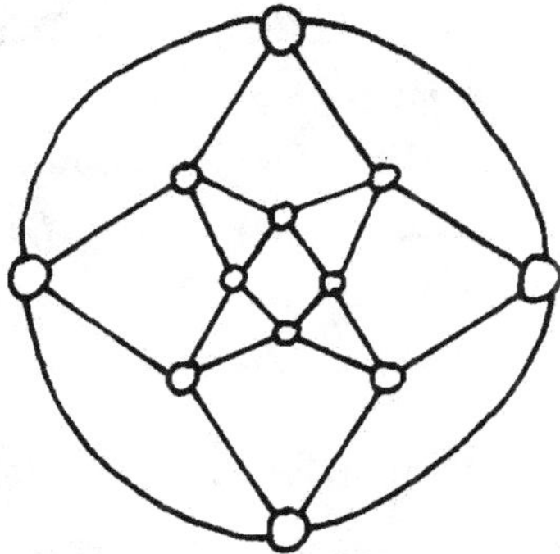

Fig. 2.17

Remark 2.5 *If D is an oriented matched diagram then one can use first $\bar{t}_4$ moves and then t_3 moves to convert D to a trivial link diagram. Namely using $\bar{t}_4$ moves D can changed to D' such that $\Psi(D')$ is an unlink.*

3 Applications

The skein (Homfly) polynomial invariant of oriented links, $P_L(a,z)$, is defined recursively by $P_0(a,z) = 1$ and $aP_{L_+} + a^{-1}P_{L_-} = zP_{L_0}$ (see [FYHLMO] or [P-T]).

The Jones polynomial invariant of oriented links, $V_L(t) \in Z[t^{\pm\frac{1}{2}}]$ is obtained from $P_L(a,z)$ by substituting $a = it^{-1}$ and $z = i(t^{1/2} - t^{-1/2})$; see [Jo].

Alternatively, one can define the Jones polynomial using the Kauffman approach [Ka]. Let L be the polynomial invariant of link diagrams preserved by the Reidemeister moves R_2 and R_3 (i.e. invariant of regular isotopy) which satisfies the following properties:

(i) $U^c = (-(A^2 + A^{-2}))^{c-1}$ where U^c is a diagram, with no crossings, of the unlink of c components.

(ii) $\langle \times \rangle = A \langle \asymp \rangle + A^{-1} \langle)(\rangle$

Let $f_L(A) = (-A^3)^{-Tait(L)}$ where $Tait(L)$ is the Tait or write number of a link diagram L obtained by summing up signs of all crossings of L ($\times$ is $+1$ and $\times$ is -1). $f_L(A)$ is an invariant of ambient isotopy of oriented links. It is a variant of the Jones polynomial; namely $V_L(t) = f_L(A)$ for $A = t^{-\frac{1}{4}}$.

Now observe that for $\varepsilon = \pm 1$ one has $P_{t_3(L)}(a, \varepsilon) = -\varepsilon a^{-3} P_L(a, \varepsilon)$. This formula follows from [P-1] and can be obtained by the simple calculation:

$$P_{t_3(L)}(a, z) = a^{-1} z P_{t_2(L)}(a, z) - a^{-2} P_{t_1(L)}(a, z) =$$

$$a^{-2} z^2 P_{t_1(L)}(a, z) - a^{-3} z P_L(a, z) - a^{-2} P_{t_1(L)}(a, z) =$$

$$a^{-2}(z^2 - 1) P_{t_1(L)}(a, z) - a^{-3} z P_L(a, z).$$

The following fact has been noted in [P-P]:

Lemma 3.1 *Let D be an unoriented link diagram and D^o be obtained from D by orienting it (of course there is no unique choice of orientation). Then one can choose the unique matched diagram D^{Ψ} in $\Psi^{-1}(D)$ such that the corresponding crossings of D^{Ψ} and D^o have the same signs. To see this, notice that the crossings of the diagrams on the left and right sides of Fig. 2.2 have opposite signs, however they go to the same crossing $\times$ under Ψ. Then we have:*

$$V_{D^o}(t) = P_{D^{\Psi}}(a, -1) \ for \ a = t^{-1/2}$$

Proof: For a trivial link of c components, T_c, one has

$$P_{T_c}(a, -1) = (-(a + a^{-1}))^{c-1} \ and \ V_{T_c}(t) = (-(t^{1/2} + t^{-1/2}))^{c-1}$$

Then one applies induction on the number of crossings using the fact that the triplet $\times$, $\asymp$, $)($ in D corresponds to the triplet $\times\!\!\times$, $\asymp$, $)($ in $\Psi^{-1}(D)$; see [P-P] for details.

Correspondence between oriented and unoriented links given by Ψ, and related correspondence between the skein polynomial of oriented links and the Jones polynomial of links, together with calculations of Murakami [Mu-2], suggest the following result which distinguishes three different types of $\bar{t}_4$ moves. □

Theorem 3.2 *Let P'_L denotes the derivative $\frac{dP_L(a,-1)}{da}$ at $a = 1$. Then*

(a)

$$P'_{\bar{t}_4(L)} = \begin{cases} 0 & \text{if } P_{\bar{t}_4(L)}(1,-1) = P_L(1,-1) \\ 3P_L(1,-1) & \text{if } -2P_{\bar{t}_4(L)}(1,-1) = P_L(1,-1) \\ 6P_L(1,-1) & \text{if } P_{\bar{t}_4(L)}(1,-1) = -2P_L(1,-1) \end{cases}$$

We say that a $\bar{t}_4$ move is of the first (respectively second or third) type if the first (respectively second or third) possibility in the equation holds.

(b) $P'_{t_3(L)} - P'_L = -3P_L(1,-1)$.

Proof: (b) follows from the formula $P_{t_3(L)}(a,-1) = a^{-3}P_L(a,-1)$. To prove (a) consider the formula 1.11(ii) for $k = 1$ from [P-1] (one can also easily derive it from the definition of the skein polynomial):

$$a^{-1}P_{[\text{diagram}]}(a,-1) - aP_{[\text{diagram}]}(a,-1) = (-a + a^{-1})P_{[\text{diagram}]}(a,-1)$$

From the formula follows $P'_{[\text{diagram}]} - P'_{[\text{diagram}]} = P_{[\text{diagram}]}(1,-1) + P_{[\text{diagram}]}(1,-1) - 2P_{[\text{diagram}]}(1,-1)$. Because any $\bar{t}_4$ move preserves $P_L(1,-1)$ therefore $P'_{[\text{diagram}]} - P'_{[\text{diagram}]} = 2P_{[\text{diagram}]}(1,-1) - 2P_{[\text{diagram}]}(1,-1)$. Now (a) follows by analyzing various possible relations between $P_{[\text{diagram}]}(1,-1)$ and $P_{[\text{diagram}]}(1,-1)$. □

Corollary 3.3 *If L_1 and L_2 are $t_3, \bar{t}_4$ equivalent links then $P'_{L_2} - P'_{L_1}$ can be interpreted as follows. Consider a family of $t_3^{\pm1}, \bar{t}_4^{\pm1}$ moves which transforms L_1 to L_2. Let C^+ (resp., C^-) be the number of t_3 (resp., t_3^{-1}) moves in the family. Let d_i^+ (resp., d_i^-) be the number of $\bar{t}_4$ (resp., $\bar{t}_4^{-1}$) moves of the i'th type in the family. Then*

$$\frac{1}{P_{L_1}(1,-1)}(P'_{L_2} - P'_{L_1}) = -3(C^+ - C_-) + 3(d_2^+ - d_2^-) + 6(d_3^+ - d_3^-).$$

Note that $C^+ - C_- \equiv \frac{P_{L_2}(1,1)}{P_{L_1}(1,1)}$ *mod* 2. *It is the case because a* $\bar{t}_4$ *move preserves* $P_L(1,1)$ *and a* t_3 *move changes its sign.*

From the formulas in the proof of Theorem 3.2, one can easily see that if $\bar{t}_4$ move $\to$ is of the first (resp., third) type then the $\bar{t}_4$ move $\to$ is of the third (resp., first) type. Therefore we have the following corollary; compare Fig.3.1.

Corollary 3.4 **(a)** *Let* L_2 *be obtained from* L_1 *by a* $\bar{t}_4^{\pm 1}$ *move of the third type. Then one can obtain* L_2 *from* L_1 *using a* $\bar{t}_4^{\mp 1}$ *move of the first type and two* $t_3^{\mp 1}$ *moves (see Fig.3.1).*

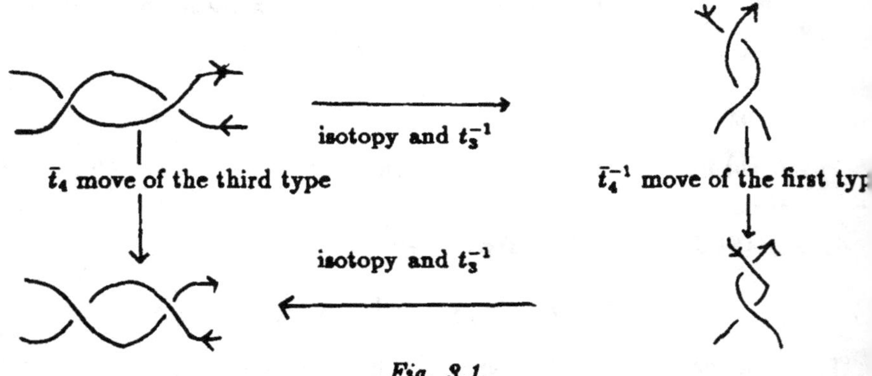

Fig. 3.1

(b) *If two links* L_1 *and* L_2 *are* $t_3, \bar{t}_4$ *equivalent, then they are equivalent by* $t_3^{\pm 1}$ *moves and* $\bar{t}_4^{\pm 1}$ *moves of the first and second types.*

There are two more observations which make use of the function Ψ^{-1}. First relates the classical signature σ of D° with the Tristram- Levine signature $\sigma(e^{2\pi i/6})$ of D^Ψ (compare [Go]).

Second translates the Kauffman polynomial of D to a polynomial invariant of $\Psi^{-1}(D)$. The regular isotopy version of this polynomial, $\widetilde{W}$, satisfies the recurrent relation

$$\widetilde{W}(\;) + \widetilde{W}(\;) = z(\widetilde{W}(\;) + \widetilde{W}(\;))$$

and additionally $\widetilde{W}$ is invariant under t_3 moves. The polynomial invariant of isotopy of links, $W(L)$, is defined by $W(L) = a^{-\frac{1}{3}Tait(L)}\widetilde{W}(L)$.

It depends on Conjecture 1.10 whether the initial condition $W(T_1) = 1$ and the above skein relation are sufficient to define $W(L)$.

To state a conjecture about $t_3, \bar{t}_8$ moves one should first define the notion of Ψ homotopic oriented links. It is the translation of the link homotopy using Ψ^{-1}.

Definition 3.5 *Two oriented links L_1 and L_2 are called Ψ homotopic if there is a finite sequence of t_3 and $\bar{t}_4$ moves of the first or third type, and isotopies which transforms L_1 onto L_2.*

The following conjecture follows from Conjectures 1.6 and 1.10; compare Proposition 3.7.

Conjecture 3.6 *If two oriented links are Ψ homotopic then they are $t_3, \bar{t}_8$ equivalent.*

Proposition 3.7

(a) *Conjecture 1.5 follows from Conjectures 1.3 and 1.10.*

(b) *Conjecture 3.6 follows from Conjectures 1.6 and 1.10.*

Proof:

(a) Let L be an oriented link. By Conjecture 1.10, L is t_3 equivalent to a link with a matched diagram, say D. By Conjecture 1.3, $\Psi(D)$ is 3-equivalent to a trivial link diagram U^n. 3 moves on an unoriented diagram E translate to $\bar{t}_6$ moves on $\Psi^{-1}(E)$, up to t_3 moves. Therefore D and L are $t_3, \bar{t}_6$ equivalent to a trivial link U^n.

(b) By Conjecture 1.10, any Ψ homotopic links L_1 and L_2 are t_3 equivalent to links with matched diagrams, say D_1 and D_2. Then $\Psi(D_1)$ and $\Psi(D_2)$ are link homotopic and by Conjecture 1.6, they are 4-equivalent. 4 moves on an unoriented diagram E translate to $\bar{t}_8$ moves on $\Psi^{-1}(E)$. Therefore D_1 and D_2 (as well as L_1 and L_2) are $t_3, \bar{t}_8$ equivalent.

□

The following conjecture was motivated by the Nash conjecture in real algebraic geometry [Nash; 1952].

Conjecture 3.8 (Akbulut) [3]

Any link with even linking numbers between any pair of components can be converted to an unlink by the Fox $t_{2,q}$ moves ([Fo-1], [P-1]), any q allowed. A $t_{2,q}$ move is described in Fig.3.2.

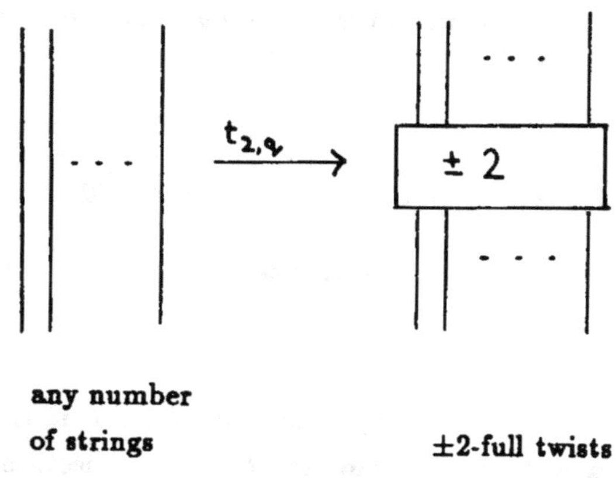

any number
of strings

±2-full twists

Fig. 3.2

The Akbulut conjecture holds for Borromean links [Na] (see Fig.3.3). One should stress here that Conjecture 1.6 (Kawauchi-Nakanishi) is still open even

[3] The Akbulut conjecture has been recently proven by Y.Nakanishi [Na-2]. Then the Nash conjecture was proven as well [A-K]

for Borromean links.

Fig. 3.3

4 Tricoloring

The idea of tricoloring has been introduced by R.Fox [Fo-2] and extensively used and popularized by J.Montesinos and L.Kauffman. We relate here tricoloring with the Jones polynomial via 3 equivalence.

Definition 4.1 *We say that a link diagram D is tricolored if every arc is colored r (red), b (blue) or y (yellow) (we consider arcs of the diagram literally so in the tunnel one arc ends and the second starts), and at any given crossing either all three colors appear or only one color appears. The number of different tricolorings is denoted by $tri(D)$. If a tricoloring uses only one color we say that it is a trivial tricoloring.*

Lemma 4.2 *If two link diagrams D and D' are 3 equivalent then $tri(D) = tri(D')$.*

Proof: We have to check that $tri(D)$ is preserved under Reidemeister moves and 3-moves. The invariance under 3 move is illustrated in Fig.4.1. The invariance under Reidemeister moves is equally simple and we leave it to the reader. □

Fig. 4.1

One can easily check that for a trivial c-component link diagram, U^c, $tri(U^c) = 3^c$ and $V^2_{U^c}(e^{\pi i/3}) = 3^{c-1}$. Furthermore 3 moves change $V^2_{U^c}(e^{\pi i/3})$ by a factor ± 1 (see [P-1]). Therefore Lemma 4.2 gives, up to Conjecture 1.3, the following result:

Theorem 4.3 $tri(L) = 3|V^2_L(e^{\pi i/3})|$

Proof: It would be nice to have a totally elementary proof of Theorem 4.3. The simplest I found, but not so elementary, goes as follows: R.Fox interprets $tri(L)$ as representations of $\pi_1(S^3 - L)$ into the permutation group S_3, and proves that the number of representations is equal to $3(3^{Dim(H_1(M_L^{(2)}, Z_3))})$, where $M_L^{(2)}$ is the branched 2-fold covering of S^3 with the branching set L (see [Fo-2]). Lickorish and Millett found the formula [L-M]: $V_L(e^{\pi i/3}) = \pm i^{c(L)-1}(i\sqrt{3})^{Dim(H_1(M_L^{(2)}, Z_3))}$ which reduces to

$$|V_L^2(e^{\pi i/3})| = 3^{Dim(H_1(M_L^{(2)}, Z_3))}.$$

This completes the proof of Theorem 4.3. □

Remark 4.4 *P.G.Tait, almost a hundred year ago, considered a similar to tricoloring concept. Namely he considered coloring edges of cubic graphs (i.e. graphs with valencies of all vertices equal to three) with three colors, so that edges with common vertex have different colors. It is now called Tait coloring. One can put both concepts in more general setting. Let $S = (V, E)$ be a setoid (set system), i.e. a set V with a family of subsets of V ($E \subset 2^V$) (compare*

[P-P]). We say that a setoid is cubic if any element of V is exactly in three elements of E. The tricoloring of S is a coloring of elements of E with 3 colors so that elements of E with common $v \in V$ have three different colors or all the same color. If we exclude the same color we get the Tait coloring.

References

[Ak] S.Akbulut, Z_2-framed link theory, Abs. of A.M.S. (Feb. 1987), Issue 49, vol.8, no2, p.183.

[A-K] S.Akbulut, H.King, Rational structures on 3-manifolds, Pacific J. Math., 150 (1991), 201-214.

[A-B,1927] J.W.Alexander, G.B. Briggs, On types of knotted curves, Ann. of Math., 28(2), (1927), 563-586.

[A-P-R] R.P.Anstee, J.H.Przytycki, D.Rolfsen, Knot polynomials and generalized mutation, Topology and Applications 32 (1989) 237-249.

[B-S] F.Bonahon, L.Siebenmann, Geometric splittings of classical knots and the algebraic knots of Conway, to appear in L.M.S. Lecture Notes Series, 75.

[B-Z] G.Burde, H.Zieschang, Knots, De Gruyter (1985).

[Ca] A.Caudron, Classification des noeuds et des enlacements, Pre-publications, Univ. Paris-Sud, Orsay 1981.

[Co] J.H.Conway, An enumeration of knots and links, and some of their algebraic properties, Computational Problems in Abstract Algebra (John Leech, ed.), Pergamon Press, Oxford and New York, 1969.

[Dun] G.W.Dunnington, Carl Friedrich Gauss, Titan of Science, Hafner Publishing Co., New York, 1955.

[Fo-1] R.H.Fox, Congruence classes of knots, Osaka Math. J., 10, 1958, 37-41.

[Fo-2] R.H.Fox, Metacyclic invariants of knots and links, Canadian J. Math., XXII(2) 1970, 193-201.

[FYHLMO] P.Freyd, D.Yetter, J.Hoste, W.B.R.Lickorish, K.Millett, A.Ocneanu, A new polynomial invariant of knots and links, Bull. Amer. Math. Soc., 12 (1985) 239-249.

[Go] C.McA.Gordon, Some aspects of classical knot theory, In: "Knot theory", L.N.M. 685 (1978) 1-60.

[Jae] F.Jaeger, On Tutte polynomials and link polynomials, Proc. Amer. Math. Soc., 103 (2) (1988) 647-654.

[Jo] V.Jones. Hecke algebra representations of braid groups and link polynomials, Ann. of Math., (1987) 335-388.

[Ka] L.H.Kauffman, State models and the Jones polynomial, Topology 26, (1987) 395-407.

[Ka-1] L.H.Kauffman, Formal knot theory, Mathematical Notes 30, Princeton University Press, 1983.

[Ka-2] L.H.Kauffman, On Knots, Princeton University Press, 1987.

[L-M] W.B.R.Lickorish, K.Millett, A polynomial invariant of oriented links, Topology 26(1987), 107-141.

[Lo] M.T.Lozano, Arcbodies, Math. Proc. Camb. Phil. Soc. 94 (1983), 253-260.

[Mo] H.R.Morton. Problems. In: "Braids", ed. J.S.Birman and A.Libgober, Contemporary Math. Vol. 78 (1988) 557-574.

[Mu] H.Murakami, Unknotting number and polynomial invariants of a link, preprint 1985.

[Mu-2] H.Murakami, On derivatives of the Jones polynomial, Kobe J. Math., 3 (1986), 61-64.

[Na-1] Y. Nakanishi, Fox's congruence classes and Conway's potential functions of knots and links, preprint, Kobe Univ., 1986.

[Na-2] Y. Nakanishi, On Fox's congruence classes of knots, II, Osaka J. Math. 27 (1990), 207-215.

[Nash] J.Nash, Real algebraic manifolds, Ann. of Math., 56 (1952), 405-421.

[P-P] T.M.Przytycka, J.H.Przytycki, Invariants of chromatic graphs, Dept. of Computer Science, The University of British Columbia, Technical Report 88-22, 1988; partially contained in "Subexponentially computable truncations of Jones-type polynomials", Contemporary Mathematics, 147 (1993).

[P-1] J.H.Przytycki, t_k-moves on links. In: "Braids", ed. J.S.Birman and A.L.Libgober, Contemporary Math. Vol. 78 (1988) 615-656.

[P-2] J.H.Przytycki, $t_3, \bar{t}_4$ moves conjecture for oriented links with matched diagrams, Math. Proc. Camb. Phil. Soc. 108 (1990) 55-61.

[P-T] J.H.Przytycki, P.Traczyk, Invariants of links of Conway type, Kobe J. Math. 4 (1987) 115-139.

[Re,1927] K.Reidemeister, Elementare Begrundung der Knotentheorie, Abh. Math. Sem. Univ. Hamburg, 5 (1927), 24-32.

[Re] K.Reidemeister, Knotentheorie, Ergebn. Math. Grenzgeb., Bd.1; Berlin: Springer-Verlag (1932) (English translation: Knot theory, BSC Associates, Moscow, Idaho, USA, 1983).

[Ro] D.Rolfsen, Knots and links, Publish or Perish, 1976.

[Tr] L.Traldi. A dichromatic polynomial for weighted graphs and link polynomials, Proc., Amer. Math. Soc., 106 (1) (1989) 279-286.

[Tu] W.T.Tutte, Graph theory, Encyclopedia of Mathematics and its Applications 21, Cambridge University Press, 1984.

Department of Mathematics and Computer Science
Odense University
Campusvej 55
DK-5230, Odense M, Denmark
e-mail: Josef@imada.ou.dk

KNOT POLYNOMIALS AS STATES OF NONPERTURBATIVE FOUR DIMENSIONAL QUANTUM GRAVITY

JORGE PULLIN
Department of Physics, University of Utah
Salt Lake City, UT 84112 USA

ABSTRACT

In this brief note we present an outline and a list of references to recent work that has brought to the forefront the role of the Jones and Kauffman polynomials as states of nonperturbative quantum gravity in the loop representation. The reader should consult the references for more details.

Traditionally, canonical quantizations of the General Theory of Relativity have been based on the use of a metric as fundamental variable. This approach leads to difficulties since the resulting constraint equations are unmanageable. Not a single solution of the resulting wave equations of the quantum theory is known. This prevents us from finding the space of physical states and further developing the theory into a stage that would enable us to make physical predictions. The situation changed when Ashtekar[1] showed that a canonical description of General Relativity in terms of an SU(2) connection A^i_a can be achieved. In this formalism, the Einstein equations bear a great resemblance with those of a Yang-Mills theory. One can construct a quantum representation in which wavefunctions of the quantum gravitational field are functionals of an SU(2) connection. Moreover one can show that the exponential of the Chern-Simons form built from this connection is actually a physical state of the quantum theory with a cosmological constant, in the sense that it is annihilated by all the quantum constraint equations[2,3].

As for any quantum theory based on a connection, one can construct a loop representation by expanding the wavefunctions of a connection in a basis of holonomies[4,5]. The wavefunctions are now functionals of loops on a three manifold $\Psi[\gamma]$. Because the theory is invariant under diffeomorphisms, the constraint equations require that the wavefunctions be functionals of loops invariant under smooth deformations of the loops. That is, they are knot invariants[6].

The constraint equations (the Einstein equations) can be written in terms of differential operators in loop space[7,8] (for instance the area derivative) and have a well defined action on wavefunctionals of loops. In order to have meaningful states, loops must have intersections[9].

Any functional of a holonomy can be expressed in terms of "loop coordinates"[10].

These are geometrical quantities that contain all the information needed from a loop to construct a holonomy. It turns out that the action of the differential operators in loop space on these coordinates is well defined.

Having developed a machinery for writing wavefunctions in the loop representation in a manageable form (loop coordinates) and for writing the constraint equations as differential operators in loop space, a natural step is to try to find wavefunctions of loops that solve the constraint equations.

As we mentioned, the exponential of the Chern-Simons form is a solution to all the constraints with cosmological constant in the representation based on connections. Its transform into the loop representation is the Kauffman bracket (even for intersecting loops[3,11]). Therefore the Kauffman bracket should be a solution of all the constraints of quantum gravity in the loop representation. An explicit check of this fact with the technology developed reveals an intriguing result[12]. In order for the Kauffman bracket to be a solution with a cosmological constant, certain portions of its expansion have to be annihilated by the vacuum constraints[13]. It turns out (at least up to second order) that these quantities are just the coefficients of the Jones polynomial. Therefore our analysis seems to reveal (up to second order) that the Jones polynomial (even for intersecting knots) is a physical state of quantum gravity in the loop representation and solves the Wheeler-DeWitt equation in terms of loops. An overview of all these techniques can be found in reference 14.

Acknowledgements: I wish to thank Lou Kauffman for inviting me to speak at the AMS meeting. This work was supported in part by grant NSF PHY92-07225 and by research funds of the University of Utah.

References:

1. A. Ashtekar, *Phys. Rev. Lett.* **57** (1986) 2244; *Phys. Rev.* **D36** (1987) 1587; *New perspectives in canonical gravity* (with invited contributions) Lecture Notes, Bibliopolis, Naples 1988; *Lectures on non perturbative quantum gravity* (notes prepared in collaboration with R. Tate), Advanced Series in Astrophysics and Cosmology Vol. 6, World Scientific, Singapore 1991.
2. H. Kodama, *Phys. Rev.* **D42** (1990) 2548.
3. B. Brügmann, R. Gambini, J. Pullin, *Nuc. Phys.* **B385** (1992) 587.
4. R. Gambini, A. Trias, *Nuc. Phys.* **B278** (1986) 436.
5. C. Rovelli, L. Smolin, *Nuc. Phys.* **B331** (1990) 80.
6. C. Rovelli, L. Smolin, *Phys. Rev. Lett.* **61** (1988) 1155.
7. R. Gambini, *Phys. Lett.* **B235** (1991) 180.
8. B. Brügmann, J. Pullin, *Nuc. Phys.* **B** (1993) (in press).
9. B. Brügmann, J. Pullin, Nuc. Phys. **B363** 221 (1991).
10. C. Di Bartolo, R. Gambini, J. Griego, A. Leal, *"Loop space coordinates, linear representations of the diffeormorphism group and knot invariants"* Montevideo Preprint (1992).

11. D. Armand-Ugon, R. Gambini, P. Mora, *"Knot invariants for intersecting loops"* Montevideo Preprint (1992); hep-th@xxx.lanl.gov:9212137.
12. B. Brügmann, R. Gambini, J. Pullin, *Gen. Rel. Grav.* (1993) (in press); hep-th@xxx.lanl.gov:9212137.
13. B. Brügmann, R. Gambini, J. Pullin, *Phys. Rev. Lett.* **68** (1992) 431; also in *"Proceedings of the XXth meeting on differential geometric methods in theoretical physics"*, S. Catto, A. Rocha, eds., World Scientific, Singapore (1992).
14. J. Pullin, *"Knot theory and quantum gravity in loop space: a primer"*, to appear in *"Proceedings of the Vth Mexican school of particles and fields"*, J. Lucio, ed. World Scientific, Singapore (1993); hep-th@xxx.lanl.gov:9301028.

ON INVARIANTS OF 3-MANIFOLDS DERIVED FROM ABELIAN GROUPS

JOSEF MATTES, MICHAEL POLYAK*, NIKOLAI RESHETIKHIN**

ABSTRACT. In this note we describe a way to construct invariants of links and 3-manifolds from 3-cocycles on abelian groups. We discuss the relation between $\mathbf{Z}_n$-invariants and $U(1)$-Chern-Simons theory. The invariant related to $\mathbf{R}^1$ exists for rational homology spheres and can be regarded as a perturbative invariant for $U(1)$-Chern-Simons theory.

1. INTRODUCTION

Topological quantum field theories [A] have become an interesting source of topological invariants, in particular, of invariants of 3-dimensional manifolds [W]. As is usual in quantum field theories, the phenomenological definition of these invariants was given in terms of a functional integral, viz.

$$Z_{k,G}(M,g) = \int \exp(ik\,\mathrm{CS}(A))\mathcal{D}[A] \tag{1.1}$$

where CS denotes the Chern-Simons functional, G is a compact semisimple simply-connected Lie group and the integral has the symbolic meaning of integration over all gauge equivalence classes of connections in a principal G-bundle over M.

There are two standard ways in which physicists avoid the fact that (1.1) is not a mathematically correctly defined object.

The first standard way of proper definition of functional integrals is via perturbation theory. In (1.1) this would be the limit $k \to \infty$. In this limit one can try to compute the integral by the stationary phase approximation method. Witten gave the following formula for the large k-limit of (1.1):

$$Z_{k,G}(M,g) = e^{i\pi \dim G\left(\frac{\eta_{grav}}{2} + \frac{1}{12}\frac{CS(A^g)}{2\pi}\right)} \sum_{[A^{(0)}]} e^{i(k+\frac{c_2(G)}{2})CS(A^{(0)})} T_{A^{(0)}} \tag{1.2}$$

where the sum is taken over gauge equivalence classes of flat connections in the principal G-bundle over M, g is a metric on M, A^g is the corresponding Levi-Civita connection, η_{grav} is the η-invariant of the operator $*D^g + D^g*$ (D^g being the exterior derivative twisted by A^g), $c_2(G)$ is the value of the quadratic Casimir

* Supported by Rotshild fellowship

** Supported by Sloan fellowship and NSF grant DMS-9015821

operator in the adjoint representation of G and $T_{A^{(0)}}$ is the Ray-Singer torsion of $A^{(0)}$. For a more refined description of this limit see [F], [FG]. The sum in (1.2) is defined if $\{[A^{(0)}]\}$ is a set of isolated points. Otherwise the sum should be replaced by an integral over classes of flat connections with a suitable measure. The coherent description of the asymptotics of (1.1) in this case is an interesting and, as far as we know, open problem.

It has been shown in [W] that the RHS should depend only on the 2-framing of 3-manifold M. Therefore (1.1) is expected to be an invariant of framed 3-manifolds. On the other hand, as was shown by Atyiah [A1], for each 3-manifold there exists a canonical framing. Choosing this canonical framing for M in (1.1) one should expect to obtain an invariant of 3-manifolds, denoted further by $Z_{k,G}(M)$. We will assume in this paper that M is endowed with a canonical framing.

The second way is to use the phenomenological formula (1.1) for studying (or rather to find) transformation properties of (1.1) under certain natural transformations. In most cases these properties fix uniquely the object in the LHS of (1.1) and give an independent rigorous definition of it. The definition of 3-manifold invariants via surgery on a link in S^3 given in [RT] can be regarded as the realization of this program for (1.1).

Formula (1.2) shows that if we really want to find the limit of (1.1) for $k \to \infty$ we at least have to sum over all classes of flat connections. On the other hand, it is clear that each individual term in (1.2) should be an invariant of the pair $(M, [A^{(0)}])$. After appropriate normalization it becomes an element of $\mathbb{C}[[\frac{1}{k}]]$, which we denote by $Z_k(M, [A^{(0)}])$. It was shown by Axelrod and Singer [AS] and Kontsevich [Ko] that this power series exists if M is a rational homology sphere and the de Rham complex twisted by $A^{(0)}$ is acyclic (in [Ko] $A^{(0)} = 0$) and that the coefficients are linear combinations of integrals $\int_{M\times\cdots\times M} \omega$ for some suitable forms ω and that they are invariants of pairs $(M, [A^{(0)}])$. The term $Z_k(M, 0) \in \mathbb{C}[[\frac{1}{k}]]$ (when it exists) is an invariant of 3-manifolds. Apparently this invariant is different from $Z_{k,G}(M)$ with $k \in \mathbb{N}$ and exists for rational homology spheres.

In this note we construct an invariant of 3-manifolds for any abelian group G if we are given an abelian 3-cocycle and a suitable invariant measure on G. In particular, we consider two examples of such invariants, corresponding to $G = \mathbb{Z}_n$ and to $G = \mathbb{R}^1$. In the $\mathbb{Z}_n$ case we compute the invariant for some manifolds and we argue that it has to coincide with appropriately defined $U(1)$-Chern-Simons topological field theory. The invariant for $G = \mathbb{R}^1$ presumably coincides with $Z_k(M, 0)$ for $U(1)$-Chern-Simons theory.

The paper is organized in the following way. In the second and third sections we recall the notions of framed 3-valent graphs, their diagrams and define G-colorings of such diagrams. Systems of weights for G-colored diagrams are introduced in sections 4, 5. They are defined via normalized abelian $\mathbb{C}^*$-valued 3-cocycles for an abelian group G. Such a system determines a certain $\mathbb{C}$-linear abelian braided balanced monoidal category [S], [GKR]. In the physics literature such weight systems are known as braiding-fusion matrices (see [MS]). A description of the corresponding invariants of framed links and 3-manifolds is given in section 6 assuming the existence of a suitable measure on G. This construction is an example of a more general approach via braided balanced monoidal categories. In sections 7-9 we present some explicit formulas for invariants of 3-manifolds related to $G = \mathbb{Z}_n$ and $G = \mathbf{R}^1$ and discuss their relation with $U(1)$-Chern-Simons theory.

We would like to thank D. Freed, R. Kirby, L. Rozansky and O. Viro for stimulating discussions.

2. Graphs and diagrams

We briefly recall the notion of 3-valent ribbon (or framed) graph. See [RT] for details. Let Y_n be the set of points $(i, 0)$, $1 \leq i \leq n$, on the plane $\mathbf{R}^2 \subset \mathbf{R}^3$. A (3-valent, oriented, ribbon) *(l,k)-graph* Γ is a closed one-dimensional cell complex embedded in $\mathbf{R}^2 \times [0,1]$ such that the boundary $\partial N(\rho)$ of a regular neighbourhood of a point $\rho \in \operatorname{int}(\Gamma)$ intersects Γ in exactly 2 or 3 points and $\partial\Gamma = Y_l \times \{0\} \cup Y_k \times \{1\}$. By a *vertex* of Γ we mean a point $\eta \in \Gamma$ such that $\partial N(\eta) \cap \Gamma$ consists of exactly 3 points. Denote by Γ_v the union of all vertices of Γ. Connected components of $\Gamma \setminus \Gamma_v$ are called *edges* of Γ. We will denote by Γ_e the set of all edges of Γ. Each edge is assumed to be oriented in such a way that in each vertex there is at least one incoming and one outgoing edge. By a *framing* of Γ we mean a trivialization of the normal bundle on each of its edges. We require that framings of three edges meeting in a common vertex coincide there (as illustrated in Figure 1a) and the framing of the boundary $\partial\Gamma$ is induced by standard xy-structure of the planes $\mathbf{R}^2 \times \{0, 1\}$ being, say, in the positive direction of the y-axis. Further on we consider graphs up to isotopy. One may define a product $\Gamma_2 \cdot \Gamma_1$ of the (l,k)-graph Γ_1 with the (k,m)-graph Γ_2 (for compatible orientations on the boundary) by gluing the top of $\Gamma_1 \subset \mathbf{R}^2 \times [0,1]$ with the bottom of the shifted copy $\Gamma_2 \subset \mathbf{R}^2 \times [1,2]$ and rescaling the z-coordinate. The tensor product $\Gamma_1 \otimes \Gamma_2$ can be also defined by juxtaposition of the corresponding graphs.

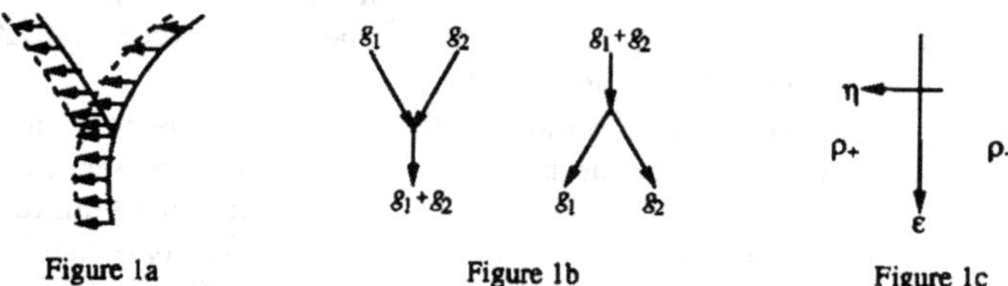

Figure 1a Figure 1b Figure 1c

It is convenient to encode graphs by diagrams. A *diagram* D_Γ of Γ is a projection of Γ to the plane $\mathbf{R}^1 \times [0,1]$ (with over- and underpasses indicated in a usual way) such that the framing of Γ is orthogonal to the plane of projection. Connected components of $\mathbf{R}^1 \times [0,1] \setminus D_\Gamma$ are called *regions* of the diagram. Two graphs are isotopic iff their diagrams are regularly isotopic, i.e. may be obtained from one another by a sequence of local moves depicted in Figure 2.

3. G-colorings of graphs and diagrams

Let G be an abelian group. A *G-coloring* of a graph Γ is a map $\Gamma_e \to G$, i.e. an assignment of elements of G to edges of Γ. It is required that the colors $g(\epsilon_i)$ of any 3 edges ϵ_1, ϵ_2, ϵ_3 with a common vertex η should satisfy the condition

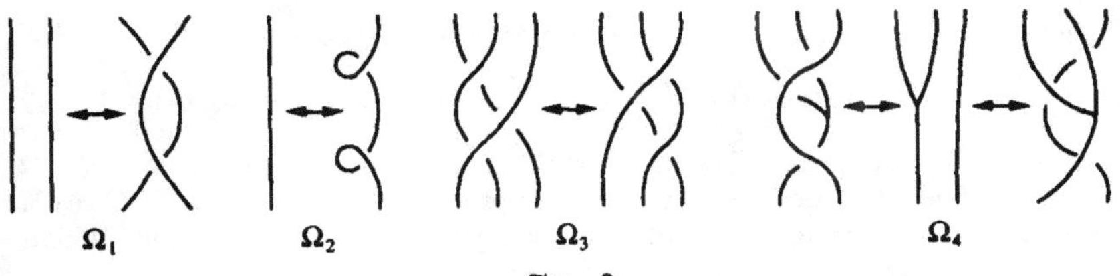

Figure 2

$\sum_{i=1}^{3} \pm g(\epsilon_i) = 0$, where $g(\epsilon_i)$ is counted with a positive sign if ϵ_i is oriented towards η and with a negative sign otherwise, see Figure 1b.

A G-coloring of Γ determines a coloring of regions for any diagram D_Γ in the following way. An orientation of an edge ϵ induces an orientation of the normal ν to ϵ. Let ρ_+ and ρ_- be the regions adjacent to ϵ in the positive and negative directions of ν respectively, as shown in Figure 1c. We require that $g(\rho_+) = g(\rho_-) + g(\epsilon)$. Let us now color the rightmost region of D_Γ by 0. It is easy to verify that this determines a well-defined coloring of all regions of D_Γ.

An important remark should be made at this stage. Note that any framed link may be considered as a $(0,0)$-graph without vertices, so in particular we can speak about colored (framed) links and their colored diagrams.

4. Cohomology of abelian groups and weights of colored diagrams

We will further use some results of the cohomology theory of abelian groups introduced by Eilenberg-MacLane. The reader is referred to [E], [EML], [ML] for details and motivation.

A *normalized abelian* 3-*cocycle* (F, Ω) for an abelian group G with coefficients in $\mathbb{C}^*$ is a pair of functions

$$F : G \times G \times G \to \mathbb{C}^*$$

$$\Omega : G \times G \to \mathbb{C}^*$$

such that F is a normalized 3-cocycle, i.e. for any $g_i \in G$

$$F(g_1, g_2, 0) = F(g_1, 0, g_3) = F(0, g_2, g_3) = 1 \tag{4.1}$$

$$F(g_1,g_2,g_3)F^{-1}(g_1,g_2,g_3+g_4)F(g_1,g_2+g_3,g_4)F^{-1}(g_1+g_2,g_3,g_4)F(g_2,g_3,g_4) = 1 \tag{4.2}$$

and F, Ω satisfy

$$F(g_1,g_2,g_3)\Omega(g_1+g_2,g_3)F(g_3,g_1,g_2) = \Omega(g_2,g_3)F(g_1,g_3,g_2)\Omega(g_1,g_3) \tag{4.3}$$

$$F^{-1}(g_1,g_2,g_3)\Omega(g_1,g_2+g_3)F^{-1}(g_2,g_3,g_1) = \Omega(g_1,g_2)F^{-1}(g_2,g_1,g_3)\Omega(g_1,g_3) \tag{4.4}$$

It is convenient for our purposes to define $B : G \times G \times G \to \mathbb{C}^*$ by

$$B(g_1,g_2,g_3) = F^{-1}(g_2,g_1,g_3)\Omega(g_1,g_2)F(g_1,g_2,g_3) \tag{4.5}$$

Remark 4.1. By (4.1), (4.2), (4.3) one may obtain

$$\Omega(g_1,0)=\Omega(0,g_2)=B(g_1,g_2,0)=B(g_1,0,g_3)=B(0,g_2,g_3)=1\ . \tag{4.6}$$

Given an abelian group G equipped with a normalized abelian 3-cocycle (F,Ω), one may define $\mathbf{C}^*$-valued weights $< >$ of colored diagrams $< >: D_\Gamma \to \mathbf{C}^*$ as the product of weights corresponding to elementary fragments of a diagram. Weights for elementary fragments are shown in Figure 3.

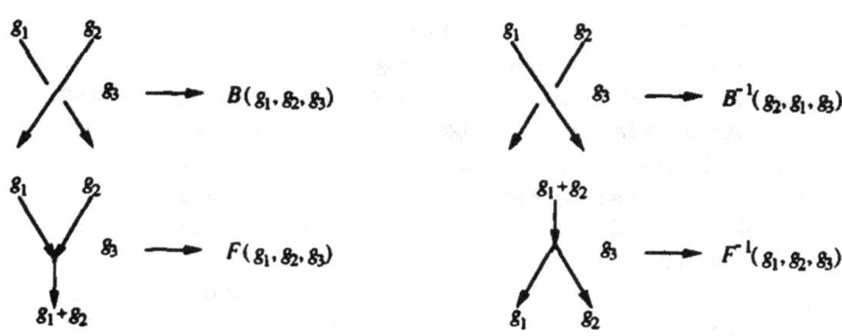

Figure 3

5. Graphical form of equations on F, B

Let us obtain some useful equalities for the functions F, B and Ω. By (4.2), (4.3) and the definition (4.5) of B it is easy to verify that the following holds:

$$B(g_1+g_2,g_3,g_4)F(g_1,g_2,g_3+g_4)=F(g_1,g_2,g_4)B(g_1,g_3,g_2+g_4)B(g_2,g_3,g_4) \tag{5.1}$$

similarly (by (4.4) instead of (4.3)) one has

(5.2)

$$B^{-1}(g_3,g_1+g_2,g_4)F(g_1,g_2,g_3+g_4)=F(g_1,g_2,g_4)B^{-1}(g_3,g_1,g_2+g_4)B^{-1}(g_3,g_2,g_4)$$

now by (4.5), (5.1), (5.2) we immediately obtain

$$\begin{aligned} &B(g_2,g_3,g_1+g_4)B(g_1,g_3,g_4)B(g_1,g_2,g_3+g_4)= \\ &\qquad B(g_1,g_2,g_4)B(g_1,g_3,g_2+g_4)B(g_2,g_3,g_4) \end{aligned} \tag{5.3}$$

$$\begin{aligned} &B^{-1}(g_3,g_2,g_1+g_4)B^{-1}(g_3,g_1,g_4)B^{-1}(g_2,g_1,g_3+g_4)= \\ &\qquad B^{-1}(g_2,g_1,g_4)B^{-1}(g_3,g_1,g_2+g_4)B^{-1}(g_3,g_2,g_4)\ . \end{aligned} \tag{5.4}$$

By the definition of weights and (4.1)-(5.4) one easily derives equalities for weights of the diagrams shown in Figure 4 (the diagrams are assumed to coincide except where shown).

As follows from the equalities in Figure 4 (compare to Figure 2), the weight $< D_\Gamma >$ does not depend on the choice of diagram D_Γ; hence it is an invariant $<\Gamma>$ of Γ.

Exercise 5.1. Let K_ϵ be a g-colored ϵ-framed unknotted circle. Then one has $< K_\epsilon >= B^\epsilon(g,g,-g)$. In particular, $< K_0 >= 1$.

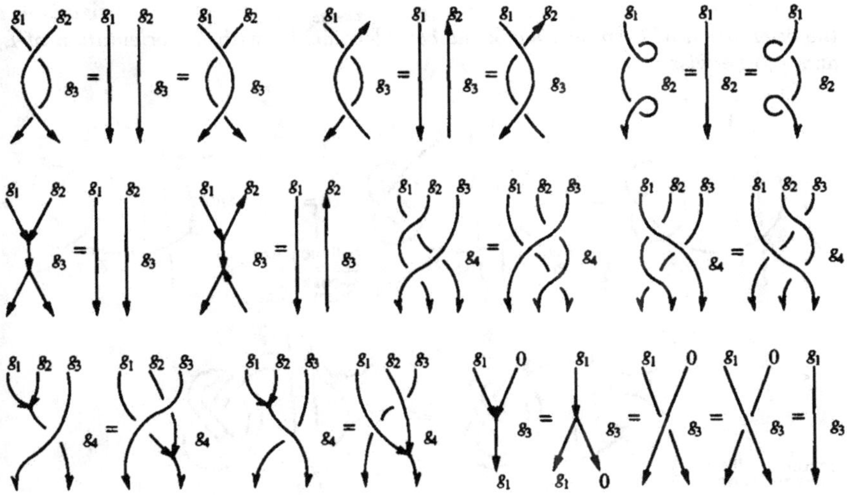

Figure 4

6. Invariant $Z(L \subset M)$ of links in 3-manifolds

It is well-known that any orientable closed 3-manifold M may be obtained by Dehn surgery on a framed link $L \subset S^3$ [Li] and $M_L = M_{L'} = M$ iff one can pass from L to L' by isotopy and a sequence of Kirby moves K_1, K_2 [K]. K_1 consists of addition (or deletion) of an unlinked unknotted component with ± 1 framing. K_2 changes a component L_1 of L by a band connected sum $L_1 \#_b L'_2$, where L'_2 is obtained by pushing another component L_2 off itself along the framing. These moves are illustrated in Figure 5.

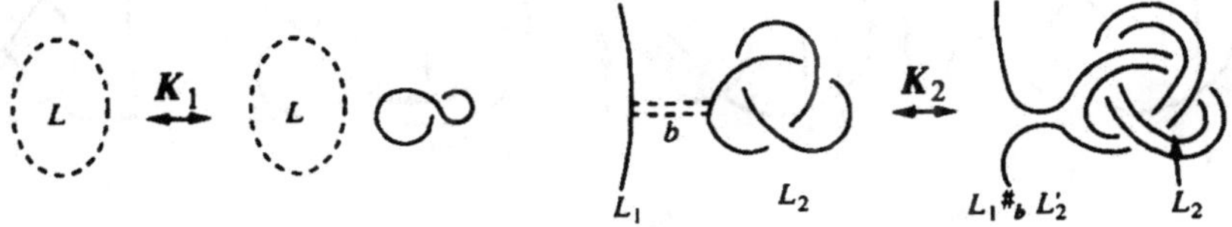

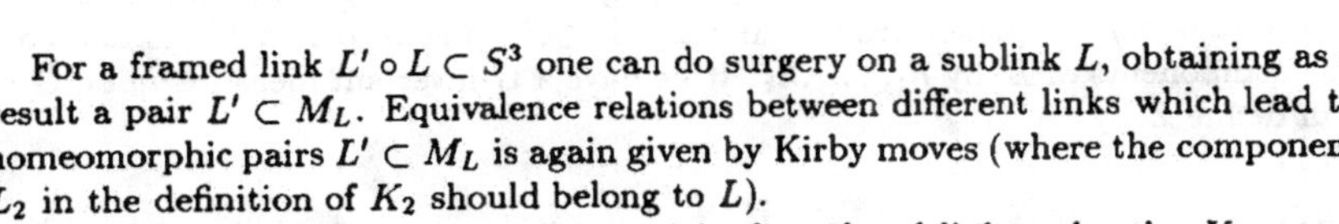

Figure 5

For a framed link $L' \circ L \subset S^3$ one can do surgery on a sublink L, obtaining as a result a pair $L' \subset M_L$. Equivalence relations between different links which lead to homeomorphic pairs $L' \subset M_L$ is again given by Kirby moves (where the component L_2 in the definition of K_2 should belong to L).

Let us verify the behaviour of the weight for colored link under the K_2 move. Denote by $L_{g_1, g_2, \ldots, g_n}$ a framed oriented link endowed with a coloring $g_1, g_2, \cdots, g_n$ of its components $L_1, L_2, \ldots, L_n$ and let L' be a link obtained from L by application of K_2 to the components L_1, L_2 as above. Then the following holds:

Proposition 6.1. $< L_{g_1,g_2,\ldots,g_n} >=< L'_{g_1,g_1\pm g_2,g_3,\ldots,g_n} >$, *where the sign is + if the orientation of L_1 pulled along the band b coincides with the orientation of L_2 and − otherwise.*

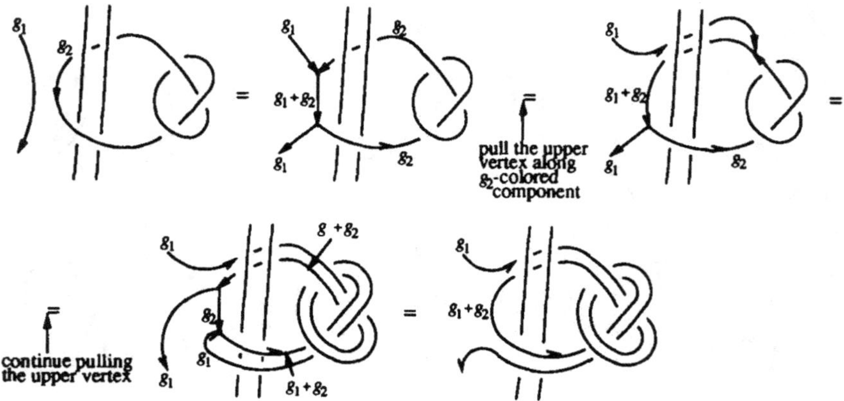

Figure 6

Proof. The proof is illustrated in Figure 6 for the case of coinciding orientations. It uses the ideas of [V] and is based on applications of equalities depicted in Figure 4. The second case is completely similar.□

Remark 6.2. Now it is easy to deduce that the weight $< L_{g_1,\ldots,g_n} >$ does not change if we reverse an orientation of L_i and simultaneously change its color g_i to $-g_i$, see Figure 7.

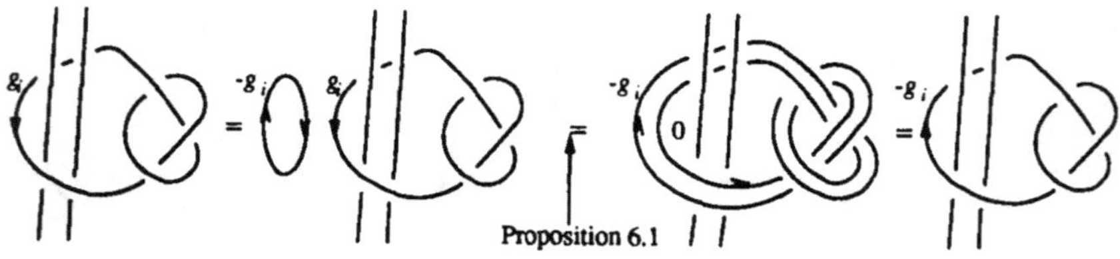

Figure 7

Let a pair $L' \subset M_L$ be obtained by surgery in S^3 on an m-component sublink L of the $(k+m)$-component framed link $L' \circ L \subset S^3$. Let us orient $L' \circ L$ and color the components of L' by $h_1, \ldots, h_k$. If we have a G-invariant measure dg on G we can define Z by

(6.1)

$$Z(L'_{h_1,\ldots,h_k} \subset M_L) = a_+^{-\frac{m+\sigma}{2}} a_-^{-\frac{m-\sigma}{2}} \int_G \cdots \int_G < L'_{h_1,\ldots,h_k} \circ L_{g_1,\ldots,g_m} > dg_1 \ldots dg_m$$

where $\sigma = \sigma(L)$ is the signature of L and $a_\pm = \int_G B^{\pm 1}(g,g,-g)dg$.

Then, if the integrals in (6.1) are well defined, the following holds:

Theorem 6.3. *$Z(L'_{h_1,\dots,h_k} \subset M_L)$ does not depend on the choice of surgery link L and is an invariant $Z(L'_{h_1,\dots,h_k} \subset M)$ of the pair: oriented colored framed link L' in 3-manifold M.*

Proof. Invariance of Z under the change of orientation for L follows from Remark 6.2. Invariance under K_1 is assured by Exercise 5.2, and under K_2 by Proposition 6.1 and G-invariance of the measure.□

Remark 6.4. For $L' = \varnothing$ the invariant $Z(M)$ is an invariant of 3-manifolds.

7. The case $G = \mathbb{Z}_n$

Let us consider the example of the cyclic group of order n. An example of a normalized abelian 3-cocycle for $\mathbb{Z}_n$ is provided by

$$F(g_1, g_2, g_3) = 1\ , \qquad \Omega(g_1, g_2) = q^{g_1 g_2}\ , \text{ where } q^n = 1. \tag{7.1}$$

Let $\mathcal{L}_L$ be the linking matrix of a framed link L (with self-linking numbers on the diagonal). For notation and basic facts concerning integer matrices we refer to [MH], in particular, $< n >$ denotes the 1×1 matrix with entry $n \in \mathbb{Z}$. If we have a link $L' \circ L$ with two distinguished complementary sets of components L' and L we will write the linking matrix $\mathcal{L}_{L' \circ L}$ in the block form:

$$\mathcal{L}_{L' \circ L} = \begin{pmatrix} \mathcal{L}_{L'L'} & \mathcal{L}_{L'L} \\ \mathcal{L}_{LL'} & \mathcal{L}_{LL} \end{pmatrix}$$

where $\mathcal{L}_{L'L'} = \mathcal{L}_{L'}$, $\mathcal{L}_{LL} = \mathcal{L}_L$ and $(\mathcal{L}_{LL'})_{ij} = (\mathcal{L}_{L'L})_{ji}$ is the linking number of i-th component of L with the j-th component of L'.

Then for the invariant (6.1) corresponding to $G = \mathbb{Z}_n$ equipped with the 3-cocycle (7.1) we have:

$$Z_{\mathbb{Z}_n}(L'_{h_1,\dots,h_k} \subset M_L) = a_+^{-\frac{m+\sigma}{2}} a_-^{-\frac{m-\sigma}{2}} \sum_{g \in \mathbb{Z}_n^m} q^{<g, \mathcal{L}_L g> + 2 <g, \mathcal{L}_{LL'} h> + <h, \mathcal{L}_{L'} h>} \tag{7.2}$$

where $< g, \mathcal{L}h >= \sum_{ij} g_i \mathcal{L}_{ij} h_j$. Here we choose $\int_{\mathbb{Z}_n} f(g) dg = \sum_{g \in \mathbb{Z}_n} f(g)$ and $a_\pm$ are given by Gaussian sums $a_\pm(q) = \sum_{g \in \mathbb{Z}_n} q^{\pm g^2}$. Note that $a_-(q) = a_+(q^{-1}) = \overline{a_+((q)}$.

Set $q = e^{\frac{2\pi i}{n}}$. Then the value of a_+ is well known to be

$$a_+(e^{\frac{2\pi i}{n}}) = \frac{(1 + i^{-n})\sqrt{n}}{(1 - i)} = \begin{cases} (1+i)\sqrt{n}, & n \equiv 0 \bmod 4 \\ \sqrt{n}, & n \equiv 1 \bmod 4 \\ 0, & n \equiv 2 \bmod 4 \\ i\sqrt{n}, & n \equiv 3 \bmod 4 \end{cases}$$

Let $q = e^{\frac{2\pi i d}{n}}$ where d and n are relatively prime, and n is odd. Then we have

$$a_+(q) = \left(\frac{d}{n}\right) a_+(e^{\frac{2\pi i}{n}})$$

where $(\frac{d}{n})$ is the Jacobi symbol [HST], [La].

Similarly, when n is even we have $a_+(q) \equiv 0$ if $n \equiv 2 \bmod 4$. Therefore the invariant (7.2) is defined only if $n \not\equiv 2 \bmod 4$.

For a framed m-component link L with a linking matrix $\mathcal{L}$ and $n \not\equiv 2 \bmod 4$ denote by $Z_n(\mathcal{L}, q)$ the value of $Z_{\mathbb{Z}_n}(M_L)$ for $q = e^{\frac{2\pi i d}{n}}$. By (7.2) we have

$$Z_n(\mathcal{L}, q) = a_+^{-\frac{m+\sigma}{2}} a_-^{-\frac{m-\sigma}{2}} \sum_{j \in \mathbb{Z}_n^m} e^{\frac{2\pi i d}{n} <j, \mathcal{L} j>}.$$

By Theorem 6.3, this is an invariant of the 3-manifold M_L obtained by surgery along L. For $Z_n(\mathcal{L}, q)$ it can be easily seen explicitly as follows:

(i) Invariance under isotopy of L is trivial.

(ii) The equality

$$\sum_{j \in \mathbb{Z}_n^{(m+1)}} e^{\frac{2\pi i d}{n} <j, (\mathcal{L} \oplus <\pm 1>) j>} = a_\pm \sum_{j \in \mathbb{Z}_n^m} e^{\frac{2\pi i d}{n} <j, \mathcal{L} j>}$$

implies invariance under the K_1 move;

(iii) invariance under K_2 follows immediately from

$$\sum_{j \in \mathbb{Z}^m / n\mathbb{Z}^m} e^{\frac{2\pi i d}{n} <j, A^t \mathcal{L} A j>} = \sum_{j \in A(\mathbb{Z}^m / n\mathbb{Z}^m)} e^{\frac{2\pi i d}{n} <j, \mathcal{L} j>}$$

It turns out that $Z_n(\mathcal{L})$ is not a very sensitive invariant:

Proposition 7.1. *For any (integer) homology sphere M_L one has $Z_n(\mathcal{L}, q) = 1$.*

Proof. Due to the invariance of $Z_n(\mathcal{L}, q)$ under K_1 we may assume $\mathcal{L}$ to be odd, indefinite and unimodular. Then $\mathcal{L}$ is equivalent over the integers to $\oplus_1^m < \pm 1 >$ (see e.g. [MH]). For $\oplus_1^m < \pm 1 >$ the result is obvious. □

Example 7.2. Let p be a prime. Then for rational homology spheres we obtain

(i) For $\mathcal{L} = < a >$, $a > 0$ one has $Z_p(a, q) = \begin{cases} (\frac{a}{p}), & p \nmid a \\ \sqrt{p}(\frac{d}{p}), & p | a, \ p \equiv 1 \bmod 4 \\ -i\sqrt{p}(\frac{d}{p}), & p | a, \ p \equiv 3 \bmod 4 \end{cases}$

(ii) For p an odd prime not dividing $\det \mathcal{L}$, $\det \mathcal{L} \neq 0$ one has $Z_{p^m}(\mathcal{L}, q) = (\frac{|\det \mathcal{L}|}{p})^m$.

Proof. (i) Follows from the well known values of Gauss sums, see e.g. [HST], [La].

(ii) By invariance under K_1 we may assume that $m = \dim \mathcal{L}$ is odd. Then it follows from [Wa, Theorem 29] that $\mathcal{L}$ is equivalent to the matrix

$$\begin{pmatrix} 1 & & & \\ & \ddots & & \\ & & 1 & \\ & & & \det \mathcal{L} \end{pmatrix}$$

modulo p^t for any $t \in \mathbf{N}$. Thus we obtain that $Z_{p^m}(\mathcal{L}) = Z_{p^m}(< \det \mathcal{L} >) = \left(\frac{|\det \mathcal{L}|}{p}\right)^m$, which finishes the proof. □

Remark 7.3. For N odd, $Z_N(\mathcal{L}, q)$ agrees with the invariant $Z_N(M, q)$ of [MOO], for N even we have $Z_{2N}(\mathcal{L}, q^2) = Z_N(M, q)$ where $q^{\frac{N}{2}} = 1$ (compare in particular with Section 7 of [MOO]). Example 7.2 (iii) reproves Corollary 4.8 of [MOO]. A general formula for $Z_N(M, q)$ is given by [MOO, Theorem 4.5]. In particular they show that these invariants are, in fact, determined by the first Betti number of M and the linking pairing on Tor $H_1(M, \mathbf{Z})$. The absolute value of $Z_N(M, q)$ turns out to be $|H^1(M, \mathbf{Z}_N)|^{\frac{1}{2}}$.

8. The invariant $Z_{\mathbf{Z}_n}$ and $U(1)$ Chern-Simons theory

In this section we will present some arguments why the invariant $Z_n(\mathcal{L})$ should be interpreted as $U(1)$ Chern-Simons theory. A phenomenological reason for that goes back to standard speculations from rational conformal field theory. It is known [MS] that $U(1)$ Chern-Simons theory must correspond to a WZW-model based on the group $U(1)$. The later has braiding and fusion matrices which are determined by an abelian 3-cocycle (F, Ω) (see [MS] for more details).

First let us clarify the definition of $U(1)$-Chern-Simons theory in terms of the functional integral. In contrast to the principal G-bundle over M which is trivial for a compact simple simply-connected Lie group G, there exist nontrivial $U(1)$-bundles over M. So the most natural modification of (1.1) is

$$Z_{k,U(1)}(M, g) = \sum_P \int \exp(ik\,\mathrm{CS}(A))\mathcal{D}[A] \tag{8.1}$$

where we take a sum over all $U(1)$-bundles over M and then we "integrate" over all gauge equivalence classes of connections in each $U(1)$-bundle P. Again, as in (1.1), we expect that (8.1) depends only on the framing of M induced by a metric g on M. Notice that in the $U(1)$ case the Chern-Simons action is quadratic with respect to A, so we can expect the stationary phase approximation to be exact. Therefore, for the integral (8.1) we obtain

$$Z_{k,U(1)}(M, g) = \sum_P e^{i\pi\left(\frac{\eta_{grav}}{2} + \frac{1}{12}\frac{CS(A^g)}{2\pi}\right)} \int_{[A^{(0)}]} e^{ikCS(A^{(0)})} T_{A^{(0)}} d\mu(A^{(0)}) \tag{8.2}$$

where we integrate over the moduli space of flat connections on a $U(1)$ bundle P. The determinant $T_{A^{(0)}}$ has the same quasiclassical origin as in (1.2) and the measure $\mu(A^{(0)})$ has to be defined.

If M is a rational homology sphere, there is at most one gauge class of flat connections in each $U(1)$ bundle over M, so in this case (8.2) simplifies to

$$Z_{k,U(1)}(M, g) = \sum_P e^{i\pi\left(\frac{\eta_{grav}}{2} + \frac{1}{12}\frac{CS(A^g)}{2\pi}\right)} e^{ikCS(A^{(0)})} T_{A^{(0)}} \tag{8.3}$$

Since the the Chern-Simons action for $U(1)$ is quadratic we expect that this formula is exact.

Let us compare this formula with the expression for $Z_{\mathbf{Z}_n}(M)$ using the reciproticity law for Gauss sums [J].

Theorem 8.1. *Let $\mathcal{L}$ be a symmetric nondegenerate integer matrix acting in $\mathbf{Z}^m$. Then the following holds:*

$$(8.4)\qquad \sum_{n\in\mathbf{Z}_r^m} e^{\frac{\pi i}{r}<n,\mathcal{L}n>} = \left(\det(\frac{\mathcal{L}}{i})\right)^{-\frac{1}{2}} r^{\frac{m}{2}} \sum_{l\in\mathbf{Z}^m/\mathcal{L}\mathbf{Z}^m} e^{-\pi i r<l,\mathcal{L}^{-1}l>}$$

We will prove this theorem later in the more general case when $\mathcal{L}$ is not necessarily nondegenerate.

Comparing (8.3) and (8.4) we see that the number of terms coincides since $\#(P) = |H_1(M,\mathbf{Z})|$ and this is equal to $\det\mathcal{L}$. Thus, we expect the term-by-term equality $Z_{k,U(1)}(M,g) = Z_k(\mathcal{L})$. For lens spaces this follows from [J], [G], [R].

Now consider the general case in which the linking matrix $\mathcal{L}$ can be degenerate. First we establish a generalization of the reciproticity law for Gauss sums [J] to arbitrary symmetric integer matrices $\mathcal{L}$. To formulate the result we need to introduce some notation.

For a symmetric $(m\times m)$ integer matrix $\mathcal{L}:\mathbf{Z}^m\to\mathbf{Z}^m$ denote by P_I and P_K the projections from $\mathbf{Z}^m$ to $\mathrm{Im}\,\mathcal{L}\otimes\mathbf{R}$ and $\mathrm{Ker}\,\mathcal{L}\otimes\mathbf{R}$ respectively. Let $(\mathrm{Im}\,\mathcal{L})^\#$ be a lattice dual to $\mathrm{Im}\,\mathcal{L}$, i.e. $(\mathrm{Im}\,\mathcal{L})^\# = \{\alpha\in\mathrm{Im}\,\mathcal{L}\otimes\mathbf{R}|<\alpha,j>\in\mathbf{Z},\forall j\in\mathrm{Im}\,\mathcal{L}\}$. Note that since $\mathcal{L}$ is symmetric, $\mathcal{L}^{-1}$ makes sense on $\mathrm{Im}\,\mathcal{L}\otimes\mathbf{R}$ and in particular on $(\mathrm{Im}\,\mathcal{L})^\#$.

Let $\mathrm{d}_I=\dim(\mathrm{Im}\,\mathcal{L})$, $\mathrm{d}_K=\dim(\mathrm{Ker}\,\mathcal{L})$ and $\mathcal{L}^{reg}$ be the restriction of $\mathcal{L}$ to $\mathrm{Im}\,\mathcal{L}$. Denote also by v_I and v_K the volume of fundamental domains for $\mathrm{Im}\,\mathcal{L}$ and $\mathrm{Ker}\,\mathcal{L}$ respectively.

Fix some set S of representatives for $\mathbf{Z}^m/(\mathrm{Ker}\,\mathcal{L}\oplus\mathrm{Im}\,\mathcal{L})$ of cardinality $|S|$. Let r be an arbitrary positive integer such that $r<j,\mathcal{L}j>\in 2\mathbf{Z}$ for any $j\in\mathbf{Z}^m$; thus $e^{\pi i r<j,\mathcal{L}j>}=1$.

Theorem 8.1. *The following relation holds:*

$$(8.5)\qquad \sum_{n\in\mathbf{Z}_r^m} e^{\frac{\pi i}{r}<n,\mathcal{L}n>} = r^{\mathrm{d}_K}\left(\det(\frac{\mathcal{L}^{reg}}{ir})\right)^{-\frac{1}{2}} \sum_{l\in(\mathbf{Z}^m\cap(\mathrm{Im}\,\mathcal{L})^\#)/\mathrm{Im}\,\mathcal{L}} e^{-\pi i r<l,\mathcal{L}^{-1}l>}$$

in particular the expression on the right is independent on the choice of ψ.

Proof. For a symmetric invertible matrix T over $\mathbf{C}^n$ with positive definite imaginary part, a vector $v\in\mathbf{C}^n$ and a lattice Λ with dual lattice $\Lambda^\#$ one has the following identity for Θ-functions:

$$(8.6)\qquad \sum_{n\in\Lambda} e^{\frac{\pi i}{r}<n+v,T(n+v)>} = \left(\det(\frac{T}{ir})\right)^{-\frac{1}{2}}\frac{1}{\mathrm{vol}\,\Lambda}\sum_{l\in\Lambda^\#} e^{-\pi i r<l,T^{-1}l>+2\pi i<v,l>}$$

This equality can be easily verified using the Poisson summation formula [H, formula (7.2.1)′]. For purely imaginary matrices see also [C, chapter XI §2]. To handle our case, note that the matrix $\mathcal{L}_\epsilon=\mathcal{L}+i\epsilon\,\mathrm{Id}$ for small positive ϵ satisfies the hypothesis of (8.6). Since $e^{\pi i r<j,\mathcal{L}j>}=1$ by assumption, the LHS of (8.6) for $v=0$ and $T=\mathcal{L}_\epsilon$ can be rewritten as

$$\sum_{j\in\mathbf{Z}^m} e^{\frac{\pi i}{r}<j,\mathcal{L}_\epsilon j>} = \sum_{j\in\mathbf{Z}^m} e^{\frac{\pi i}{r}<j,\mathcal{L}n>}e^{-\frac{\pi\epsilon}{r}<j,j>} =$$

(8.7)

$$\sum_{n\in\mathbf{Z}_r^m} e^{\frac{\pi i}{r}<n,\mathcal{L}n>}\sum_{j\in\mathbf{Z}^m} e^{-\frac{\pi\epsilon}{r}<n+rj,n+rj>} = \sum_{n\in\mathbf{Z}_r^m} e^{\frac{\pi i}{r}<n,\mathcal{L}n>}\sum_{j\in\mathbf{Z}^m} e^{\frac{\pi i}{r}<j+\frac{n}{r},ir^2\epsilon(j+\frac{n}{r})>}$$

Applying (8.6) to the last sum in (8.7) one derives that (8.7) equals

$$\sum_{n\in\mathbb{Z}_r^m} e^{\frac{\pi i}{r}\langle n,\mathcal{L}n\rangle}(\det(r\epsilon\,\mathrm{Id}))^{-\frac{1}{2}}\sum_{j\in\mathbb{Z}^m} e^{-\pi i r\langle j,\frac{j}{ir^2\epsilon}\rangle+2\pi i\langle j,\frac{n}{r}\rangle} =$$

$$\frac{1}{\sqrt{r\epsilon}^m}\sum_{n\in\mathbb{Z}_r^m} e^{\frac{\pi i}{r}\langle n,\mathcal{L}n\rangle}\sum_{j\in\mathbb{Z}^m} e^{-\frac{\pi}{r\epsilon}\langle j,j\rangle+2\pi i\langle j,\frac{n}{r}\rangle}$$

As $\epsilon\to 0$, the second sum approaches 1, so asymptotically

$$\sum_{j\in\mathbb{Z}^m} e^{\frac{\pi i}{r}\langle j,\mathcal{L}_\epsilon j\rangle} \overset{\epsilon\to 0}{\sim} \frac{1}{\sqrt{\epsilon}^m}\frac{1}{\sqrt{r}^m}\sum_{n\in\mathbb{Z}_r^m} e^{\frac{\pi i}{r}\langle n,\mathcal{L}n\rangle} \tag{8.8}$$

On the other hand we can rewrite the sum in the LHS of (8.8) using the orthogonality of Ker $\mathcal{L}$ and Im $\mathcal{L}$ as

$$\sum_{j\in\mathbb{Z}^m} e^{\frac{\pi i}{r}\langle j,\mathcal{L}_\epsilon j\rangle} = \sum_{j\in S,n\in\mathrm{Ker}\,\mathcal{L},k\in\mathrm{Im}\,\mathcal{L}} e^{\frac{\pi i}{r}\langle n+k+j,i\epsilon n+\mathcal{L}_\epsilon(l+j)\rangle} =$$

$$\sum_{j\in S}\sum_{n\in\mathrm{Ker}\,\mathcal{L}} e^{-\frac{\pi\epsilon}{r}\langle n,n\rangle+2\langle n,\mathrm{P}_K(j)\rangle}\sum_{l\in\mathrm{Im}\,\mathcal{L}} e^{\frac{\pi i}{r}\langle l+j,\mathcal{L}_\epsilon(l+j)\rangle} =$$

$$\sum_{j\in S}\sum_{n\in\mathrm{Ker}\,\mathcal{L}} e^{-\frac{\pi\epsilon}{r}\langle n+\mathrm{P}_K(j),n+\mathrm{P}_K(j)}\sum_{l\in\mathrm{Im}\,\mathcal{L}} e^{\frac{\pi i}{r}\langle l+\mathrm{P}_I(j),\mathcal{L}_\epsilon(l+\mathrm{P}_I(j))\rangle}$$

Using (8.6) for the sum over Ker $\mathcal{L}$ we see that the last expression equals

$$\sum_{j\in S}\left(\det(\frac{\epsilon}{r}|_{\mathrm{Ker}\,\mathcal{L}})\right)^{-\frac{1}{2}}\frac{1}{\mathrm{v}_K}\sum_{n\in(\mathrm{Ker}\,\mathcal{L})^\#} e^{-\frac{\pi\epsilon}{r}\langle n,n\rangle+2\pi i\langle n,\mathrm{P}_K(j)\rangle}\times$$

$$\times\sum_{l\in\mathrm{Im}\,\mathcal{L}} e^{\frac{\pi i}{r}\langle l+\mathrm{P}_I(j),\mathcal{L}_\epsilon(l+\mathrm{P}_I(j))\rangle}$$

which in turn is, asymptotically as $\epsilon\to 0$,

$$\sum_{j\in S}(\frac{r}{\epsilon})^{\frac{d_K}{2}}\frac{1}{\mathrm{v}_K}\sum_{l\in\mathrm{Im}\,\mathcal{L}} e^{\frac{\pi i}{r}\langle l+\mathrm{P}_I(j),(\mathcal{L}+i\epsilon)(l+\mathrm{P}_I(j))\rangle}$$

Applying (8.6) to the inner sum we see that the above is equal to

$$(\frac{r}{\epsilon})^{\frac{d_K}{2}}\frac{1}{\mathrm{v}_K}\sum_{j\in S}\frac{1}{\mathrm{v}_I}\left(\det(\frac{\mathcal{L}_\epsilon^{reg}}{ir})\right)^{-\frac{1}{2}}\sum_{l\in(\mathrm{Im}\,\mathcal{L})^\#} e^{-\pi i r\langle l,\mathcal{L}_\epsilon^{-1}l\rangle+2\pi i\langle l,\mathrm{P}_I(j)\rangle} =$$

$$(\frac{r}{\epsilon})^{\frac{d_K}{2}}\frac{1}{\mathrm{v}_K}\frac{1}{\mathrm{v}_I}\left(\det(\frac{\mathcal{L}_\epsilon^{reg}}{ir})\right)^{-\frac{1}{2}}\sum_{j\in S}\sum_{l\in(\mathrm{Im}\,\mathcal{L})^\#} e^{-\pi i r\langle l,(\mathcal{L}^{-1}-i\epsilon\mathcal{L}^{-1}\mathcal{L}_\epsilon^{-1})l\rangle+2\pi i\langle l,\mathrm{P}_I(j)\rangle} =$$

$$(\frac{r}{\epsilon})^{\frac{d_K}{2}}\frac{1}{\mathrm{v}_K}\frac{1}{\mathrm{v}_I}\left(\det(\frac{\mathcal{L}_\epsilon^{reg}}{ir})\right)^{-\frac{1}{2}}\sum_{j\in S}\sum_{l\in(\mathrm{Im}\,\mathcal{L})^\#/\mathrm{Im}\,\mathcal{L}} e^{-\pi i r\langle l,\mathcal{L}^{-1}l\rangle+2\pi i\langle l,\mathrm{P}_I(j)\rangle}\times$$

$$\times\sum_{n\in\mathrm{Im}\,\mathcal{L}} e^{-\pi i r\langle l+n,-i\epsilon\mathcal{L}^{-1}\mathcal{L}_\epsilon^{-1}(l+n)\rangle} \tag{8.9}$$

where we used that $<\mathcal{L}x, \mathrm{P}_I(j)>=<x, \mathcal{L}\mathrm{P}_I(j)>=<x, \mathcal{L}(\mathrm{P}_I+\mathrm{P}_K)(j)>=<x, \mathcal{L}j>\in \mathbf{Z}$ for $x \in \mathbf{Z}^k$.

Using (8.6) again we calculate the asymptotic behaviour of the inner sum in (8.9) as $\epsilon \to 0$:

$$\sum_{n\in \mathrm{Im}\, \mathcal{L}} e^{-\pi i r<l+n,-i\epsilon\mathcal{L}^{-1}\mathcal{L}_\epsilon^{-1}(l+n)>} =$$

$$\frac{1}{\mathrm{v}_I}\big(\det(r\epsilon\mathcal{L}^{-1}\mathcal{L}_\epsilon^{-1}\mathcal{L}|_{\mathrm{Im}\,\mathcal{L}\otimes\mathbf{R}}\big)^{-\frac{1}{2}} \sum_{n\in\mathrm{Im}\,\mathcal{L}} e^{-\frac{\pi}{r\epsilon}<n,\mathcal{L}_\epsilon\mathcal{L}n>+2\pi i<n,l>} \overset{\epsilon\to 0}{\sim}$$

$$\overset{\epsilon\to 0}{\sim} \Big(\frac{1}{\sqrt{r\epsilon}}\Big)^{\mathrm{d}_I} \mathrm{v}_I{}^{-1}\sqrt{(\det\mathcal{L})^2} = \Big(\frac{1}{\sqrt{r\epsilon}}\Big)^{\mathrm{d}_I}$$

Substituting this expression into (8.9) we derive that, as $\epsilon \to 0$,

$$\sum_{j\in\mathbf{Z}^m} e^{\frac{\pi i}{r}<j,\mathcal{L}_\epsilon j>} \overset{\epsilon\to 0}{\sim}$$

$$\frac{1}{\sqrt{\epsilon}^m}\frac{\sqrt{r}^{\mathrm{d}_K-\mathrm{d}_I}}{\mathrm{v}_I\,\mathrm{v}_K}\Big(\det\Big(\frac{\mathcal{L}_\epsilon^{reg}}{ir}\Big)\Big)^{-\frac{1}{2}}\sum_{j\in S}\ \sum_{l\in(\mathrm{Im}\,\mathcal{L})^\#/\mathrm{Im}\,\mathcal{L}} e^{-\pi i r<l,\mathcal{L}^{-1}l>+2\pi i<l,\mathrm{P}_I(j)>}$$

Combining this with (8.8) we obtain:

$$\frac{1}{\sqrt{r}^m}\sum_{n\in\mathbf{Z}_r^m} e^{\frac{\pi i}{r}<n,\mathcal{L}n>} \overset{\epsilon\to 0}{\sim}$$

$$\frac{\sqrt{r}^{\mathrm{d}_K-\mathrm{d}_I}}{\mathrm{v}_I\,\mathrm{v}_K}\Big(\det\Big(\frac{\mathcal{L}_\epsilon^{reg}}{ir}\Big)\Big)^{-\frac{1}{2}}\sum_{j\in S}\ \sum_{l\in(\mathrm{Im}\,\mathcal{L})^\#/\mathrm{Im}\,\mathcal{L}} e^{-\pi i r<l,\mathcal{L}^{-1}l>+2\pi i<l,\mathrm{P}_I(j)>}$$

which implies

$$\sum_{n\in\mathbf{Z}_r^m} e^{\frac{\pi i}{r}<n,\mathcal{L}n>} = \frac{r^{\mathrm{d}_K}}{\mathrm{v}_I\,\mathrm{v}_K}\Big(\det\Big(\frac{\mathcal{L}_\epsilon^{reg}}{ir}\Big)\Big)^{-\frac{1}{2}} \sum_{l\in(\mathrm{Im}\,\mathcal{L})^\#/\mathrm{Im}\,\mathcal{L}} e^{-\pi i r<l,\mathcal{L}^{-1}l>}\sum_{j\in S} e^{2\pi i<l,j>}$$

by the definition of P_I. Note that $\sum_{j\in S} e^{2\pi i<l,j>}$ equals $|S|$ if $k \in \mathbf{Z}^m$ and 0 otherwise, so finally we obtain

$$\sum_{n\in\mathbf{Z}_r^m} e^{\frac{\pi i}{r}<n,\mathcal{L}n>} = \frac{r^{\mathrm{d}_K}|S|}{\mathrm{v}_I\,\mathrm{v}_K}\Big(\det\Big(\frac{\mathcal{L}_\epsilon^{reg}}{ir}\Big)\Big)^{-\frac{1}{2}} \sum_{l\in(\mathbf{Z}^m\cap(\mathrm{Im}\,\mathcal{L})^\#)/\mathrm{Im}\,\mathcal{L}} e^{-\pi i r<l,\mathcal{L}^{-1}l>}$$

which proves the theorem, since $|S| = \mathrm{v}_I\,\mathrm{v}_K$. □

Remark 8.2. For $\det\mathcal{L} \neq 0$ we get

$$\sum_{n\in\mathbf{Z}_r^m} e^{\frac{\pi i}{r}<n,\mathcal{L}n>} = r^{\frac{m}{2}}\Big(\det\Big(\frac{\mathcal{L}}{i}\Big)\Big)^{-\frac{1}{2}} \sum_{l\in\mathbf{Z}^m/\mathcal{L}\mathbf{Z}^m} e^{-\pi i r<l,\mathcal{L}^{-1}l>}$$

which also follows from [J, Proposition 4.3].

We expect that applying this theorem to the sum $Z_k(\mathcal{L})$ we will have term-by-term correspondence between $Z_k(\mathcal{L})$ and the sum over all components of the moduli space of flat connections in (8.2).

9. The case $G = \mathbf{R}^1$

For $G = \mathbf{R}^1$ we will choose the normalized abelian cocycle as

$$F(g_1, g_2, g_3) = 1\ , \qquad \Omega(g_1, g_2) = e^{ig_1 g_2} \tag{9.1}$$

This choice leads to the following expression for the invariant (6.1):

$$\begin{aligned} Z_{\mathbf{R}^1}(L'_{h_1,\dots h_k} \subset M_L) &= a_+^{-\frac{m+\sigma}{2}} a_-^{-\frac{m-\sigma}{2}} \times \\ &\times \int \dots \int_{\mathbf{R}^m} \exp(i < g, \mathcal{L}_L g > + 2i < g, \mathcal{L}_{LL'} h > + < h, \mathcal{L}_{L'} h >) dg_1 \dots dg_m \end{aligned} \tag{9.2}$$

where σ is the signature of the m-component link L, $a_\pm = \int_{\mathbf{R}^1} e^{\pm i x^2} dx = e^{\frac{i\pi}{4}} \sqrt{\pi}$ and dx is the usual Lebesgue measure on $\mathbf{R}^1$.

If the matrix $\mathcal{L}_L$ is nondegenerate, the integral (9.2) exists as a number. If the matrix $\mathcal{L}_L$ is degenerate, but $\mathrm{Ker}(\mathcal{L}_L) \subset \mathrm{Im}(\mathcal{L}_{LL'})$ where $\mathcal{L}_{LL'}$ is regarded as a linear map $\mathcal{L}_{LL'} : \mathbf{R}^k \to \mathbf{R}^m$, then the integral exists as a distribution on $\mathbf{R}^k$. In other cases the integral (9.2) does not exist and the invariant is not defined. Let us first consider the nondegenerate case $\det(\mathcal{L}_L) \neq 0$. In this case we obtain

$$\begin{aligned} &Z_{\mathbf{R}^1}(L'_{h_1,\dots h_k} \subset M_L) = \\ &(e^{\frac{i\pi}{4}} \sqrt{\pi})^{m-\sigma} \det(\mathcal{L}_L)^{-\frac{1}{2}} \exp(-i < h, \mathcal{L}_{L'L} \mathcal{L}_L^{-1} \mathcal{L}_{LL'} h > + i < h, \mathcal{L}_{L'} h >) \end{aligned} \tag{9.3}$$

Here we used the following formula for the Gaussian integral for a nondegenerate symmetric linear map $\mathcal{L}$:

$$\int_{\mathbf{R}^m} \exp(i < x, Ax > + 2i < x, y >) dx = (e^{\frac{i\pi}{4}} \sqrt{\pi})^m \det(A)^{-\frac{1}{2}} \exp(-i < y, \mathcal{L}^{-1} y >)$$

where $< x, y >= \sum_{i=1}^m x_i y_i$ is the standard bilinear form on $\mathbf{R}^m$.

Thus, for rational homology spheres we have

$$Z_{\mathbf{R}^1}(M_L) = (e^{\frac{i\pi}{4}} \sqrt{\pi})^{m-\sigma} \det(\mathcal{L}_L)^{-\frac{1}{2}}$$

Notice that this invariant coincides with the $k = 0$ term in the sum (8.5) when we apply it to $Z_n(\mathcal{L})$. On the other hand it is exactly the term corresponding to the trivial connection $A^{(0)}$ in the quasiclassical limit (8.3).

Consider now the case $\det(\mathcal{L}_L) = 0$, $\mathrm{Ker}(\mathcal{L}_L) \subset \mathrm{Im}(\mathcal{L}_{LL'})$. The linking matrix $\mathcal{L}_L$ regarded as a linear map $\mathcal{L}_L : \mathbf{R}^m \to \mathbf{R}^m$ is a symmetric linear operator and can be diagonalized by an orthogonal transformation. $\mathbf{R}^m$ splits into a direct sum $\mathbf{R}^m = \mathrm{Ker}(\mathcal{L}_L) \oplus \mathrm{Ker}(\mathcal{L}_L)^\perp$. Denote by P the projector onto $\mathrm{Ker}(\mathcal{L}_L)$ and decompose $x, y \in \mathbf{R}^m$ as $x = x' \oplus x''$ and $y = y' \oplus y''$, where $x', y' \in \mathrm{Ker}(\mathcal{L}_L)$, $x'', y'' \in \mathrm{Ker}(\mathcal{L}_L)^\perp$. Then we have

$$\begin{aligned} &\int_{\mathbf{R}^m} \exp(i < x, \mathcal{L}_L x > + 2i < x, y >) dx = \\ &(e^{\frac{i\pi\pi}{4}} \sqrt{\pi})^{m - \dim(\mathrm{Ker}(\mathcal{L}_L))} \det(\mathcal{L}_L^{reg})^{-\frac{1}{2}} \exp(-i < y'', (\mathcal{L}_L^{reg})^{-1} y'' >) \delta(y') \end{aligned} \tag{9.4}$$

where $\mathcal{L}_L^{reg}$ is the action of $\mathcal{L}_L$ on $\mathrm{Ker}(\mathcal{L}_L)^\perp \subset \mathbf{R}^m$. Substituting (9.4) into (9.2) we obtain

$$Z_{\mathbb{R}^1}(L'_{h_1,\dots h_k} \subset M_L) = (e^{\frac{i\pi}{4}}\sqrt{\pi})^{m-\sigma-\dim(\mathrm{Ker}(\mathcal{L}_L))}\times$$
$$\times \det(\mathcal{L}_L^{reg})^{-\frac{1}{2}}\exp(-i < h, \mathcal{L}_{L'L}(\mathcal{L}_L^{reg})^{-1}\mathcal{L}_{LL'}h > +i < h, \mathcal{L}_{L'}h >)\delta(h')$$

References

[A] M. F. Atiyah, *The geometry and physics of knots*, Cambridge Univ. Press, 1990.

[A1] M. F. Atiyah, *On framings of 3-manifolds*, Topology **29** (1990), 1-7.

[AS] S. Axelrod, I. Singer, *Perturbation theory for Chern-Simons theory*, Proc. XX Int. Conf. on Diff. Geom. Methods in Phys., W. Sci. (1991), 3-45.

[C] K. Chandrasekharan, *Elliptic functions*, Springer GL281, 1985.

[E] S. Eilenberg, *Homotopy groups and algebraic homology theories*, Proc. Intl. Congress of Mathematicians **I** (1950), 350-353.

[EML] S. Eilenberg, S. MacLane, *On the groups $H(\pi, n)$ I*, Annals of Math. **58** (1953), 55-106; *On the groups $H(\pi, n)$ II*, Annals of Math. **70** (1954), 49-137.

[F] D. Freed, *Classical Chern-Simons theory*, preprint.

[FG] D. Freed, R. Gompf, *Computer calculations of Witten's 3-manifold invariants*, Comm. Math. Phys. **141** (1991), 79-117.

[G] S. Garoufalidis, *Relations among 3-manifold invariants*, preprint of Univ. of Chicago.

[GKR] S. Gelfand, D. Kazhdan, N. Reshetikhin, in preparation.

[HST] E. Hlawka, J. Schoissengeier, R. Taschner, *Geometric and analytic number theory*, Springer Universitext, 1991.

[H] L. Hörmander, *The analysis of linear partial differential operators I*, 2nd ed., Springer, 1990.

[J] L. Jeffrey, *Chern-Simons-Witten invariants of lens spaces and torus bundles and the semi-classical approximation*, Comm. Math. Phys. **147** (1992), 563-604.

[K] R. C. Kirby, *A calculus for framed links in S^3*, Invent. Math **45** (1978), 35-56.

[Ko] M. Kontsevich, *Feynman diagrams and low-dimensional topology*, preprint (1992).

[La] S. Lang, *Algebraic number theory*, Springer GTM110, 1986.

[Li] W.B.R. Lickorish, *A representation of orientable combinatorial 3-manifolds*, Ann. Math. **76** (1962), 531-540.

[ML] S. MacLane, *Cohomology theory of abelian groups*, Proc. Intl. Congress of Mathematicians **II** (1950), 8-14.

[MH] J. Milnor, D. Husemoller, *Symmetric bilinear forms*, vol. 73, Springer Ergeb. Math. ser. 2, 1973.

[MS] G. Moore, N. Seiberg, *Classical and quantum conformal field theory*, Comm. Math. Phys. **123** (1989), 177-254.

[MOO] H. Murakami, T. Ohtsuki, M. Okada, *Invariants of three-manifolds derived from linking matrices of framed links*, Osaka J. Math. **29** (1992), 545-572.

[RT] N. Yu Reshetikhin, V. G. Turaev, *Invariants of 3-manifolds via link polynomials and quantum groups*, Invent. Math. **103** (1991), 547-597.

[R] L. Rozansky, *A large k asymptotics of Witten's invariant of Seifert manifolds*, preprint of Univ. of Texas, Austin.

[S] M. C. Shum, *Tortile tensor categories*, preprint (1989).

[V] O. Viro, private communication.

[Wa] G. Watson, *Integral quadratic forms*, vol. 51, Cambridge Tracts in Math., Cambridge UP, 1960.

[W] E. Witten, *Quantum field theory and the Jones polynomial*, Comm. Math. Phys. **121** (1989), 351-399.

Department of Mathematics, University of California, Berkeley, CA 94720

Some Knots Not Determined by Their Complements

Yongwu Rong

Department of Mathematics, George Washington University, Washington, D.C. 20052, U.S.A.

Abstract

We consider the problem of whether knots in a closed 3-manifold are determined by their complements or not. The main result determines all knots whose complements are Seifert fibered spaces, with the property that the unoriented knot complements do not determine the knot types. We also describe a way to produce such examples whose complements are not Seifert fibered spaces.

1. Introduction. The well-known knot complement conjecture, solved by Gordon and Luecke ([1]), states that knots in S^3 are determined by their complements. In this paper we consider the analogous problem for knots in any closed 3-manifold, or, more precisely, Q1 and Q2 below. Our main result determines all knots whose complements are Seifert fibered spaces, with the property that the unoriented knot complements do not determine the knot types.

First we need some notations. Let X be a 3-manifold with $\partial X \cong T^2$, r be a slope on ∂X, i.e. an unoriented isotopy class of an essential simple closed curve on ∂X. Let $X(r)$ denote the closed 3-manifold obtained by doing r-Dehn filling to X, that is, $X(r) = X \cup_\partial S^1 \times D^2$ so that r is identified with a meridian. Let K be a knot in a closed 3-manifold M, let $X(K)$ denote its exterior. Doing Dehn surgery on K is the same as doing Dehn filling to $X(K)$. Denote $X(K)(r)$ by $K(r)$, where r is a slope on $\partial X(K)$. For manifolds M and N, $M \cong N$ means M and N are unoriented homeomorphic manifolds, $M \cong +N$ (resp. $-N$) means M and N are homeomorphic by an orientation preserving (resp. reversing) homeomorphism.

The following question is a natural generalization of the knot complement problem in S^3.

Q1. *Let K, K' be two knots in a closed 3-manifold M. Does $X(K) \cong X(K')$ imply that there is a self homeomorphism of M sending K to K' ?*

It is easy to see that a positive answer to Q1 is equivalent to a positive answer to Q2:

Q2: *Let X be a compact 3-manifold with $\partial X \cong T^2$, let r, s be two slopes of ∂X. Does $X(r) \cong X(s)$ imply that there is a homeomorphism of X sending r to s ?*

The answer to Q1 and Q2 is in fact no if we ignore the orientation. Mathieu ([3]) has found examples of knots in S^3 such that each knot admits two different surgeries that give homeomorphic manifolds (with opposite orientation), and such that the corresponding slopes are not related by a self homeomorphism of the knot exterior. In other words, there is a closed orientable 3-manifold M and two knots K and K' in M, such that $X(K) \cong -X(K')$, but there is no self homeomorphism of M sending K to K'. What if we insist that everything be in the oriented category ?

The answer again turns out to be no. The following example shows that there exist knots K and K' in a 3-manifold M with $X(K) \cong +X(K')$, but there is no self homeomorphism of M sending K to K'.

Example: Let X be the solid torus, l and m be its longitude-meridian pair. Let $r_{p,q} = pl + qm$ be a slope. Then $X(r_{p,q})$ is the lens space $L(p,q)$. Now if q, q' satisfy $qq' \equiv 1 \pmod p$, then it is well known that $L(p,q) \cong +L(p,q')$ (such a homeomorphism can be obtained by "switching" the two sides of the standard genus one Heegaard splitting of the lens space). Thus $X(r_{p,q}) \cong X(r_{p,q'})$. However, any orientation preserving self homeomorphism f of X sends l to $\epsilon(l + km)$, and m to ϵm, where $\epsilon = \pm 1, k \in Z$. This easily implies that if $q \not\equiv q' \pmod p$, then there is no such f sending $r_{p,q}$ to $r_{p,q'}$. (If, in addition, $q \not\equiv -q' \pmod p$, then there is no orientation reversing homeomorphism of X sending $r_{p,q}$ to $r_{p,q'}$ either. An example is $p = 7, q = 2, q' = 4$). Hence this gives a negative answer to Q2 in the oriented category. This example can be translated into the setting of Q1. Let $L(p,q) = V \cup V'$ be the standard Heegaard splitting, K and K' be the cores of V and V' respectively. Then $X(K) \cong +X(K')$. But there is no orientation preserving self homeomorphism of $L(p,q)$ sending K to K' if $q^2 \not\equiv 1 \pmod p$ (no such orientation reversing homeomorphism exists either if in addition $q^2 \not\equiv -1 \pmod p$). The classification of lens spaces implies that these are the only examples which arise from doing Dehn fillings to a solid torus.

Note that, in the above example, K and K' are rather trivial knots in the sense that $X(K)$ and $X(K')$ are ∂-compressible. It must be pointed out that if we insist that $\partial X(K)$ (resp. ∂X) be incompressible, and everything be in the oriented category, then Q1 (resp. Q2) remains open.

Back to Mathieu's example. We will determine all the examples of Mathieu's type which arise from doing Dehn fillings to Seifert fibered spaces. Our result will cover all of Mathieu's example, since all his examples are torus knots. Our basic tool is the unnormalized Seifert invariant ([4]), and the rational number Seifert pairing notation. These we review in section 2. In section 3 we prove our main result. In section 4 we describe a way to produce more examples of knots not determined by their complements.

2. Unnormalized Seifert invariants. A Seifert fibered space (abbreviated by SFS from now on) is a compact 3-manifold which is a disjoint union of circles, called fibers, such that each fiber has a regular neighborhood which is homeomorphic to a standard fibered solid torus via a fiber-preserving homeomorphism. Here a standard fibered solid torus is a fibered solid torus V whose fibers consist of disjoint (μ,ν) torus knots and the core of V for some coprime integer pair (μ,ν). If we identify each fiber to a point, the quotient space is a 2-manifold with an orbifold structure, denoted by $O(M)$. Let $\pi : M \to O(M)$ be the projection map. The triple $(M, \pi, O(M))$ can be regarded as a generalized version of a bundle over an orbifold. Up to fiber-preserving homeomorphisms, each SFS belongs to exactly one of the following classes ([5]):

$o_1, o_2, n_1, n_2, n_3, n_4$, where $O(M)$ is orientable for o_1 and o_2, and nonorientable otherwise. The fiber orientations are preserved in the o_1 and n_1 cases. For o_2 and n_2 all the generators reverse the fiber orientation. For n_3 exactly one generator preserves the fiber orientation and $g > 1$. For n_4 exactly two generators preserve the fiber orientation and $g > 2$.

Let M be a SFS, $O_1, \cdots, O_s$ be a non-empty collection of disjoint fibers in M, including all the singular fibers. Let $T_1, \cdots, T_s$ be disjoint vertical tubular neighborhoods of $O_1, \cdots, O_s$. Let $M_0 = M - \text{Int}(T_1 \cup \cdots \cup T_s)$ and $F_0 = \pi(M_0)$. Since $\pi|M_0 : M_0 \to F_0$ is an S^1-bundle over a surface with boundary, it admits a section R. Let $c_i = R \cap \partial T_i$. Let h_i be a regular fiber on ∂T_i, and m_i be a meridional curve of T_i. Since c_i, h_i form a basis for $H_1(\partial T_i)$, $m_i = \alpha_i c_i - \beta_i h_i$ in $H_1(\partial T_i)$ for some coprime integers α_i, β_i. A suitable orientation convention guarantees that all h_i's are parallel in M if the regular fibers can be consistently oriented in M (o_1 or n_1 case), all c_i's have the induced orientation from R if R can be oriented (o_1 or o_2 case), and (c_i, h_i) is the orientation of T_i induced from M if M is oriented (o_1 or n_2 case). The orientation of m_i can always be chosen so that $\alpha_i > 0$. Thus β_i is a well-defined integer if M is oriented and is well-defined up to sign otherwise.

The unnormalized Seifert invariant of M is

$$I(M) = \{\epsilon, g; (\alpha_1, \beta_1), \cdots, (\alpha_s, \beta_s)\}$$

where $\epsilon = o_1, ..., n_4$ is as above, and g is the genus of $O(M)$.

Since $(\alpha_i, \beta_i) = 1$ and $\alpha_i \geq 1$, we can denote (α_i, β_i) by $r_i = \frac{\beta_i}{\alpha_i}$ without losing any information. Thus $I(M) = \{\epsilon, g; r_1, \cdots, r_s\}$

The following propositions characterize the nonuniqueness of the unnormalized Seifert invariants. Proposition 1 appeared in [4]. A similar argument proves Proposition 2.

Proposition 1 *Let M and M' be closed oriented SFS with unnormalized Seifert invariants $I(M) = \{\epsilon, g; r_1, \cdots, r_s\}$ and $I(M') = \{\epsilon', g'; r'_1, \cdots, r'_t\}$, Then M and M' are* *orientation preservingly homeomorphic by a fiber-preserving homeomorphism if and only if, after adding* 0*'s to* $\{r_1, \cdots, r_s\}$ *or* $\{r'_1, \cdots r'_t\}$ *so that* $s = t$*, and reindexing the* r_i*'s if necessary,*

1. $\epsilon = \epsilon'$,
2. $g = g'$,
3. $r_i = r'_i + k_i$, *where* $k_i \in Z$ *for* $i = 1, \cdots, s$,
4. $\sum k_i = 0$

Proposition 2 *Let* M *and* M' *be closed nonorientable SFS with unnormalized Seifert invariants* $I(M) = \{\epsilon, g; r_1, \cdots, r_s\}$ *and* $I(M') = \{\epsilon', g'; r'_1, \cdots, r'_t\}$*, Then* M *and* M' *are homeomorphic by a fiber-preserving homeomorphism if and only if, after adding* 0*'s to* $\{r_1, \cdots, r_s\}$ *or* $\{r'_1, \cdots r'_s\}$ *so that* $s = t$*, and reindexing the* r_i*'s if necessary,*

1. $\epsilon = \epsilon'$,
2. $g = g'$,
3. $r_i = \pm r'_i + k_i$, *where* $k_i \in Z$ *for* $i = 1, \cdots, s$,
4. $\sum k_i \equiv 0 \pmod 2$

Remark 1: Let $R = \{r_1, \cdots, r_s\}, R' = \{r_1, \cdots, r_t\}$. It is easy to check that 3 and 4 in Proposition 1 are equivalent to: R and R' are related by a sequence of the following moves:

a) permute the indices;

b) add or delete 0;

c) replace r_1, r_2 by $r_1 + 1, r_1 - 1$.

Similarly, 3 and 4 in Proposition 2 are equivalent to: R and R' are related by a sequence of the following moves:

a) permute the indices;

b) add or delete 0;

c) replace r_1 by $r_1 + 2$.

d) replace r_1 by $-r_1$;

e) replace r_1, r_2 by r_1+1, r_1-1.

Remark 2: It follows from Proposition 1 that if M is oriented, then $e(M) = -\sum r_i$, called the Euler number, is an invariant of the Seifert fibration of M.

Remark 3: Let $-M$ denote M with opposite orientation; then $-M$ has Seifert invariant $I(-M) = \{\epsilon, g; -r_1, \cdots, -r_s\}$.

3. The main result. First some notations for numbers. For each rational number $x \in Q$, define $\overline{x} = x \pmod 1 \in Q/Z$, and $\overline{\overline{x}} = [x] \in Q/\sim$, where $\sim$ is the equivalence relation (which descends to Q/Z) on Q defined by $a \sim b$ iff $a \equiv \pm b \pmod Z$.

The proof of the following lemma is elementary and is implicitly contained in the proof of Proposition 1.

Lemma 1 *Let O_1, O_2 be two fibers in a SFS M, r_1, r_2 be the corresponding rational numbers in $I(M)$.*

(1) if M is oriented, then there exists a fiber preserving and orientation preserving homeomorphism $f : M \to M$ sending O_1 to O_2 if and only if $r_1 \equiv r_2 \pmod 1$

(2) if M is nonorientable, then there exists a fiber preserving homeomorphism $f : M \to M$ sending O_1 to O_2 if and only if $r_1 \equiv \pm r_2 \pmod 1$.

Now let X be a SFS with $\partial X \cong T^2$. As before, we delete all the singular fibers and let R be a section of the remaining bundle. Let $I(X) = \{\epsilon, g; r_1, \cdots, r_t\}$ be the resulting Seifert invariant, and $c = R \cap \partial M$ be the section restricted to ∂X, with a similar orientation convention as in section 1. We say that c is compatible with $I(X)$. Changing the sections R changes the r_i's by integers, and changes the compatible section c into $c + kh$ for some integer k. Once we fix c, each slope r on ∂X is uniquely determined by a rational number $\frac{\beta}{\alpha}$, where $[r] = \alpha c - \beta h$. We will identify r with the rational number $r = \frac{\beta}{\alpha}$. If $\alpha \neq 0$, $X(r)$ is a SFS with $I(X(r)) = \{\epsilon, g; r_1, \cdots, r_t, r\}$.

Lemma 2 *Let $X \not\cong S^1 \times D^2$ be a SFS with $\partial X \cong T^2$. Let h be the regular fiber slope on ∂X and let r be a different slope. If $X(h) \cong X(r)$, then $X \cong B \times S^1$, where B is the Möbius band. Furthermore, there is a self homeomorphism of X sending h to r.*

Proof. Since $r \neq h$, $X(r)$ is a SFS. Hence $X(h)$ is homeomorphic to a SFS. Let $O(X)$ denote the orbifold of X, then we have $\pi_1(X(h)) \cong \pi_1(X)/ < h > \cong \pi_1(O(X))$. Since $X \not\cong S^1 \times D^2$, $O(X)$ is not a disk with at most one cone point. Thus $\pi_1(O(X))$ is either Z or a free product. Since it is also isomorphic to π_1 of a Seifert fibered space $X(r)$, the only possibilities for $\pi_1(O(X))$ are $Z_2 * Z_2$ and Z.

If $\pi_1(X(h)) \cong \pi_1(O(X)) \cong Z_2 * Z_2$, then $X(h) \cong RP^3 \# RP^3$, and $O(X)$ is a disk with 2 cone points both with order 2. Thus $X(r)$ is a SFS over S^2 with at most three cone points. Such a manifold cannot be homeomorphic to $RP^3 \# RP^3$.

If $\pi_1(X(h)) \cong \pi_1(O(X)) \cong Z$, then $X(h)$ is $S^2 \times S^1$ or $S^2 \tilde{\times} S^1$, and $O(X)$ is the Mōbius band B (the annulus is ruled out since it has 2 boundary components). Thus X is either $B \tilde{\times} S^1$ or $B \times S^1$. If $X \cong B \tilde{\times} S^1$, then $X(h) \cong S^2 \times S^1$. But $X(r)$ is a SFS over P^2 with at most one cone point. Such a SFS cannot be homeomorphic to $S^2 \times S^1$. The only possibility remaining is $X \cong B \times S^1$ and so $X(h) \cong S^2 \tilde{\times} S^1$. The last statement of the lemma follows from the following

Lemma 3 *Let $X = B \times S^1$, let h be a regular fiber on ∂X, and let $c = \partial B$. On the set of slopes on ∂X, define an equivalence relation by $r \sim s$ iff there is a self homeomorphism $f : X \to X$ sending r to s. Then for each slope $r = xc + yh, r \sim c$ iff y is even, $r \sim h$ iff y is odd. In particular, $X(r) \cong P^2 \times S^1$ if y is even, $X(r) \cong S^2 \tilde{\times} S^1$ if y is odd.*

Proof. Let d be the center circle of B; then (d, h) form a basis for $H_1(X)$. Let a be a spanning arc of B, let $A = a \times S^1$ be a vertical annulus. Consider two self homeomorphisms α and β of X, where α is a Dehn twist along A, and β is a "Dehn twist" along B. Here a Dehn twist along a Mōbius band is defined similarly to a Dehn twist along an annulus except that you rotate along the center line twice. The actions of these two homeomorphisms on $H_1(X)$ are $\alpha_*(d\ h) = (d\ h)\begin{pmatrix} 1 & 0 \\ 1 & 1 \end{pmatrix}$ and $\beta_*(d\ h) = (d\ h)\begin{pmatrix} 1 & 2 \\ 0 & 1 \end{pmatrix}$. The actions on $H_1(\partial X)$ are $\alpha_*(c\ h) = (c\ h)\begin{pmatrix} 1 & 0 \\ 2 & 1 \end{pmatrix}$ and $\beta_*(c\ h) =$

$(c\ h)\begin{pmatrix}1 & 1\\ 0 & 1\end{pmatrix}$. Thus the actions on slopes $r = \begin{pmatrix}x\\ y\end{pmatrix}$ are $\begin{pmatrix}x\\ y\end{pmatrix} \xrightarrow{\alpha_*} \begin{pmatrix}1 & 0\\ 2 & 1\end{pmatrix}\begin{pmatrix}x\\ y\end{pmatrix}$ and $\begin{pmatrix}x\\ y\end{pmatrix} \xrightarrow{\beta_*} \begin{pmatrix}1 & 1\\ 0 & 1\end{pmatrix}\begin{pmatrix}x\\ y\end{pmatrix}$. Simple linear algebra checks that the two matrices $\begin{pmatrix}1 & 0\\ 2 & 1\end{pmatrix}$ and $\begin{pmatrix}1 & 1\\ 0 & 1\end{pmatrix}$ generate a subgroup G of $SL_2(Z)$ of index two (namely, all the 2 by 2 matrices (a_{ij}) with a_{11}, a_{22} odd, and a_{21} even), and $\begin{pmatrix}x\\ y\end{pmatrix} \sim \begin{pmatrix}1\\ 0\end{pmatrix}$ if y is even, $\begin{pmatrix}x\\ y\end{pmatrix} \sim \begin{pmatrix}0\\ 1\end{pmatrix}$ if y is odd, where $u \sim v$ means $u = Av$ for some $A \in G$. This proves the "if" part. To prove the "only if" part, note that h is a primitive element in $H_1(X)$, but c is not. Thus there cannot be a self homeomorphism f sending c to h. □

We are concerned with Dehn fillings to a SFS which yield homeomorphic manifolds so that the slopes are not related by a self homeomorphism of the SFS. The result says such a SFS must be orientable, and the surgered manifolds must be orientation reversingly homeomorphic. Furthermore, at most two different slopes yield homeomorphic manifolds, and there are an infinite number of such pairs provided one such pair exists.

Theorem 1 *Let X be an orientable SFS with $\partial X \cong T^2$ and ∂X incompressible. Let r, s be two slopes on ∂X such that $X(r) \cong X(s)$, but there is no homeomorphism $f : X \to X$ sending r to s. Then*

(1) $X(r) \cong -X(s)$, $X(r) \not\cong +X(s)$.

(2) Under some choice of the section c on ∂X, the Seifert invariant of X can be written as

$$I(X) = \{\epsilon, g; \frac{1}{2}, \cdots, \frac{1}{2}, r_2, -r_2, \cdots, r_k, -r_k, r_1\} \text{ where } r_i \not\equiv 0, \frac{1}{2} \pmod 1$$

Furthermore $r = -r_1 + m$, $s = -r_1 - n - m$, where n is the number of $\frac{1}{2}$'s in $I(M)$ and $m \neq -\frac{n}{2}$ is an integer.

Conversely, if $I(X), r, s$ are as in (2) above, then $X(r) \cong -X(s) (\not\cong +X(s))$, and there is no homeomorphism $f : X \to X$ sending r to s.

Proof. First by Lemma 2, neither r nor s is a regular fiber of X. Thus $X(r)$ and $X(s)$ both have a natural Seifert fibration induced from X. We claim that there exists a homeomorphism $f : X(r) \to X(s)$ which preserves the above natural Seifert fibrations. To prove the claim, suppose $X(r)$ and $X(s)$ are two different Seifert fibrations of the same manifold. By [2] [VI.17], the only possibilities are: (1) $O(X(r))$ is S^2 with three cone points of orders $2, 2, \alpha$, and $O(X(s))$ is RP^2 with at most 1 cone point; or (2) $O(X(r))$ is S^2 with four cone points all of order 2, and $O(X(s))$ is the Klein bottle. On the other hand, however, $O(X(r))$ and $O(X(s))$ are both obtained from $O(X)$ by adding a disk with possibly one cone point. Thus they differ only by a cone point. Neither (1) nor (2) satisfies this. Thus the claim must be true.

Now we fix a section c on ∂X, and let $I(X) = \{\epsilon, g; r_1, \cdots, r_t\}$ be a Seifert invariant of X compatible with c. Under the basis (c, h) of $H_1(\partial X)$ we identify r, s with rational numbers. If $X(r) \cong +X(s)$, then comparing the Euler numbers gives $r = s$, contradicting the assumption that $r \neq s$. Thus $X(r) \cong -X(s)$. Now $X(r), -X(s)$ have Seifert invariants

$$I(X(r)) = \{\epsilon, g; r_1, \cdots, r_t, r\}, \text{ and } I(-X(s)) = \{\epsilon, g; -r_1, \cdots, -r_t, -s\}$$

Let $R_1 = \{r_1, \cdots, r_t, r\}, R_2 = \{-r_1, \cdots, -r_t, -s\}$. Then R_1 and R_2 are related as in Proposition 1. It follows that

(a). $\{\overline{r_1}, \cdots, \overline{r_t}, \overline{r}\} = \{\overline{-r_1}, \cdots, \overline{-r_t}, \overline{-s}\}$, and

(b). $\{\overline{\overline{r_1}}, \cdots, \overline{\overline{r_t}}, \overline{\overline{r}}\} = \{\overline{\overline{-r_1}}, \cdots, \overline{\overline{-r_t}}, \overline{\overline{-s}}\}$.

as unordered sets counted with multiplicity.

Since $\overline{\overline{r_i}} = \overline{\overline{-r_i}}$, equation (b) implies $\overline{\overline{r}} = \overline{\overline{-s}}$. Thus $\overline{r} = \overline{\pm s}$.

If $\overline{r} = \overline{-s}$, then by Lemma 1, there exists a homeomorphism $f : X(r) \to -X(s)$ sending K_r to K_s, where K_r, K_s are the cores of the added tori. Thus $f|X$ is a homeomorphism sending r to s, contradicting our assumption. Thus $\overline{r} = \overline{s}$. Since $\overline{r} \neq \overline{-s}, \overline{r} \neq \overline{0}, \overline{\frac{1}{2}}$.

Now equation (a) gives $\overline{r} = \overline{-r_i}$ for some r_i, say $\overline{r} = \overline{-r_1}$, where $\overline{r_1} \neq \overline{0}, \overline{\frac{1}{2}}$. Then $\overline{-s} = \overline{-r} = \overline{r_1}$. Thus in equation (a) we can delete $\overline{r_1}, \overline{r}$ from the left, and $\overline{-r_1}, \overline{-s}$ from the right, giving the equation $\{\overline{r_2}, \cdots, \overline{r_t}\} = \{\overline{-r_2}, \cdots, \overline{-r_t}\}$. Hence for each i, either $\overline{r_i} = \overline{-r_j}$ for some $j \neq i$ or $\overline{r_i} = \overline{\frac{1}{2}}$. It follows that after reindexing $\{\overline{r_2}, \cdots, \overline{r_t}\} = \{\overline{\frac{1}{2}}, \cdots, \overline{\frac{1}{2}}, \overline{r_2}, \overline{-r_2}, \cdots, \overline{r_k}, \overline{-r_k}\}$, where $r_i \not\equiv 0, \frac{1}{2}$ (mod 1). At the expense of using a different section c of ∂X, we can change r_i by a suitable integer by using a different section R so that

$$I(X) = \{\epsilon, g; \frac{1}{2}, \cdots, \frac{1}{2}, r_2, -r_2, \cdots, r_k, -r_k, r_1\}$$

Such a change of c changes r and s by integers. Thus we still have $\overline{r} = \overline{-r_1} = \overline{s}$. Let $r = -r_1 + m, s = -r_1 + l$, where m and l are integers. Comparing the Euler numbers of $X(r)$ and $-X(s)$, we have $\frac{1}{2}n + m = -\frac{1}{2}n - l$, where n is the number of $\frac{1}{2}$'s in $I(X)$. This gives $l = -n - m$. So $s = -r_1 - n - m$. Since $r \neq s, m \neq -\frac{n}{2}$.

Conversely, given $I(X), r, s$, as in (2) of the theorem, then

$$I(X(r)) = \{\epsilon, g; \frac{1}{2}, \cdots, \frac{1}{2}, r_2, -r_2, \cdots, r_k, -r_k, r_1, -r_1 + m\}, \quad \text{and}$$

$$I(-X(s)) = \{\epsilon, g; -\frac{1}{2}, \cdots, -\frac{1}{2}, -r_2, r_2, \cdots, -r_k, r_k, -r_1, r_1 + n + m\}$$

It is clear that after some moves described in Remark 1 of section 1, the two invariants are identical. Hence $X(r) \cong -X(s)$.

Since $X \not\cong S^1 \times D^2, X(r)$ is not a lens space. The Euler number $e(X(r)) = \frac{n}{2} + m \neq 0$. Hence by [4], $X(r) \not\cong -X(r)$. Thus $X(r) \not\cong +X(s)$.

If there is a homeomorphism $f : X \to X$ sending r to s, then by [6, Theorem 3.9], after a homotopy f is a fiber preserving homeomorphism. This extends to a homeomorphism from $X(r)$ to $X(s)$ (still denoted by f). It must be orientation reversing since $X(r) \not\cong +X(s)$. Thus we have a fiber preserving orientation preserving homeomorphism $f : X(r) \to -X(s)$ sending K_r to K_s, where K_r and K_s are the cores of the added tori. By Lemma 1, $\overline{r} = \overline{-s}$, i.e. $\overline{-r_1 + m} = \overline{r_1 + n + m}$. Thus $\overline{r_1} = \overline{\frac{1}{2}}$ or $\overline{0}$, a contradiction. □

Theorem 2 *Let X be a nonorientable SFS with $\partial X \cong T^2$ and ∂X incompressible, r, s be two slopes on ∂X. If $X(r) \cong X(s)$, then there is a homeomorphism of X sending r to s.*

Proof. If $X \cong B \times S^1$, the theorem follows from Lemma 3.

If $X \not\cong B \times S^1$, the theorem follows from Lemma 1 and the following lemma.

Lemma 4 *Let X, r, s be as in Theorem 2. If $X \not\cong B \times S^1$, then $X(r) \cong X(s)$ if and only if $r \equiv \pm s \pmod{Z}$.*

Proof. First, similar to the proof of Theorem 1, we can show that $X(r), X(s)$ have natural Seifert fibrations induced from X, and there is a homeomorphism $f : X(r) \to X(s)$ that preserve the fibrations. The Seifert invariants of $X(r)$ and $X(s)$ are given by $I(X(r)) = \{\epsilon, g; r_1, \cdots, r_t, r\}$ and $I(X(s)) = \{\epsilon, g; r_1, \cdots, r_t, s\}$ respectively. Thus $\{r_1, \cdots, r_t, r\}$ and $\{r_1, \cdots, r_t, s\}$ are related as in Proposition 2. It follows that $\{\overline{\overline{r_1}}, \cdots, \overline{\overline{r_t}}, \overline{\overline{r}}\} = \{\overline{\overline{r_1}}, \cdots, \overline{\overline{r_t}}, \overline{\overline{s}}\}$. This gives $\overline{\overline{r}} = \overline{\overline{s}}$ and hence the lemma follows. □

To summarize, the following is the complete list of knots whose complements are SFS, and which are not determined by their complements:

1. M is an oriented SFS with $I(M) = \{\epsilon, g; \frac{1}{2}, \cdots, \frac{1}{2}, r_1, -r_1, \cdots, r_k, -r_k, m\}$, where $r_i \not\equiv 0, \frac{1}{2} \pmod 1), m \neq -\frac{n}{2}, n$ is the number of $\frac{1}{2}$'s, and K, K' are the singular fibers corresponding to the Seifert pairing $r_1, -r_1$ respectively. We have $X(K) \cong -X(K')$ but there is no self homeomorphism of M sending sending K to K'.

2. M is the lens space $L(p, q)$, K and K' are the cores of the two solid tori in the standard Heegaard splitting of M. We have $X(K) \cong +X(K)$. But there is no orientation preserving self homeomorphism of M sending K to K' if $q^2 \not\equiv 1 \pmod p$. Nor is there any self homeomorphism of M sending K to K' if $q^2 \not\equiv \pm 1 \pmod p$.

4. Further examples. In this section we describe a way to produce more examples of knots not determined by their complements. The knot complements here are no longer Seifert fibered spaces, instead they are various Haken manifolds with torus splittings. Examples with hyperbolic complements were also found by Bleiler, Hodgson, and Weeks (personal communication). Again, all these examples are in the unoriented category.

Our construction starts with a given example of a knot not determined by its complement. Let M be a closed orientable 3-manifold, let K and K' be two disjoint knots in M such that there is a homeomorphism $f : X(K) \to X(K')$ but there is no homeomorphism $g : (M, K) \to (M, K')$. Let us also assume that $X(K)$ is irreducible so that is has a torus splitting. Suppose that there is an oriented knot H in M, disjoint from K and K', that is invariant under f. Also assume that $X(H) = M - N(H)$ is ∂-incompressible, where $N(H)$ denotes an open tubular neighborhood of the knot H. Let Y be a compact orientable irreducible 3-manifold with $\partial Y \cong T^2$. We form a closed 3-manifold $W = (M - N(H)) \cup_\theta Y$, and think of K and K' as knots in the closed 3-manifold W.

Proposition 3 *One can choose Y and the gluing map such that K and K' are inequivalent knots in W, but $X(K) \cong +X(K)$ if f is orientation preserving, and $X(K) \cong -X(K')$ if f is orientation reversing.*

Proof. First we consider the case when f is orientation preserving. We can isotope f so that $f|N(H)$ is the identity map for a tubular neighborhood of $N(H)$. Thus f can be extended over Y to become a homeomorphism from $W - N(K)$ to $W - N(K')$. Thus $X(K) \cong +X(K')$, where K and K' are regarded as knots in W. Next we show that K and K' are generally not equivalent knots in W. To do this, let us choose Y so that ∂Y is incompressible. Then W has a natural torus splitting induced from that of Y and $M - N(H)$. The uniqueness of the torus splitting implies that any homeomorphism $g : (W, K) \to (W, K')$ must send Y to Y, and $M - N(H)$ to $M - N(H)$. Furthermore, for a homological reason, g must fix a slope on ∂Y and a slope on $\partial(M - N(H))$, namely their longitudes (By the longitude of Y, we mean the unoriented simple loop that generates $\ker\{i_* : H_1(\partial Y, Q) \to H_1(Y, Q)\}$, and the same for $M - N(H)$). Thus as long as our gluing map does not identify these two slopes, g must fix two different slopes on the common boundary $\partial Y = \partial(M - N(H)) = T$. It follows that g fixes all the slopes, and in particular, the meridian of the knot H. Thus g induces a homeomorphism from M to itself sending K to K', contradicting to the assumption that K and K' are not equivalent. Hence the proposition is proved if f is orientation preserving.

Before we move to the orientation reversing case, we would like to argue that even if ∂Y is compressible, i.e. Y is the solid torus, we still have the conclusion very often. In this case, W is obtained from M by doing a (p,q) Dehn surgery to the knot H. Let $H_{p,q}$ be the core of the added torus. If (p,q) is sufficiently large then the knot $H_{p,q}$ is often rigid in W. By this we mean any self homeomorphism of W can be homotoped to preserve $H_{p,q}$. (For example, if $M - N(K)$ is hyperbolic, then by for large (p,q), W is hyperbolic, and $H_{p,q}$ is the shortest geodesic. Thus by Mostow's rigidity theorem, any self homeomorphism of W is homotoped to an isometry, which must preserve $H_{p,q}$. Similar results can be proved for most Seifert fibered spaces using π_1.) Thus a similar argument as above shows that K and K' are knots with homeomorphic complement but not equivalent.

Next we consider the case when f is orientation reversing. After an isotopy, we may assume that f maps a tubular neighborhood $N(H)$ of H to itself, and $f(l) = l, f(m) = -m$, where (l,m) is a longitude-meridian pair for $N(H)$. Let Y be a compact orientable 3-manifold with $\partial Y \cong T^2$, and there is an orientation reversing map $f_1 : Y \to Y$ that sends l_1 to $-l_1$, and m_1 to m_1, where (l_1, m_1) form a basis for $H_1(\partial Y)$, and l_1 is a generator for $\ker\{i_* : H_1(\partial Y, Q) \to H_1(Y, Q)\}$. An example of such an Y is the exterior of the figure eight knot. Now form a closed 3-manifold $W = (M - N(H)) \cup_\partial Y$, where l is glued to m_1 and m is glued to l_1. As before, we can show that K and K' are knots in W with $W - N(K) \cong -(W - X(K'))$ but there is no homeomorphism sending (W,K) to (W,K'). $\square$

Examples of such a knot H can be obtained in the following two different ways:

1. Let M, K, K' be the example of the SFS in Theorem 1, H be a singular fiber

with Seifert invariant $\frac{1}{2}$. By Lemma 1 we can choose f that leaves H invariant.

2. Let $M = L_{p,q} = L_{16,5}$ with a standard Heegaard splitting $M == V \cup V'$, let K and K' be the cores of V and V' respectively. Since $5^2 = 25 \not\equiv \pm 1 \pmod{16}$, K and K' are not equivalent. Let $T = \partial V$ be the Heegaard surface, (l, m) (resp. (l', m')) be a longitude-meridian pair of V (resp. V'). Define $f : T \to T$ by $f(l \quad m) = (l \quad m)\begin{pmatrix} b & p \\ d & q \end{pmatrix} = (l \quad m)\begin{pmatrix} -3 & 16 \\ -1 & 5 \end{pmatrix}$. Since $f(m) = pl + qm = m'$, f extends to a map $f : V \to V'$, i.e. $X(K) \to X(K')$. This map is orientation reversing since $f|T$ is orientation preserving and f switches the sides of T. Let $H = 4l + m$ be a torus knot on T. Since $f(H) = (l \quad m)\begin{pmatrix} -3 & 16 \\ -1 & 5 \end{pmatrix}\begin{pmatrix} 4 \\ 1 \end{pmatrix} = (l \quad m)\begin{pmatrix} 4 \\ 1 \end{pmatrix} = H$, H is invariant under f.

More generally, if $q + b = 2$, then the map $f|T$ is reducible since its characteristic polynomial is $x^2 - (q + b)x + 1 = (x - 1)^2$. Thus there is an invariant torus knot H. This is always possible as long as $q(2 - q) \equiv 1 \pmod{p}$.

Unfortunately it is not clear whether such a knot H could exist if f is orientation preserving. Thus we do not obtain examples of knots not determined by their oriented complements if requiring the exteriors being ∂-incompressible.

Acknowledgements

I wish to thank Cameron Gordon for his helpful comments.

References

[1] C. McA. Gordon and J. Luecke. Knots are determined by their complements. *Journ. of the Amer. Math. Soc.* Vol. 2, No. 2 371-415 (1989).

[2] W. Jaco, "Lectures on Three-Manifold Topology," CBMS Lecture Series, Number 43, Amer. Math. Soc. Providence, R.I. (1980).

[3] Y. Mathieu, Seminar notes at University of Texas (1990)

[4] W. D. Neumann and F. Raymond, "Seifert manifolds, plumbing, μ-invariant and orientation reversing maps," Lecture Notes in Mathematics 664, 162-165 Springer, Berlin, (1978)

[5] P. Orlik, "Seifert Manifolds," Lecture notes in Mathematics, No. 291, Springer-Verlag, (1972)

[6] P. Scott, "The Geometries of 3-manifolds," Bull. London Math. Soc. 15, 401-487 (1983)

Triangulations and TQFT's

David N. Yetter

Department of Mathematics
Kansas State University
Manhattan, KS 66506-2602, U.S.A.

Abstract We present a general method for the construction of topological quantum field theories (TQFT's) using triangulations of manifolds and cobordisms, discuss examples (notably Turaev/Viro theory [TV] and a (3+1)-dimensional TQFT recently constructed by Crane and Yetter [CY]), and suggest further avenues for investigation using the general method.

1. Introduction

Since the study of topological quantum field theories (TQFT's) was initiated by Atiyah [A] and Witten [Wi], one of the main motivations has been the study of smooth 4-manifolds. In particular, an object of considerable speculation and work has been the quest for "Donaldson/Floer theory", a TQFT which would extend and unify the work of Donaldson on invariants of smooth simply-connected 4-manifolds [Do], and Floer's instanton-homology [F]. Thus far this quest has not been fulfilled.

The notion of TQFT, has, however, already borne fruit in the study of 3-manifolds, by providing a unifying framework for invariants of 3-manifolds constructed in work of Reshetikhin and Turaev [RT] from surgery descriptions of 3-manifolds (cf. Kirby [K]) and the Jones polynomial [J], with other new invariants of 3-manifolds arising in the work of Crane [C] (using Heegaard splittings) and Turaev/Viro [TV] (using triangulations).

Also for 3-manifolds, in the particularly simple case of "finite gauge-group Chern-Simons theory", the functional integral approach to TQFT's has been made rigorous by Freed and Quinn [FQ]. Unfortunately, attempts to extend the functional integral approach to more interesting theories appear to meet foundational difficulties.

On the other hand, a Turaev/Viro-style approach has been used to provide another rigorous construction of finite gauge-group Chern-Simons theory (Wakui [Wak], Yetter [Y1]).

Recent work of Piunikhin [Pi], showing that the invariants of [RT] and [C] coincide, together with the easy classical construction relating triangulations of 3-manifolds to Heegaard diagrams, show that the known examples of (2+1)- dimensional TQFT's not originally constructed from triangulations, can nonetheless be viewed as arising from constructions on triangulations.

The work of Walker [Wal] on factorization-at-a-corner suggests that the connection between triangulations and TQFT's may be perfectly general.

Finally, the coincidence for 4-manifolds between smooth and PL structures suggests the possibility of a PL approach to invariants of smooth 4-manifolds.

In most of what follows, we will deal with ordered triangulations, that is, triangulations for which the set of vertices has been equipped with a linear ordering. For the discussion in section 2., we will simply refer to triangulations, but will implicitly include the case where the triangulations are equipped with some auxiliary structure (like an ordering of the vertices), provided this auxiliary structure can be made well-behaved under subdivision. Finally, the reader should be warned that we follow the diagrammatic ordering when writing composition of maps; thus fg means f followed by g.

2. Generalities

Atiyah's original definition of TQFT [A] was given essentially in the Dirac bras and kets formulation [Di]. When, again following Dirac, the induced operators on kets are constructed, his definition is seen to be equivalent to

Definition 1 *A* $(d+1)$-dimensional TQFT *is a monoidal functor, Z, from* $d-$ **cobord** *to* **VECT**, *where* $d-$ **cobord** *is the category whose objects are compact oriented smooth d-manifolds, whose maps are cobordisms, equipped with the monoidal category structure given on both objects and maps by disjoint union (with $I =$ empty manifold), and* **VECT** *is the category of complex vector spaces.*[1] *The vectorspace $Z(M)$ will be called* the state space of M, *the linear map $Z(X)$: $Z(M) \to Z(N)$ associated to a cobordism X will be called* the evolution operator of X.

This definition can, of course, be generalized by replacing **VECT** with some (other) category of modules or of some type of topological vector spaces, or by including tangles in the cobordisms.

Now, the (ordered) triangulations of a d-manifold (or cobordism) form a directed set when ordered by refinement, and thus provide a good indexing set for some sort of approximation procedure. To outline such a procedure we need

Definition 2 *If (M,T) and (N,S) are triangulated d-manifolds, a* triangulated cobordism *from (M,T) to (N,S) is a cobordism X from M to N equipped with a triangulation $\mathcal{T}$ whose restriction to M (resp. N) is T (resp. S).*

Suppose we have a way of assigning to each triangulated d-manifold (M,T) a vector space $Z(M,T)$, and to each triangulated cobordism $(X,\mathcal{T})$ from (M,T) to (N,S) a map $Z(X,\mathcal{T})$ from $Z(M,T)$ to $Z(N,S)$. Thus for any cobordism we have

[1] A point often overlooked here is that the non-degenerate bilinear pairing between $Z(M)$ and $Z(-M)$ is obtained "for free" by regarding $M \times [0,1]$ as a cobordism from $M \coprod -M$ to $\emptyset$. (Non-degeneracy follows from the "drainpipe identity" version of the triangle identity for adjunctions familiar from tangle-theory (cf. [FY1,FY2,S]).)

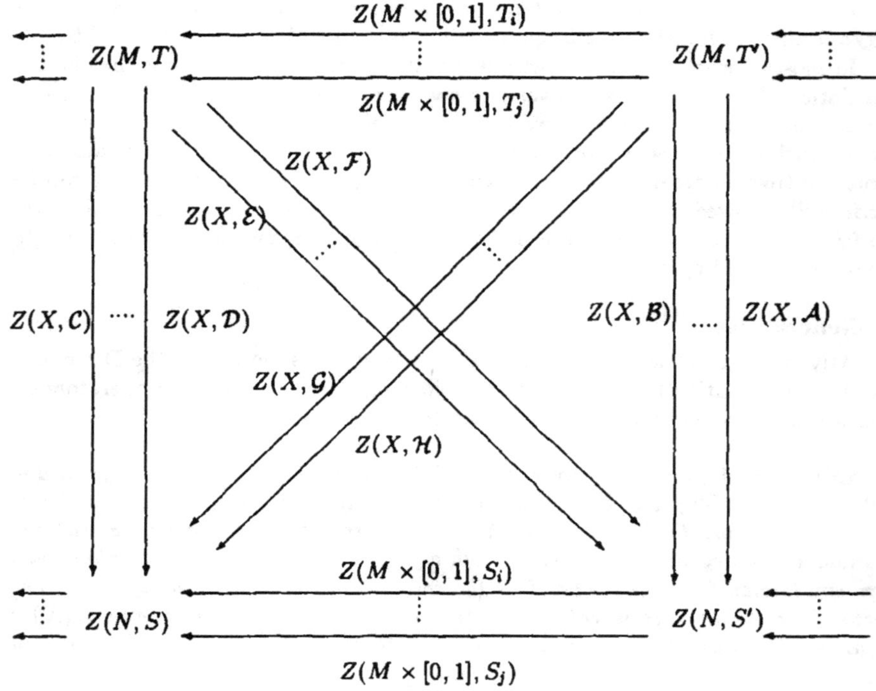

Figure 1: The Basic Diagram

a large, usually non-commutative diagram of the form shown in Figure 1, in which the horizontal maps correspond to triangulated cobordisms between $Z(M,T)$ and $Z(M,T')$ for T' a refinement of T, with the identity cobordism from M to M as underlying cobordism.

For now, the only requirements we wish to place on the assignments is that they echo the monoidal functoriality of a TQFT: we want $Z(X,\mathcal{T})Z(Y,\mathcal{S}) = Z(XY,\mathcal{T}\cup\mathcal{S})$ whenever $(X,\mathcal{T})$ and $(Y,\mathcal{S})$ are composable triangulated cobordisms, $Z(\emptyset,\emptyset)\cong C$, and $Z(X\coprod Y, S\cup T)\cong Z(X,S)\otimes Z(Y,T)$ (this last should hold both for triangulated d-manifolds, in which case $\cong$ is linear isomorphism, and for triangulated cobordisms, in which case $\cong$ is "conjugation" by the linear isomorphisms for the source and target d-manifolds).

Although the details of how such an assignment is constructed are for the present unimportant, all examples currently in the literature [TV, CY, Wak, Y1, Y2, Y3] use state-sum techniques: a pure state of a triangulated d-manifold (resp. cobordism) is an assignment of "spins" to simplices of certain dimensions in the triangulation, satisfying some local constraints. $Z(M,T)$ is then the free vector space on the basis of pure states. A product of "interactions", numbers defined as contributions of local configurations of spins, is then used to define a linear map $Z(X,T) : Z(M,T) \to Z(N,S)$ by

$$Z(X,T)(\lambda) = \sum_{\substack{\text{states } \mu \text{ on } T \text{ re-} \\ \text{stricting to } \lambda \text{ on } T}} \left(\prod \text{interactions}\right) \cdot (\mu \text{ restricted to } S).$$

Constructions via state-sums of this sort give the requirements of the previous paragraph automatically.

Unfortunately, for almost any naive construction, the parallel maps are unequal, and none of the possible triangles in the diagram commute. One must now find a renormalization procedure to replace the parallel maps with a single map and make the diagram commute.

Again in all examples currently in the literature [TV, CY, Wak, Y1, Y2, Y3], there is a single way of doing this: the diagram is already *projectively* commutative, and the ratio between two parallel maps is a number conveniently computed from coarse data associated to the triangulation and its refinement and the initial data for the state-sums. If it is suspected that the diagram is projectively commutative, the proof of projective commutativity, the discovery of correction factors needed to make the diagram projectively commutative (if it is not already so), and the discovery of correction factors needed to make it actually commutative can all be carried out by considering the local contributions of parts of triangulations related by the combinatorial moves of Pachner [Pa1, Pa2]. Pachner's moves on an n-manifold may be described as follows: Consider the triangulation $\mathcal{S}$ of an n-sphere realized as the boundary of an $(n+1)$-simplex. It will consist of $n+2$ n-simplices. A k-for-$(n+2-k)$ Pachner move consists of replacing an n-ball triangulated by a copy of the triangulation of k of the n-simplices in $\mathcal{S}$ by a copy of the triangulation of the other $n+2-k$ (agreeing on the boundary).

Figure 2 illustrates the Pachner move for dimensions 1, 2, and 3.

Theorem 3 (Pachner [Pa2]) *If T and T' are triangulations of a PL n-manifold, then there is a finite sequence of k-for-$(n+2-k)$ moves (for $k=1,\ldots,n+1$) which transforms T into T'.*

If the ratio of local contributions from the "before" and "after" parts of the triangulation is independent of the (fixed) spins on the boundary of the region modified by the Pachner move, the diagram is already projectively commutative. Moreover, by "attributing" the ratio to the difference in the number of simplices in the "before" and "after" triangulations the correction factor needed to make parallel maps in Figure 1 equal can be expressed as a product over simplices interior to the cobordism of factors depending only on the dimension of the simplex.

If the ratio of local contributions from the "before" and "after" parts of the triangulation is not independent of the spins on the boundary region modified by

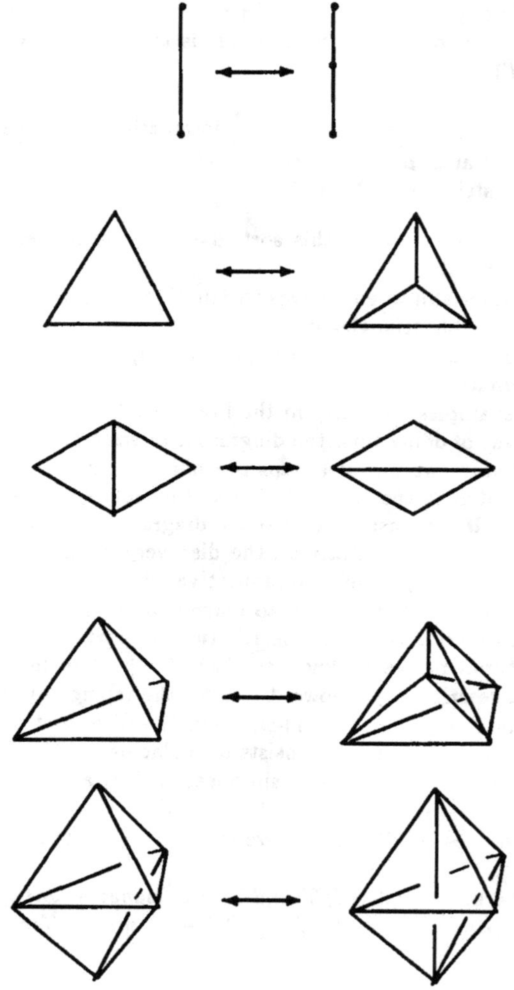

Figure 2: Pachner's Moves in Dimensions 1, 2, and 3

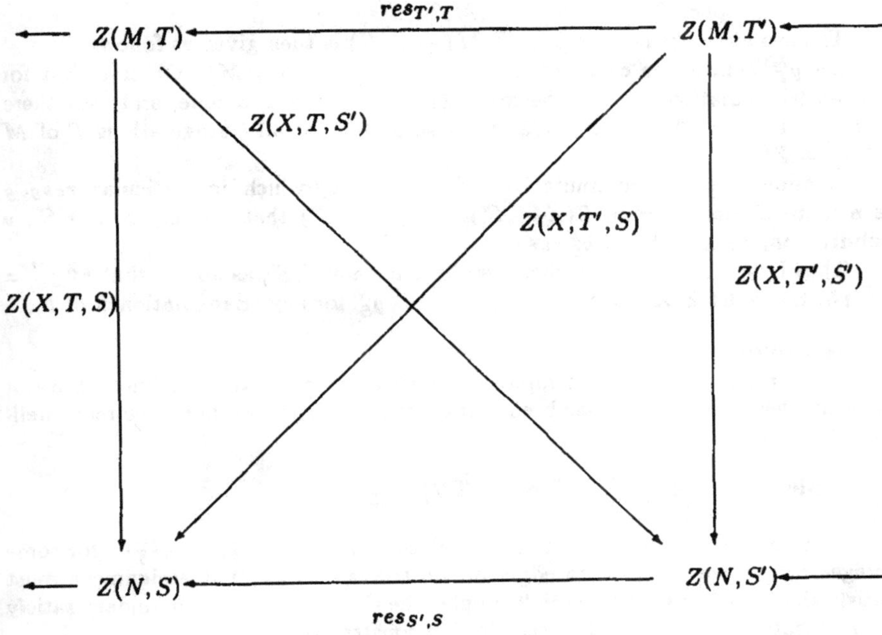

Figure 3:

the Pachner move, a more subtle correction factor depending on the spins occurring inside the region modified must be found to make the diagram projectively commutative. In the examples below, this step was necessary in Examples 1, 2, and 6. Regrettably, this step is presently a matter of art rather than a rigorous procedure. In the examples, it is carried out by ensuring the possibility of using "Schur's lemma" in the setting of a suitable monoidal category (in Examples 1 and 6, specifically the "small" representations of $U_q(sl_2)$ for q a root of unity).

If one has been able to renormalize, one now has a projectively commutative diagram of the form shown in Figure 3.

This diagram will not in general be commutative, and the preservation of composition of cobordisms will have been destroyed in the last step. However, the diagram can be made commutative and the composition property restored by multiplying each of the maps by another correction factor. This new factor is the product over simplices on the boundary of the triangulated cobordism of the square-root of the factor associated to a simplex of the same dimension in the previous step.

At last, we have for each cobordism a *commutative* diagram of the form shown in Figure 3. From this diagram, we can then define the state-spaces and evolution operators.

The state-space $Z(M)$ is $\operatorname{colim}(Z(M,T), res_{T',T})$.

The evolution operator $Z(X): Z(M) \to Z(N)$ is then given as follows:

Let g_T^M denote the canonical map from $Z(M,T)$ to $Z(M)$. Observe that for a fixed triangulation S of N, the maps $Z(X,T,S)$ form a cocone, and thus there is a unique map $k_S^X : Z(M) \to Z(N,S)$ such that for all triangulations T of M $g_T^M k_S^X = Z(X,T,S)$.

It follows from the commutativity of Figure 3 (by which, in particular, $res_{S',S}$ is a map of cocones from $Z(X,-,S')$ to $Z(X,-,S)$) that for any S and S', a subdivision, we have $k_S^X = k_{S'}^X res_{S',S}$.

Thus by a trivial diagram chase, we have for any S, S', as above, that $k_S^X g_S^N = k_{S'}^X g_{S'}^N$. So we let $Z(X) : Z(M) \to Z(N)$ be $k_S^X g_S^N$ for any triangulation S of N.

3. Examples

There are a number of examples in which the construction outlined above or an equivalent construction has been carried out by various authors, the most well-known being

Example 1: Turaev/Viro Theory [TV]

$(2+1)$-dimensional

Here spins (no quotes in this example) chosen from $\{0, \frac{1}{2}, 1, ..., \frac{r-2}{2}\}$ for some integer $r > 3$ are assigned to edges of the triangulation. The assignment must satisfy the condition that for each 2-simplex the three spins on the boundary satisfy a modification of the usual Clebsch-Gordan constraints:

$$|j-k| \le l \le j+k \text{ if } j+k \le \tfrac{r-2}{2},\ |j-k| \le l \le r-2-j-k \text{ if } j+k > \tfrac{r-2}{2}$$

$$\text{and } \ 2l \equiv 2j + 2k \pmod 2.$$

The interactions are then defined on tetrahedra of the triangulation by the quantum 6j-symbol associated to the six spins (in the correct normalization).

Following the procedure outlined in section 2 will then rediscover the correction factors: the product over edges of the quantum dimension of the spin, regarded as naming a representation of $U_q(sl_2)$ at q a primitive r^{th} root of unity, times the product over vertices of the reciprocal of the sum of the squares of the quantum dimensions. (This sum is denoted w^2 by Turaev and Viro [TV]–to see this, compute w^2 using Turaev and Viro's formula with 0 as the spin in the denominator.)

In the preceding example, the construction is simplified by the fact that the initial data came from an artinian semisimple C-linear abelian tortile category (a.k.a. an artinian semisimple ribbon quasi-tensor category over C) in which

a. all objects are self-dual

and

b. the fusion rules are multiplicity free.

The analogous construction can be carried out using any artinian semisimple K-linear abelian tortile category $\mathcal{X}$ (for K a field), but the absence of a. and b. complicates the construction.

Example 2: Generalized Turaev/Viro Theory [Y3]
$(2+1)$-dimensional

Here the use of an ordered triangulation becomes important. "Spins" assigned to oriented edges are chosen from a set of representatives for the isomorphism classes of simple objects in $\mathcal{X}$, while "spins" (really intertwiners, rather than spins) assigned to 2-simplices are chosen from the union of fixed bases for the hom-spaces $\mathcal{X}[a \otimes b \otimes c, I]$. The assignment must satisfy the condition that if an edge with one orientation is labelled a, the edge with the orientation reversed must be labelled a^* (where a^* is the representative of the isomorphism class of the dual object to a), and for each 2-simplex the label is an element of $\mathcal{X}[a \otimes b \otimes c, I]$ for a, b, c the labels on the bounding edges in the order induced by the ordering on the triangulation.

Now, given a choice of basis element for $\mathcal{X}[a \otimes b \otimes c, I]$ there is a dual basis of splittings for $\mathcal{X}[I, a \otimes b \otimes c]$. We denote the splitting of ϕ by $\overline{\phi}$. The interactions are then defined on tetrahedra as the coefficient of the multiple of the identity on I given in diagrammatic notation in the first (resp. second) part of Figure 4 when the orientation on the 3-simplex induced by the ordering of the vertices agrees with (resp. is opposite) the orientation on the cobordism. (In the figures, A, B, D, E are the labels on the 2-simplices obtained by deleting the last, third, second, first vertex in the ordering restricted to the vertices of the tetrahedron.)

Again following the procedure outlined in section 2 will discover the required correction factors: the "quantum dimension" (i.e. trace of the identity map in the internal categorical sense–cf. [Y4]) of the label on each edge, and the reciprocal of the sum of the squares of the "quantum dimensions" for each vertex.

One can also use simpler structures to provide initial data for the construction:

Example 3: Finite Gauge-Group Chern-Simons Theory [Wak,Y1]
$(2+1)$-dimensional

Here again, the use of an ordered triangulation is important. "Spins" assigned to oriented edges are chosen from a finite group. The assignment must satisfy the conditions that if g is assigned to an oriented edge then g^{-1} is assigned to the edge with the orientation reversed, and the product of the edges around a 2-simplex must be trivial.

The interactions are defined on tetrahedra by

$$\alpha([g|h|k])^{\epsilon}$$

where g, h, k are the edges from the first vertex to the second, the second to the third, and the third to the fourth (with respect to the ordering of vertices restricted to the tetrahedron), $\alpha \in Z^3(BG, U(1))$ is a fixed 3-cocycle, and where

$$\epsilon = \begin{cases} 1 & \text{if the orientation induced by the ordering agrees with the orientation on the cobordism} \\ -1 & \text{if the orientations disagree} \end{cases}$$

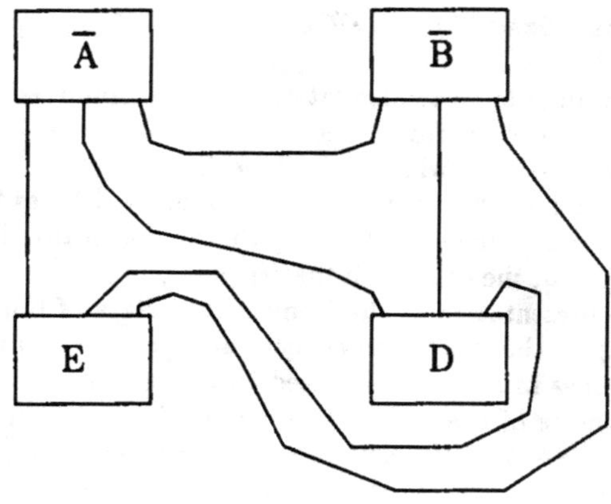

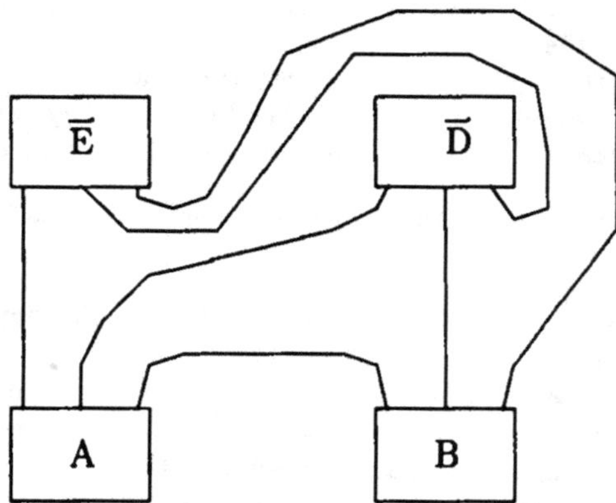

Figure 4:

Again the procedure of section 2 will uncover the required correction terms: the reciprocal of the order of the group for each vertex.

For $\alpha = 1$ (notice the cocycle is written multiplicatively), the last example extends to any dimension.

Example 4: $Hom(\pi_1, G)$**-counting Theory**
$(d+1)$-dimensional

A construction identical to that in Example 3 in the case $\alpha = 1$ gives rise to a TQFT in any dimension. For connected, closed $d+1$-manifolds regarded as self-cobordisms of the empty d-manifold the value of such a theory is given by

$$Z(M) = \frac{|Hom(\pi_1(M), G)|}{|G|}.$$

Example 5: A Theory from Homotopy 2-types [Y2]
$(d+1)$-dimensional

From the point of view of homotopy theory, the initial data in the last construction, groups, plays the role of "an algebraic model for a connected homotopy 1-type." We can use finite *categorical* groups in a similar way to construct TQFT's which measure the homotopy 2-type of manifolds.

By a categorical group we mean a group object $\mathcal{G}$ in the category of groupoids. (We denote the group law functor by $\bullet$ and the group inverse functor by †, the set of objects by $\mathcal{G}_0$, and the set of maps by $\mathcal{G}_1$.)

Given a finite categorical group $\mathcal{G}$, we use as "spins" on oriented edges, objects of $\mathcal{G}$, and as "spins" on 2-simplices, maps of $\mathcal{G}$, subject to the conditions that

a. if an oriented edge is assigned g, the edge with the orientation reversed is assigned $g^\dagger$;

b. 2-simplices are assigned maps from $g \bullet h$ to k, when g, h, k are the ordered edges obtained by omitting the third, second, first vertex in the ordering restricted to the 2-simplex;

and

c. the faces of every ordered 3-simplex form a commutative square in $\mathcal{G}$.

Interactions are taken to be identically 1.

The correction factors provided by the procedure in section 2 are then the reciprocal of $|\mathcal{G}_0|$ for each vertex, and the ratio $\frac{|\mathcal{G}_0|}{|\mathcal{G}_1|}$ for each edge.

This last example shows that the conjecture of Crane that TQFT's in $d > 4$ all arise from generalized cohomology theories is false. The conjecture must be refined to include theories which are sensitive to more refined homotopical data.

Example 6: [CY] Theory
(3 + 1)-dimensional

The initial data here is the same as for Turaev/Viro theory, but is used differently. Each 2-simplex is assigned a spin from the set $\{0, \frac{1}{2}, \ldots \frac{r-2}{2}\}$. The ordering of the vertices induces an orientation on all of the faces, allowing us to partition the 2-faces of each tetrahedron into "in" and "out" faces. Each tetrahedron is then to be labelled with an additional spin. This assignment is subject to the restriction that if j, k are the spins on the "in" (resp. "out") faces, and l is the spin on the tetrahedron, then j, k, and l satisfy the modified Clebsch-Gordan constraints (see Example 1).

We can now define the interaction on each 4-simplex as the number given as an element of $C = End(0)$ in Figure 5 using a slightly modified version of the diagrammatic notation of Reshetikhin/Kirillov [RK]. Here, minima labelled j denote $(-1)^j$ times the minima in [RK], and maxima labelled j denote $(-1)^{-j}$ times the maxima in [RK]. The verticals labelled with $\hat{\alpha}, \ldots, \hat{\varepsilon}$ are labelled with the spin assigned to the tetrahedra resulting from the deletion of the vertices of the 4-simplex in order, and the arcs connecting them are labelled with the spins assigned to the 2-simplices shared by the tetrahedra represented at the ends.

The correction factors are then discovered to be the reciprocal of the quantum dimension of the spin on tetrahedra, the quantum dimension of the spin on faces, the reciprocal of the sum of the squares of the quantum dimensions on edges, and the sum of the squares of the quantum dimension on vertices.

This last example is of particular interest because it is the first rigorous example of a (3 + 1)-dimensional TQFT without an immediate homotopical interpretation.

4. Other Ways? Negative Results and Speculations

We begin by answering a question about the general construction of section 2. which rules out some naive attempts at generalizations: Why do the maps run from finer to coarser triangulations?

The answer is "Simply for convenience." One can just as well perform the construction with a diagram similar to Figure 1 with the horizontal maps given by triangulated cobordisms from coarser to finer triangulations or even between arbitrary pairs of triangulations.

All three cases give the same result, as we will now show.

To do this, we need to recall a definition and a theorem from [M].

Definition 4 *A functor $J : \mathcal{D}' \to \mathcal{D}$ is* final *if*

(a) given any object $D \in \mathcal{D}$, there is an object $K \in \mathcal{D}'$ and a map $D \to J(K) \in \mathcal{D}$

and

(b) given any two such maps $D \to J(K_0)$ and $D \to J(K_n)$ there exists a finite commutative diagram of the form shown in Figure 6.

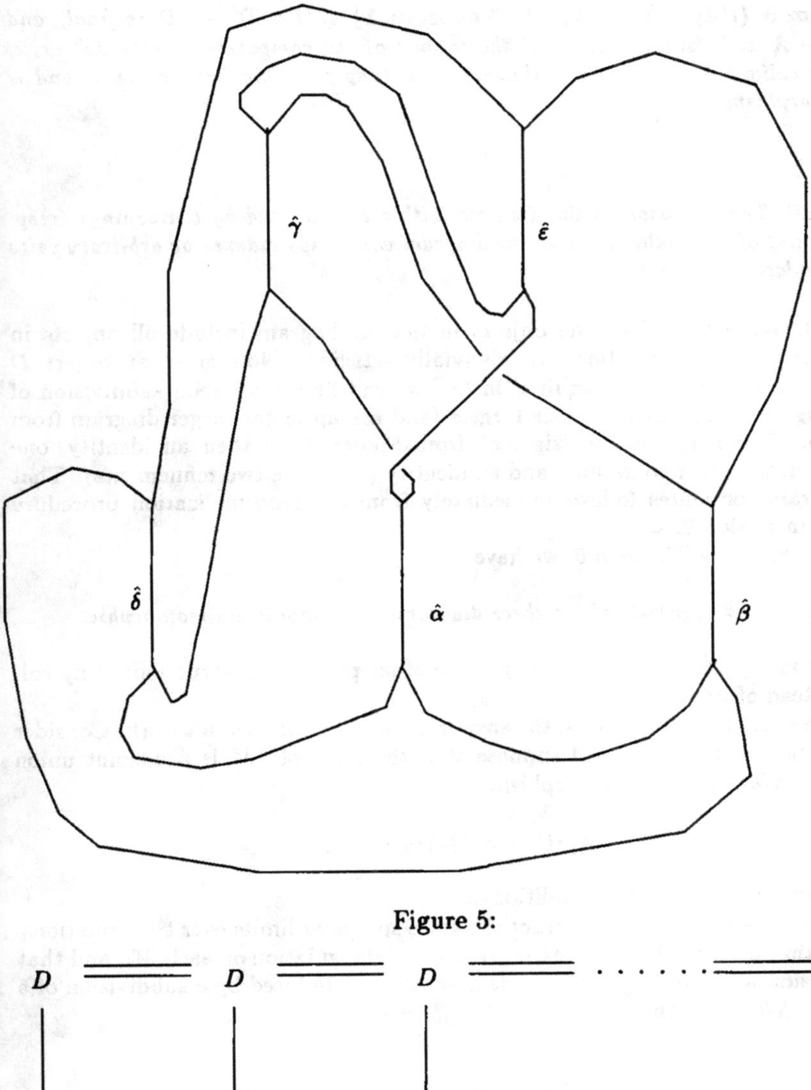

Figure 5:

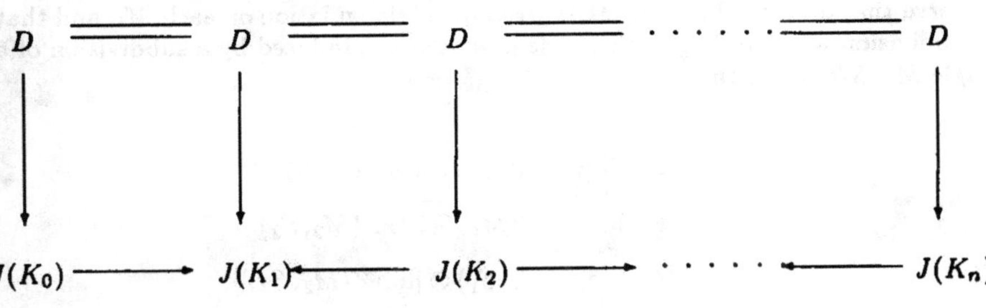

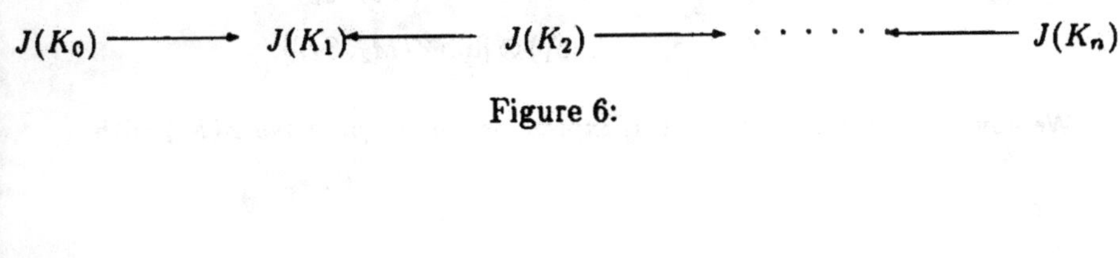

Figure 6:

Theorem 5 **([M], Ch. IX, §3, Theorem 1)** *If $J : \mathcal{D}' \to \mathcal{D}$ is final, and $F : \mathcal{D} \to \mathcal{X}$ is a functor such that the colimit of the composite functor JF exits, then the colimit of F exists and the canonical map from the first to the second is an isomorphism.*

Now

Lemma 6 *The inclusion of the diagram with maps indexed by coarsenings (resp. refinements) of triangulations into the diagram with maps indexed by arbitrary pairs of triangulations is final.*

proof: Observe that since the objects in the subdiagram include all objects in the larger diagram, condition (a) is trivially satisfied. Now given an object D and two maps of the form required in (a), we can find a common subdivision of the triangulations named by their targets (and a map in the larger diagram from D to it). The maps in the "zig-zag" from Figure 6 are then an identity, one coarsening, the other coarsening and an identity (resp. the two refinements). That the diagram commutes follows immediately from the renormalization procedure outlined in section 2. □

Thus, applying Theorem 5, we have

Theorem 7 *The colimits of the three diagrams are canonically isomorphic.*

A second question, however, suggests another possible construction: Why colimits instead of limits?

If one sticks to vector spaces, the answer is: because limits don't work. Consider the situation in Figure 3, and suppose that the manifold M is a disjoint union $M_1 \coprod M_2$. We require an isomorphism

$$Z(M) \cong Z(M_1) \otimes Z(M_2)$$

(satisfying certain coherence conditions).

Suppose we attempt to construct the state-spaces as limits over triangulations. Observe that a triangulation of M is given by a triangulation on each M_i, and that a subdivision which introduces a single new vertex is induced by a subdivision of a single M_i. We would then have

$$\begin{aligned} Z(M) &= \lim_{S_1,S_2} Z(M_1,S_1) \otimes Z(M_2,S_2) \\ &\cong \lim_{S_1} \lim_{S_2} Z(M_1,S_1) \otimes Z(M_2,S_2) \\ &\cong \lim_{S_1} [Z(M_1,S_1) \otimes [\lim_{S_2} Z(M_2,S_2)]]. \end{aligned}$$

We now wish to compare this last expression for the putative $Z(M)$ with

$$Z(M_1) \otimes Z(M_2) = [\lim_{S_1} Z(M_1, S_1)] \otimes [\lim_{S_2} Z(M_2, S_2)].$$

To carry out the comparison, choose a basis $\mathcal{L}$ for the limit over the S_2's, which we denote $Z(M_2)$ (since that's what it would be if the construction worked).

Now, suppose we have a cone $f_S : X \rightarrow Z(M_1, S) \otimes Z(M_2)$. Using our choice of basis, we can write

$$f_S(x) = \sum_{l \in \mathcal{L}} \phi_{S,l}(x) \otimes l$$

for $\phi_{S,l}$ linear functions from X to $Z(M_1, S)$. Observe that for each S this sum has only finitely many non-zero terms.

For each fixed $l \in \mathcal{L}$ we can then use the universal property of the limit to construct a map

$$\phi_l : X \longrightarrow \lim_{S} Z(M_1, S).$$

Now if only finitely many of these maps were non-zero, this would produce a map from X to $Z(M_1) \otimes Z(M_2)$, which is what we want. Unfortunately, finiteness at each S does not suffice: passing to a finer triangulation S', one will, for general cones, have non-zero $\phi_{S',\lambda}$'s for which $\phi_{S,\lambda} = 0$ (they factor through $\ker(res_{S',S})$).

Thus the tensor product of the limits is not the limit of the tensor products. This raises the question of whether one can save the situation by passing to a suitable completion. (Preferably as a Hilbert space, though more exotic possiblities might also be tried.)

Finally, there may indeed be other methods of proceding from a "triangulated TQFT" to an actual TQFT obtained not by modifying the last step, but by finding initial data with more subtle renormalization procedures. There are at least two different places to look: analysis or scheme theory.

The introduction of analysis might be accomplished by first putting a norm (or semi-norm, or inner product) on the spaces $Z(M, T)$, then using convergence of the parallel maps in the sense

$$\lim_{T} Z(X, T) = Z(X) \text{ iff } \forall \epsilon > 0 \; \exists \mathcal{F} \; T \in \mathcal{F} \Rightarrow \|Z(X, T) - Z(X)\| < \epsilon$$

where F is a filter on the set of triangulations of X ordered by refinement to define a renormalization procedure leading to a diagram of the form in Figure 3. It is my suspicion that the quest for Donaldson/Floer theory will, in the end, require some such marriage of the categorical and analytic approaches to TQFT's.

The introduction of scheme theory might proceed first by attempting to carry out Example 4 for the restriction of a group scheme to finite fields of a fixed characteristic. The problem is then to find a renormalization procedure that takes into account the global aspects of the scheme.

References

[TV] V.G. Turaev and O.Y. Viro, *State Sum Invariants of 3-Manifolds and Quantum 6j-Symbols*, Topology **31** (4) (1992) 865-902.

[CY] L. Crane and D.N. Yetter, *A Categorical Construction of 4D Topological Quantum Field Theories*, preprint (Dec. 1992).

[A] M. Atiyah, *New Invariants for Three and Four Manifolds* in The Mathematical Heritage of Hermann Weyl, AMS (1988).

[Wi] E. Witten, *Topological Quantum Field Theory*, Comm. Math. Phys. **177** (1988) 353-386.

[Do] S. Donaldson, *An Application of Gauge Theory to Four Dimensional Topology*, J. Diff. Geom. **18** 269-316.

[F] A. Floer, *An Instanton-Invariant for 3-Manifolds*, Comm. Math. Phys. *118* (1988) 215-240.

[RT] N.Yu. Reshetikhin and V.G. Turaev, *Invariants of 3-Manifolds via Link Polynomials and Quantum Groups*, Invent. Math. **103** (1991) 547-597.

[K] R. Kirby, *A Calculus for Links in* S^3, Invent. Math. **45** (1978) 35-56.

[J] V.F.R. Jones, *A Polynomial Invariant of Knots via von Neumann Algebras*, Bull. AMS **12** (1985) 103-111.

[C] L. Crane, *2D Physics and 3D Topology*, Comm. Math. Phys. **135** 615-640.

[FQ] D. Freed and F. Quinn, *Chern-Simons Theory with Finite Gauge Group*, preprint (1991).

[Wak] M. Wakui, *On the Dijkgraaf-Witten Invariant for 3-Manifolds*, Osaka J. Math. **29** (4) (1992) 675-696.

[Y1] D.N. Yetter, *Topological Quantum Field Theories Associated to Finite Groups and Crossed G-Sets*, J. Knot Th. and R., **1** (1) (1992) 1-20.

[Pi] S. Piunikhin, *Reshetikhin-Turaev and Kontsevich-Kohno-Crane 3-Manifold Invariants Coincide*, J. Knot Th. and R., (to appear).

[Wal] K. Walker, *On Witten's 3-Manifold Invariant*, preprint (Feb. 1991).

[Di] P.A.M. Dirac, *The Principles of Quantum Mechanics*, (4th ed.) Oxford Univ. P., Oxford (1958).

[FY1] P.J. Freyd and D.N. Yetter, *Braided Monoidal Categories with Applications to Low-Dimensional Topology*, Adv. in Math. **77** (1989) 156-182.

[FY2] P.J. Freyd and D.N. Yetter, *Coherence Theorems via Knot Theory*, J. Pure and App. Alg. **78** (1992) 49-76.

[S] M.-C. Shum, *Tortile Tensor Categories*, Ph.D. Thesis, Macquarie Univ. (1989).

[Y2] D.N. Yetter, *TQFT's from Homotopy 2-Types*, J. Knot Th. and R., (to appear).

[Y3] D.N. Yetter *State-Sum Invariants of 3-Manifolds Associated to Artinian Semisimple Tortile Categories*, preprint (Aug. 1992).

[Pa1] U. Pachner, *P.L. Homeomorphic Manifolds are Equivalent by Elementary Shelling*, Eur. J. Comb. **12** (1991) 129-145.

[Pa2] U. Pachner, *Konstruktionsmethoden und das Kombinatorische Homöomorphieproblem für Triangulationen Kompakter Semilinearer Mannigfaltigkeiten*, Abh. Math. Sem. Hamb. **57** (1987) 69-86.

[Y4] D.N. Yetter *Framed Tangles and a Theorem of Deligne on Braided Deformations of Tannakian Categories* in Deformation Theory and Quantum Groups with Applications to Mathematical Physics, AMS Contemp. Math **134** (M. Gerstenhaber and J. Stasheff, eds.) (1992) 325-349.

[RK] N.Yu. Reshetikhin and A.N. Kirillov, *Representations of the Algebra $U_q(sl(2))$, q-orthogonal Polynomials and Invariants of Links* in Infinite Dimensional Lie Algebras and Groups (V.G. Kac, ed.), World Scientific, Singapore (1989).

[M] S. Mac Lane, *Categories for the Working Mathematician*, Springer, New York (1971).

List of Contributed Papers with Authors and Author's Affiliations:

1. ***Introduction to Quantum Topology***
   Louis H. Kauffman, University of Illinois, Chicago, Illinois

2. ***Knot Theory, Exotic Spheres and Global Gravitational Anomalies***
   Randy A. Baadhio, University of California, Berkeley, California

3. ***A Diagrammatic Theory of Knotted Surfaces***
   J. Scott Carter, University of South Alabama, Mobile, Alabama
   Mashahico Saito, University of Texas, Austin, Texas

4. ***Four Dimensional TQFT; A Triptych***
   Louis Crane, Kansas State University, Manhattan, Kansas

5. ***A Categorical Construction of 4D Topological Quantum Field Theories***
   Louis Crane and David Yetter, Kansas State University, Manhattan, Kansas

6. ***Evaluating the Crane-Yetter Invariant***
   Louis Crane, David Yetter, Kansas State University, Manhattan, and Louis H. Kauffman, University of Illinois, Chicago

7. ***Canonically Quantized Lattice Chern-Simon Theory***
   Dave Eliezer, Lawrence Livermore National Laboratory, Livermore, California;
   G. W. Semenoff, University of British Columbia, Vancouver, Canada

8. ***Extended Structures in Topological Quantum Field Theory***
   Daniel S. Freed, University of Texas, Austin, Texas

9. ***A Method for Computing the Arf Invariants of Links***
   Patrick Gilmer, Louisiana State University, Baton Rouge, Louisiana

10. ***Differential Equations for Loop Expectations in Quantum Gauge Theories***
    Christopher King, Northeastern University, Boston, Massachusetts

11. ***Triangulations, Categories and Extended Topological Field Theories***
    Ruth J. Lawrence, Harvard University, Cambridge, Massachusetts

12. ***A Geometric Construction of the First Pontrjagin Class***
    Dennis McLaughlin, Princeton University, New Jersey
    Jean-Luc Brylinski, Pennsylvania State University, University Park, Pennsylvania

13. ***The Casson Invariant for Two-Fold Branched Covers of Links***
    David Mullins, Division of Natural Sciences, New College of USF Sarasota, Florida

14. ***D-Algebras, The D-Simplex Equation and MultiDimensional Integrability***
    F. P. Michielsen & F. W. Nifhoff, Clarkson University, Potsdam, New York

15. ***The Chern-Simons Character of a Lattice Gauge Field***
    Anthony V. Phillips, State University of New York, Stony Brook, New York
    David A. Stone, Brooklyn College, City University of New York

16. ***Elementary Conjectures in Classical Knot Theory***
    Jozef H. Przytycki, Odense University, Odense M, Denmark

17. ***Knot Polynomials As States of Nonperturbative Four Dimensional Quantum Gravity***
    Jorge Pullin, University of Utah, Salt Lake City, Utah

18. ***On Invariants of 3-Manifolds Derived From Abelian Groups***
    J. Mattes, M. Polyak & N. Reshetikin, University of California, Berkeley, California

19. ***Some Knots Not Determined By Their Complements***
    Yongwu Rong, George Washington University, Washington D.C.

20. ***Triangulations and TQFT's***
    David N. Yetter, Kansas State University, Manhattan, Kansas

List of Participants

1. *Dr. Randy A. Baadhio*
Theory Group, Physics Division,
Lawrence Berkeley Laboratory and
Department of Physics
University of California at Berkeley
Berkeley, California 94720

2. *Dr. Scott Carter*
Department of Mathematics and Statistics,
University of South Alabama
Mobile, Alabama 36688

3. *Dr. Louis Crane*
Mathematics Department
Kansas State University
Manhattan, Kansas 66506-2602

4. *Dr. David Eliezer*
Department of Physics, L-412
Lawrence Livermore Laboratories, P.O. Box 808
Livermore, California 94551

5. *Dr. John Fischer*
Department of Mathematics
Yale University
New Haven, Connecticut 06520

6. *Dr. Daniel S. Freed*
Department of Mathematics
University of Texas at Austin
Austin, Texas 78712

7. *Dr. Stavros Garoufalidis*
Mathematical Sciences Research Institute
1000 Centennial Drive
Berkeley, California 94720

8. *Dr. Patrick Gilmer*
Mathematics Department
Louisiana Sate University
Baton Rouge, Louisiana 70803-4918

9. *Dr. Jay Goldman*
School of Mathematics
Vincent Hall, University of Minnesota
Minneapolis, Minnesota 55455

10. *Dr. Joanna Kania-Bartoszynska*
Mathematics Department
University of Iowa
Iowa City, Iowa 52242

11. *Dr. Louis H. Kauffman*
Department of Mathematics, Statistics and Computer Sciences (M/C 249), University of Illinois at Chicago
Box 4348, Chicago, Illinois 60680

12. *Dr. Christopher King*
Mathematics Department
Lake Hall, Northeastern University
Boston, Massachussetts 02115

13. *Dr. Ruth Lawrence*
Department of Mathematics
Harvard University
Cambridge, Massachussetts 02139

14. *Dr. Sam Lomonaco*
Department of Mathematics
University of Maryland
Elicott City, Maryland 21042

15. *Dr. Dennis A. McLaughlin*
Mathematics Department
Princeton University
Princeton, New Jersey 08544

16. *Dr. David Mullins*
Division of Natural Sciences
New College of USF
Sarasota, Florida 34243

17. *Dr. Kunio Murasugi*
Department of Mathematics
University of Toronto
Toronto, Ontario M5S, 1A1, Canada

18. *Dr. Frank Nijhoff*
Mathematics and Computer Sciences Department
Clarkson University
Potsdam, New York 13676

19. *Dr. Anthony V. Phillips*
Department of Mathematics
State University of New York
Stony Brook, New York 11974-3651

20. *Dr. Josef Przytycki*
Mathematics Department
University of California
Riverside, California 92521

21. *Dr. Jorge Pullin*
Physics Department
University of Utah
Salt Lake City, Utah 84112

22. *Dr. Nicolai Reshetikhin*
Department of Mathematics
University of California
Berkeley, California 94720

23. *Dr. Yongwu Rong*
Mathematics Department
George Washington University
Washington, District of Columbia 20052

24. *Dr. Masahiko Saito*
Department of Mathematics
University of Texas, Box 78
Austin, Texas 78712

25. *Dr. David Yetter*
Mathematics Department
Kansas State University
Manhattan, Kansas 66506-2602

www.ingramcontent.com/pod-product-compliance
Lightning Source LLC
Chambersburg PA
CBHW061617220326
41598CB00026BA/3796

* 9 7 8 9 8 1 0 2 2 5 7 5 9 *